D0199607

Weyerhaeuser Environmental Books

William Cronon, Editor

Weyerhaeuser Environmental Books

Weyerhaeuser Environmental Books explore human relationships with natural environments in all their variety and complexity. They seek to cast new light on the ways that natural systems affect human communities, the ways that people affect the environments of which they are a part, and the ways that different cultural conceptions of nature profoundly shape our sense of the world around us.

The Natural History of Puget Sound Country
Arthur R. Kruckeberg

Forest Dreams, Forest Nightmares:
The Paradox of Old Growth in the Inland West
Nancy Langston

Landscapes of Promise: The Oregon Story, 1800–1940
William G. Robbins

The Dawn of Conservation Diplomacy:
U.S.–Canadian Wildlife Protection Treaties in the Progressive Era
Kurkpatrick Dorsey

Irrigated Eden:
The Making of an Agricultural Landscape in the American West
Mark Fiege

Weyerhaeuser Environmental Classics

The Great Columbia Plain: A Historical Geography, 1805–1910
D. W. Meinig

Mountain Gloom and Mountain Glory:
The Development of the Aesthetics of the Infinite
Marjorie Hope Nicolson

Cycle of Fire *by Stephen J. Pyne*

World Fire: The Culture of Fire on Earth

Vestal Fire: An Environmental History, Told through Fire, of Europe and Europe's Encounter with the World

Fire in America: A Cultural History of Wildland and Rural Fire

Burning Bush: A Fire History of Australia

The Ice: A Journey to Antarctica

Irrigated Eden

The Making of an Agricultural Landscape in the American West

MARK FIEGE

Foreword by William Cronon

UNIVERSITY OF WASHINGTON PRESS
SEATTLE & LONDON

Irrigated Eden by Mark Fiege has been published with the assistance of a grant from the Weyerhaeuser Environmental Books Endowment, established by the Weyerhaeuser Company Foundation, members of the Weyerhaeuser family, and Janet and Jack Creighton.

Copyright © 1999 by the University of Washington Press
Printed in the United States of America

All rights reserved. No part of this publication may be reproduced or transmitted in any form or by any means, electronic or mechanical, including photocopy, recording, or any information storage or retrieval system, without permission in writing from the publisher.

Library of Congress Cataloging-in-Publication Data

Fiege, Mark.
Irrigated Eden : the making of an agricultural landscape in the American West / Mark Fiege ; foreword by William Cronon.
p. cm. — (Weyerhaeuser environmental books)
Includes bibliographical references (p.) and index.
ISBN 0-295-97757-4 (alk. paper)
1. Irrigation farming—Snake River Valley (Wyo.-Wash.) 2. Irrigation farming—Idaho. 3. Irrigation—Snake River Valley (Wyo.-Wash.) 4. Irrigation—Idaho. 5. Water-supply—Snake River Valley (Wyo.-Wash.) 6. Water-supply—Idaho. I. Title. II. Series: Weyerhaeuser environmental book.
S616.U6F54 1999
333.91'3'097961—dc21 98-50247
CIP

The paper used in this publication meets the minimum requirements of American National Standard for Information Sciences—Permanence of Paper for Printed Library Materials, ANSI Z39.48-1984. ♾

The paper used in this publication is acid-free and recycled from 10 percent post-consumer and at least 50 percent pre-consumer waste.

For Janet and Alexandra

Contents

MAPS

Foreword

Paradise Lost or Gained?

William Cronon

Ordinarily, a book on the history of irrigation in Idaho might seem like pretty dull stuff to readers encountering it for the first time, an all-too-typical academic monograph on a narrow subject likely to be of interest only to scholars with decidedly eccentric tastes for obscure topics in remote places. I'm sure at least a few library and bookstore browsers will be inclined to pass up this volume with just such thoughts in mind. But they will be wrong to do so, having no idea what they are missing. Not only is Mark Fiege's *Irrigated Eden* one of the best histories we yet have of a form of agriculture which has literally transformed the American West and become a mainstay of much of the American diet. Not only is this an elegantly written and subtle meditation on the complex interplay of human dreams and natural realities in reshaping our national landscape. Its ambitions and achievements in fact reach out toward a much bigger quarry, for the book asks us to look anew at the boundaries we draw—often with remarkably little reflection or self-consciousness on our part—between the places and things we call "natural" and those we call "unnatural." With Mark Fiege as our guide, both of these categories become much more difficult to define . . . and therefore much richer and more intriguing as sources of new insight and understanding.

Fiege's starting point for this inquiry is deceptively simple. It has become an unquestioned article of faith for many of us who care about the natural environment that the trajectory of environmental change in the nineteenth and twentieth centuries has involved a long-term decline in which a "natural" landscape has been replaced by an "artificial" one. We think of the free-flowing rivers, ancient forests, and vast grasslands of the western wilderness as "authentic" and "natural" in a way that no human landscape could ever hope to be. Conversely, we imagine that the changes people have imposed on such places—the dams in the rivers, the clearings in the forests, the fences in the grasslands—involve an assertion of human control that has become ever more absolute with time. Wandering amid the potato fields and irrigation ditches of southern Idaho today, and remembering the wilderness that preceded them, we imagine we are encountering a landscape so completely subordinated to the human will that nature has here been utterly overwhelmed and replaced by artifice. Whether we celebrate or lament this conquest—whether we see this place as a garden blooming in the desert or as the sad remnant of a lost and fallen world—we imagine that the human triumph is so complete that nature in its most authentic essence has been banished far away.

It is precisely here that Mark Fiege asks us to stop and look more closely, not only at our own assumptions about what is and is not natural, but also at the seemingly unnatural landscape that appears to lie so prostrate before the conquering hand of humanity. Over and over again in this remarkable book, he manages to demonstrate in all sorts of unexpected ways that the human conquest of nature is far less complete than we imagine. By forcing us to walk Idaho's irrigation ditches and really look at what they are doing, by making us examine potato fields more closely than we would ever have done without his help, he calls into question our easy assumption that nature is entirely absent from such places. Under his guidance, we come to see that the irrigation ditches in many ways mimic the underlying terrain, and that when they violate that natural order, their builders get themselves into trouble. When we look more closely at the plants growing in the ditches and the animals living along their banks, we suddenly see that the conquest going on in this landscape is more multidirectional than we at first realized: yes, human beings have worked hard to impose their ordering will on this environment, but many many creatures are working just as hard to recolonize these human structures to reclaim them for nonhuman nature. If one of the things we mean by "nature" is the ability of the nonhuman to resist the human in all sorts of autonomous and unexpected ways, then nature is alive and well even in so "controlled" and "artificial" a place as this.

Fiege's search for persistent and resistant nature, his project of blurring the boundaries between the natural and the artificial, operates at many levels, from entire watersheds to individual farms all the way down to the genetic machinery buried deep in the cells of an individual potato plant. Throughout the book, he also attends to the myths and perceptions that have tempted human beings to imagine themselves more in control of this environment than they really are. The result is a book that genuinely alters the way we see the world, giving us new tools for questioning our own assumptions and thinking in subtler ways about what it means to inhabit and use a particular landscape. Fiege is no apologist for those who wish to remake nature in their own image; it is far from his purpose in this book to argue that "anything goes," that one can do whatever one wants in altering landscapes and nature will still survive just fine. Quite the contrary. What he demonstrates in myriad ways is that no matter how we seek to alter the natural world, unexpected consequences of our own actions will always come back to haunt us. To say that nature will survive no matter what we do is very far from saying that we will be pleased with what survives, or that it will conform to our particular vision of the good, the true, and the beautiful. Fiege offers no easy solutions or quick fixes to the many dilemmas he poses in this book; rather, his goal is always to remind us that the world is a far subtler and more complicated place than we usually imagine it to be, and that an indispensable tool for understanding the puzzles it poses is to reflect long and hard on its history.

Why bother to read a book about the history of irrigation in southern Idaho? Because it has surprisingly interesting things to say about irrigation ditches and potato plants. Because it has great stories to tell about people doing remarkable and sometimes self-defeating things in the service of their dreams. Because it asks good questions about what is natural and what is not. And perhaps most of all because it offers new ways of thinking about the past which may, just possibly, offer new ways of thinking about the future as well.

Acknowledgments

Numerous institutions and people assisted me in my effort to research and write this history. The Idaho Historical Society, the Idaho Department of Water Resources, the Idaho Falls and Twin Falls public libraries, the Aberdeen-Springfield Canal Company, the Enterprise Irrigation District, and the Twin Falls Canal Company deserve special thanks, as do the National Archives, the Family History Library of the Church of Jesus Christ of Latter-day Saints, Marriott Library at the University of Utah, Merrill Library at Utah State University, Oboler Library at Idaho State University, and Suzzallo and Allen Libraries at the University of Washington. Equally important were the Idahoans who gave generously of their time and help. I am especially grateful to Gary Bettis, Tim Collett, Gary Domitz, Jack Eakin, Guila Ford, Lonnie Freeman, Inez Hopkins, Jake Isaak, Elizabeth Jacox, Tomas Jaen, Naida Kelleher, Pam Muirbrook, Alan Robertson, Kevin Ryan, James Siddoway, Carol Silvers, Sharon Walker, Tim Wehrkamp, Merle Wells, and Chuck Yost.

I could not have completed my work without financial assistance, and I would like to thank the institutions that provided it: the Aspen Institute for Humanistic Studies and the Woodrow Wilson National Fellowship Foundation, the University of Utah Humanities Center, Colorado State University, and the CSU Department of History.

The Institute for Environmental Studies at the University of Washington and the Wildlife Management Institute published early versions of my work in their journals; I thank them for their permission to reprint portions of these essays.

A number of outstanding scholars and editors offered advice, criticism, and encouragement at various stages. I especially would like to thank Tom Carter, Ellen Chu, Ed Davies, Thomas Dunlap, Paul Johnson, Jim Lehning, Donald Pisani (who read and critiqued the entire manuscript), John Rogers, and Virginia Scharff for their help. Bill Cronon, my editor, helped me a great deal to improve the quality of the manuscript's prose. I'm grateful for the lessons he taught me. Julidta Tarver, a well of patience, understanding, and good sense, shepherded the book to completion.

This project would have been infinitely more difficult without friends. I am proud to count Ruth Alexander, Mark Harvey, Don Jackson, Dale Martin, Roger Roper, Greg Silkensen, and Louis Warren among them. These scholars possess enormous energy, creativity, integrity, good humor, and intelligence; their example has been an inspiration to me. I am especially grateful to Dale, my mapmaker, who first introduced me to the historical study of common landscapes and people of the American West. Our many hours of conversation and our excursions together have enriched my life and work. In ways that only Dale can appreciate, this book really began years ago, during late-night talks over chocolate milk, on hikes along railroad tracks in the Palouse, and on a memorable train trip across the Cascade mountains and the Columbia plateau.

I would like to thank members of my family for their love and support: my mother, Phyllis Eckstein Fiege, and my sister, Gale Fiege Mann; my parents-in-law, Florence and Wes Ore; my brothers-in-law, Roger and Stephen Ore; and my sister-in-law, Anne Hepburn Ore. All of my family—immediate, extended, and in-law—have experienced a full measure of life's tribulations; their ability to overcome tragedy and adversity has motivated me to press on.

My daughter, Alexandra Fiege Ore, deserves special mention. Her birth gave me a renewed sense of purpose at a crucial moment and made me understand why people believe in angels. Her love and laughter have sustained me through bleak days and dark nights.

I could not have written this book, much less completed any constructive project, without the guidance and unflagging support of two people. Richard White opened my mind and helped me to see the world and the past in ways that I never thought possible. If this book accomplishes anything, it is because of his influence. My wife and yokemate,

Janet Ore, has been my greatest supporter and teacher. Her love, encouragement, good judgment, and faith have flowed in an endless stream. Her insights into the creative process and her passion for scholarship have inspired me and brought forth the best in me. I owe my life to her, and to her and Alexandra I dedicate this book.

Irrigated
Eden

Cover of a publicity pamphlet, early 1900s. Promoters, in this case the Oregon Short Line railway, issued illustrated booklets advertising Idaho's towns and irrigated farms. Such scenes often depicted female figures symbolizing nature in a fertile, well-watered landscape. *(ISHS No. Ms 544, Box 8/Pam 3)*

Introduction

Discovering the Irrigated Landscape

My understanding of Idaho's irrigated landscape began to take shape on a summer day in 1990. I was traveling over country roads, heading west from the town of Twin Falls through the heart of a 200,000-acre tract of farms, itself located within the vast arid basin sometimes called the Snake River valley. I had come here to study irrigated agriculture, and I had been told to look closely at the land, even its most mundane features. Look at the crops, the fields, the livestock, the trees, the farmyards, the canals; look, even, at the roadkill. All had the potential to reveal something important about this place.

As I passed mile after mile of farms, what drew my attention most was the network of irrigation conduits that delivered water to the fields. The largest canals were big—one could easily swallow my automobile—and they obviously carried a tremendous quantity of water. Grass grew on their engineered banks, and gravel service roads followed alongside. There were lesser canals, too, of various sizes; and bordering the fields were the smallest ditches, only inches deep and a few feet wide at most. It was difficult for me to tell how all the conduits fit together, where one branch ended and another began. But every now and then I came across an intersection, and here stood a low concrete foundation supporting steel gates that could be raised and lowered by turning the wheel of a worm gear. Usually the gates were at least partially open, and water flowed from one canal or ditch into the next. I was deeply impressed with all that I saw; the conduits formed an ingenious, intricate, technological system.

Then, somewhere around the small town of Buhl, I began to notice

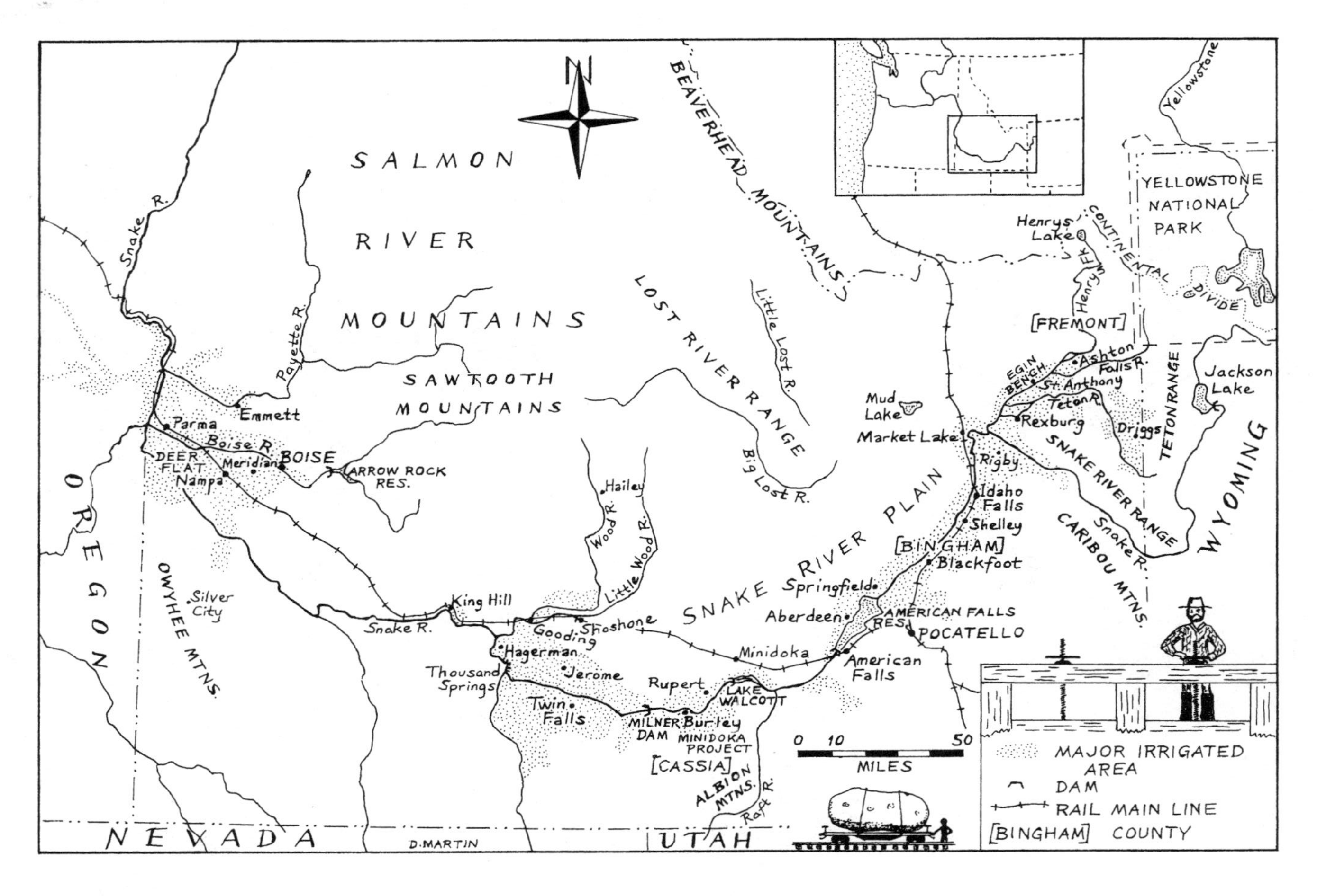

Map 1 Southern Idaho, 1930, showing major rivers, dams, reservoirs, and irrigated areas. *(Map by Dale Martin)*

something different about the waterways that coursed through this hydraulic landscape. I realized that I wasn't looking at just ditches and canals anymore—I was also seeing what appeared to be little creeks, perhaps two or three feet wide and from several inches to a foot deep. In this area the land sloped roughly south to north, and it was down this slope, through verdant pastures, that the creeks tumbled and splashed. I was puzzled. I knew that the water in the canals and ditches came from the Snake River, but what about these streams? What was their source? They must, I first thought, come from beyond the irrigated tract, no doubt somewhere in the hills to the south. But this outlying land, I reminded myself, was dry and sagebrush covered, by most standards a desert. How could these creeks rise in such an arid environment?

I had to stop and look more closely. I pulled off the road, turned off the engine, and got out. I walked to a place where one of the creeks flowed under a fence, into a culvert, and beneath the road. Standing above the culvert, I surveyed the pastoral scene that spread before me. The turf was green and lush. The sunlight was bright, the sky a deep blue. A few cows grazed leisurely, now and then looking up and staring at me with dumb bovine curiosity. My eyes followed the creek from the foreground to the background of my view. In front of me the stream was tranquil, almost languid, and in places the grass grew right up to its edge. Farther back, where hooves had trampled the ground, the banks were muddy. Beyond that, in the distance, the creek disappeared in the folds of the pasture. From my vantage I couldn't tell where the creek came from, where it originated. For the moment, at least, it would remain a hydrologic riddle. But my curiosity was aroused, and I knew that I must find an answer.

I did not get a complete answer that day, but soon enough, after a little exploration and some research, I had something. Here is what I learned. I learned that the creek was not really a creek, that it was not natural, in the conventional sense of these words. It did not originate in a natural source—it did not bubble up from a spring or flow from the distant mountains. It did not even have a name of the sort that people give to creeks. It was designated simply Lateral L, and it took its water from a conduit called the Low Line Canal. The creek, then, was part of the very irrigation system that had so fascinated me in the first place; it was a component of hydraulic technology.

And yet, to my surprise, I also found that Lateral L was not entirely artificial, either. When the Twin Falls Land and Water Company system was built in 1905–1907, engineers did not simply lay out irrigation conduits according to an abstract hydraulic geometry. They did not look at a blank environment and say, "We will put Lateral L here." In fact, Lateral L already existed in an incipient form: it was a draw, a crease made in the land by geological forces, by the earth's subterranean activity and the movement of wind and water over the surface. Indeed, it is possible that Lateral L originally carried water from rain or melting snow. Thus when engineers surveyed the irrigation system, they realized that they could install gates in a canal and divert the water through ready-made draws (which became Laterals A, B, C, D, etc.) and down to the crops and pastures. True, the irrigation water that flowed through some of these draws scoured well-defined channels where none had existed previously. No doubt, too, irrigators "improved" the channels with their shovels. But ultimately, these watercourses followed a route that the engineers had not plotted, and each still looked and flowed much like a creek. No wonder I thought Lateral L was natural. In certain respects it was.

I found more evidence of this ambiguous entangling of artifice and nature as I followed the water upstream from Lateral L. At first glance, it was obvious that the principal canals—Low Line, High Line, Main—were artifacts of hydraulic technology. They ran for about eighty miles, and their uniform dimensions and smooth shape reflected the precise survey and earth-moving techniques that had produced them. Yet, on closer examination, the canals' artificial qualities became less certain. Not just sterile technology, the conduits provided habitat for organisms. One summer evening, for example, I crossed a bridge over the High Line Canal, and there, illuminated in the rays of the setting sun, was a huge, dense, shimmering swarm of mayflies, as fine as I had seen over any river. I realized, too, that like Lateral L the canals conformed to natural topography: hydraulic engineers had situated them so that they followed the earth's contours, its bumps, hollows, and slopes. Indeed, this was the pattern that the builders had followed for the entire irrigation system. Viewed on a map, main canals and smaller laterals together thus appeared not as arrow-straight conduits but as a vast network that resembled a branching tree, veins in a leaf, or the numerous channels of a stream spreading across a delta.

When I reached the Snake, I discovered that the dividing line between nature and technology was no clearer there than in the Main Canal to which the river gave up its water. In one way, the Snake seemed so transformed by human activity that in places it could scarcely be called a river. Numerous dams along its length generated electricity and diverted huge quantities of water—sometimes nearly all of it—for irrigation. One such structure, the rock-fill Milner Dam, 86 feet high and 2,320 feet across, directed virtually the entire remaining flow of the Snake into the Twin Falls Main Canal. The view downstream was astonishing: save for a trickle, the river was empty. I came upon an equally staggering sight at Shoshone Falls some twenty-five miles below. Two hundred twelve feet high, Shoshone had not a trace of the thundering white torrent that had led local boosters to dub the falls the Niagara of the West. At places such as Milner Dam and Shoshone Falls it was easy to imagine that the Snake had become, over the past century, a pipeline with spigots that irrigators turned on and off at will.

Yet, viewed in another way, the Snake River did not seem so artificial after all. I eventually learned that the plumbing metaphor, so often applied to western American rivers, obscured more than it revealed. Nature, that part of the world not fabricated or controlled by humans, was still present here. Birds and fish inhabited the Snake River valley's reservoirs—most memorable for me were the flocks of white pelicans that circled around and around in the sky above these giant ponds, at last coming to light on the surface. But more important, the actual flow of the Snake bore little resemblance to the movement of water through a pipeline. I came to understand the Snake as a complex stream that varied according to geological formations below its bed, springs, the weather, seasons, and other factors. Dams did not squelch such processes. These massive hydraulic structures indeed disrupted and altered the flow of the river, but not in ways that irrigators could fully predict or control. In late spring, for example, when heavy accumulations of mountain snow turned liquid, the dams could not contain the entire flood—and once again Shoshone Falls appeared in all its splendor. Had the dams humanized the Snake, turned it into plumbing? Perhaps it was the other way around, I came to think; perhaps the river had naturalized the dams.

It was through this process of inquiry, a process that began at Lateral

L, that I moved deeper and deeper into Idaho's irrigated landscape. The little creeks that first caught my eye near Buhl, and the specific creek that I had stopped to look at, were my points of entry into the ecology and history of an environment that encompassed dozens of canal systems and millions of acres of land. Gradually I broadened my spatial and historical exploration into the farthest reaches of this world. I looked more closely at the early development of irrigation systems. I examined the crops and livestock, indeed the entire flora and fauna, that flourished here. I studied the methods that farmers and government officials devised to allocate the waters of the upper Snake. I looked at cropping systems and the ways that the farmers mobilized labor and marketed what they raised. My investigation even drew me into the human imagination, where the irrigators conjured vivid metaphors and compelling myths to make sense of the new world that they were helping to create. Eventually, I took all that I had discovered and fashioned it into a narrative of how the irrigated landscape came to be, how it functioned, and what significance it has for us now.

My narrative draws on analytical approaches that range from the materialist to the idealist. This is an account of the irrigated landscape as a physical entity, as a place made from earth, water, air, plants, animals, and artifacts. It is an explanation of abstract human systems: the agricultural economy or water laws, for example. It is also an assessment of ideas, of the meanings that people assigned to their lives and land. Each of these approaches is essential, indeed integral, to the others. A discussion of the physical aspects of the irrigated landscape necessarily must include a history of the human systems that rested on and shaped it. And a discussion of these necessarily must include a discussion of meaning: the values that motivated people to create and live in this environment, and the figures of speech and the stories they used to explain their actions. People often expressed meaning through metaphor, and they often built metaphor into a larger tale called myth. Every American landscape embodies a myth; to discover the physical place is to discover the great story that its inhabitants told about it. Thus the materialist and idealist approaches reinforce and complement one another. The reader must have both.[1]

At the core of my narrative is a lesson that I first learned, on a small scale, from my encounter with the little creeks around Buhl: what is human in the irrigated landscape, and what is natural, cannot be easily teased apart, if at all. This understanding, of course, runs against the way that we conventionally think of humans and nature in American history. We do not normally interpret heavily used landscapes, places of work and production, as environments that are also natural. Rather, we tend to see them as departures from nature. These places appear to be degraded or destroyed ecosystems from which nature has vanished; they seem to be environments that have been humanized and transformed by artifice into components of modern industrial society.[2]

We need to move beyond this conventional view and appraise our landscapes from the perspective of Lateral L. Like Lateral L, or the Low Line Canal, or the Snake River itself for that matter, nearly all environments are both human and natural creations. Humans alter land and water, often destroying much of what is natural and turning the land into an artifact. But nature is seldom if ever completely eliminated; often, as in the case of Lateral L, people only modify something in the biophysical environment that was already there. In turn, nature changes what humans build, often in unanticipated ways; sometimes nature comes back more powerful than before. Mayflies inhabit the canal; the Snake's hydrology fluctuates and upsets the control that the irrigators attempted to assert over the river. Again we try to arrange land and water to suit our objectives. And again nature circumvents our plans. Back and forth it goes, a process of alteration, intermingling, and layering, the result of which is landscape. Thus we should view each place that we inhabit, Idaho's irrigated farmland or any other, not simply as a departure or degradation but as a new environment, a new ecological system, that has been created and formed. An older, perhaps more natural environment was eliminated, but in its place stands a new, hybrid landscape that should be understood on its own terms.[3]

The irrigated landscape has much to tell us about the human place in nature, but it is not a landscape that today elicits much interest or admiration. Outsiders find its vast expanses too hot in summer, too cold in winter, and always too dusty, smelly, junky, bleak, and boring. The signs of intense human activity—acres of monocultures, utilitarian buildings, machinery, the odors of factory processing, animal wastes, and chemicals—assault modern aesthetic sensibilities. Thus, travelers usually react to the irrigated landscape as they do to so many places in the

American West: they drive straight through, rarely stopping at anything but roadside eateries, convenience stores, and rest areas. To be fair, though, even some of the people who are from here do not much care for it. To these natives, the irrigated landscape no longer stands as the symbol of hope and progress that it was during the early twentieth century. So they seek escape.

But the irrigated landscape refuses to be ignored. I have spent a great deal of my life living in or studying unappealing places in the American West. If there is one thing I have learned from my observations, it is that these are precisely the kinds of environments from which we can learn the most about our connections to nature. Sometimes it is difficult for me to accept this fact. Like most Americans, I desire landscapes untroubled by manifestations of work and production. I, too, want purity of surroundings and peace of mind. But as an environmental historian, I seldom ask questions that lead me to such places. Always my curiosity carries me back to the hard-used landscapes: to the mines of Butte, the wheat fields of the Palouse and the Great Plains, the test sites and grazing lands of Utah's West Desert, the suburbs of Denver and Seattle, the oil refineries of Billings, the pulp mills of Puget Sound, and the irrigated fields of the Snake River valley. It is in these apparently unnatural places, I believe, that we most directly confront the reality of our deeply tangled and problematic relationship to the natural world that we inhabit.

1/ **Genesis**

Water, Earth, and Irrigation Systems

Idaho's irrigated landscape developed during the nineteenth and early twentieth centuries from the interaction of agriculture with the Snake River valley environment. At the heart of this agriculture was a vision, shared by many people, of the human relationship to nature. Engineers, farmers, journalists, and social reformers frequently asserted that irrigation offered a method for transcending the limits that nature imposed. Most fundamentally, they said, it promised to liberate farmers from their dependence on precipitation. With irrigation, no tiller of the soil need fear drought. In 1900, Idaho state engineer D. W. Ross envisioned an environment in which the farmer could "laugh at the cloudless skies" because irrigation would allow him to have "the much needed moisture under perfect control." The *Twin Falls News,* boosting local reclamation projects, concurred. "Irrigation is the science of farming," the paper announced in 1904. "Rainfall farming is accidental farming."[1]

Irrigation advocates even claimed that farmers exercised a mastery not just over water but over nature in general. This was the message of the *Idaho Republican* in 1907 when it announced the arrival of R. H. Loomis, who had come to Idaho to settle on the Aberdeen-Springfield project near American Falls. "Loomis," the paper stated, "left a nice home in Illinois to make a new home on this manless, homeless, weedless, bugless tract where he can put everything on the land just to his liking." After journalist W. F. G. Thacher visited one of the Twin Falls irrigation projects in 1911, he similarly described farmers who worked the land with mechanistic precision. "Here," wrote Thacher, "agriculture is reduced to one of the exact sciences. The farmer knows the elements of his soil; he knows the amount of water he has to depend upon; he knows practically what the weather will be. He proceeds like a chemist in his laboratory."[2]

This rhetoric of a plastic and precisely controlled landscape appealed to farmers, especially those who had fallen victim to the hazards of "rainfall farming." The writer Hamlin Garland recalled that the promise of irrigation instilled hope in his father, nearly broken after several disastrous years of wheat farming on the sunbaked and wind-blasted Dakota prairie. "The irrigated country is the next field for development," Dick Garland told his son in 1893. "I'm going to sell out here and try irrigation in Montana. I want to get where I can regulate the water for my crops." Hamlin and his brother eventually convinced their westering father that the family's destiny lay eastward, back in humid Wisconsin, but the elder Garland's words lingered in the son's memory: with irrigation, a man could *regulate* his water.[3]

In retrospect, the Garland brothers probably judged wisely when they decided to discourage their father from going farther west. Dick Garland would only have encountered environmental problems similar to the ones he hoped to escape. In places such as Montana or Idaho's Snake River valley, the proponents of irrigation never achieved the technical mastery that they promised. Behind their extravagant claims lay a complicated, difficult landscape in which farms did not function like laboratories and irrigators did not even attain their most basic objective: freedom from drought.

Contrary to the boosters' pronouncements, the irrigated landscape was not a place in which humans mastered nature. Rather, it was the site of an ongoing interaction between people and the land, a reciprocal interplay in which irrigators seldom if ever achieved the control they desired. Through plowing, planting, and irrigating, farmers and engineers tried to impose their designs on nature. They also adjusted to certain natural conditions, such as climate and landforms. By both manipulating the land and adjusting to natural conditions, they created a productive agricultural system. Their system, however, did not function with perfect precision. Streams and climate proved erratic, soil conditions changed, reservoirs and canals leaked. And if irrigation made crops and livestock possible, it inadvertently created a habitat for an array of vexatious flora and fauna: weeds, insects, mammals, birds. Confronted with such problems, inhabitants of the irrigated landscape readjusted their agricultural systems and once again attempted to make nature bend to their will.

Gradually, this interaction between irrigators and nature created a new, complicated landscape in which human and natural systems overlapped, intermingled, and finally merged. Dynamic, ambiguous, often inscrutable, the irrigated landscape forced farmers to acknowledge that they could not always control nature as they wished. Behind their boasts they sometimes expressed frustration and doubt, and they grasped for new metaphors and images to explain their relationship to the complex environment in which they lived.

The irrigated landscape was the product, first of all, of the irrigators' encounter with the landforms and streams of the Snake River valley. This is a story of rock and soil, of changing climate and the transforming power of water, of fundamental physical processes that have shaped the world as we know it. And it is a story of people, the irrigators, interacting with these processes. Ancient forces laid down a substratum, the land; farmers and engineers then sought to incorporate this natural base into irrigation systems. From the 1860s through the 1920s, the irrigators built dams in canyons, excavated canals in the ground, and scraped fields from the soil. But in attempting to transform the Snake River valley into a tool of agricultural production, the irrigators exposed their systems to some of the very forces that had shaped the land in the first place. Thus the irrigators' efforts to change the environment resulted in unanticipated changes in their own irrigation works. From this convergence of hydraulic systems and geological forces, human and earth history, the irrigated landscape began to form.

Intense heat, pressure, and erosion, awesome powers that at once destroy and create, imparted the basic structure to the land that the irrigators would one day inhabit. Seventeen million years ago, during the Miocene, a giant meteorite smashed into the earth in what is now southeastern Oregon. The impact was so violent that it broke the earth's crust and opened a fissure in the underlying mantle. Enormous quantities of molten lava welled up and surged north into what would become Washington. But over millions of years, as the continental crust inched southwest across the opening, the "hot spot" shifted northeast, through the future Idaho. As the hot spot migrated it erupted with staggering force, producing a furrow that eventually became the Snake River plain. A thick layer of rhyolite, a pale volcanic rock, filled the furrow bottom. Later, after the hot spot moved on to the area that is now Yellowstone National Park, eruptions of a different sort added yet another layer of rock to Idaho's volcanic zone. Until 2,000 years ago, crustal stretching opened new fissures, and the earth spewed forth basalt that cooled and hardened into a black cap on the rhyolite. Volcanic ash, blown into the air, drifted on the surface, and there began to form soil.[4]

Volcanism defined the land that farmers later irrigated, but so did nearby mountains. During the Mesozoic, 70 to 80 million years ago, crustal movements shoved the Rocky Mountains into place. Later, the Oregon meteorite initiated a second phase of mountain building. Besides creating the migrating hot spot, the missile fractured the earth. As the continental crust slowly moved over the fracture, the Rockies shuddered and broke into huge blocks. This Bunyanesque stonework alternately thrust some blocks up and others down, creating a series of elongated ranges and basins that extended north-south. The Rockies and the "basin and range" would have covered all of southern Idaho, but the hot spot's eruptions left the furrow in their midst.

In the Pleistocene, water at last began to leave its mark on the land that irrigators would eventually transform: cool weather and increased moisture slowly turned the volcanic furrow into a fully developed river valley. Between 2 million and 3 million years ago, glacier-covered mountains sent streams gushing into the valley. Alluvial deposits formed at canyon mouths, and wind picked up the lightest sediments and laid them down across the Snake River plain. Water trickled into the ground and formed aquifers in gravel that lay sandwiched between separate lava flows. Eventually the water seeped into a major river, the Snake. Flowing along a westward gradient, the Snake and its tributaries slowly carved channels in the underlying volcanic rock. Then, about 15,000 years ago, a cataclysmic flood reshaped the Snake channel and gave the river its modern form. Lake Bonneville, which covered much of what is now northern Utah, absorbed so much inflow that its basin could no longer contain it. The water at last burst out, sending a vast rushing torrent into Idaho. The swirling waters scoured the Snake River canyon and created alcoves, rapids, and a series of magnificent falls.

Over the past 10,000 years, the cool, moist conditions dissipated, and the Snake River basin gradually grew warmer and drier. The glaciers retreated, and the Ice Age, the Pleistocene, came to an end. Surrounding mountain ranges blocked moist air and captured much of its precipitation, denying it to the valley. Rain and snow that did fall evaporated quickly into the crisp dry air. After several thousand years, aridity, the compelling reason for irrigation, became a predominant feature of the Snake River valley environment. To be sure, there were relatively brief periods when the climate grew cooler and wetter, but overall the valley became a dry place.

Yet within this dry environment, water remained an important agent of landscape change and formation. While profound geological forces laid down successive layers of rock and soil, an incessant exchange of moisture between earth and atmosphere—the hydrologic cycle—kept at least some water moving through the valley. As snow or rain, water fell to earth. There, plants absorbed and transpired it, and it evaporated into the air. Some water infiltrated the ground and percolated downward into gravel and the underlying volcanic rock, forming aquifers. Much water fell on mountains adjacent to the generally arid valley: the Boise, Sawtooth, Pioneer, Lost River, Lemhi, and Beaverhead ranges to the north; the Owyhee, Albion, Deep Creek, and other mountains to the south; the Caribou, Snake, and craggy Tetons to the east. Streams poured from these mountains and converged on the Snake, which flowed from the eastern ranges and followed its westward course in a huge arc across southern Idaho. The pattern never ended; the hydrologic cycle repeated itself, on and on, in perpetuity.[5]

Earth history had produced the Snake River valley and driven the hydrologic cycle; this land and the ancient natural processes that shaped it would soon attract thousands of irrigation farmers. These people first came to the lower valley in the 1860s, and they settled adjacent to the Bruneau, Boise, Payette, Owyhee, and Weiser rivers, tributaries of the Snake. Here they diverted water to their fields and raised crops for sale in nearby mining camps. Their presence helped boost Idaho's territorial population to nearly 15,000 people in 1870.[6]

The influx of farmers intensified in the late 1870s and early 1880s, when the Oregon Short Line and the Utah and Northern, subsidiaries of the Union Pacific Railroad, penetrated the valley. Idaho's doors opened to the wider world; outside markets beckoned, and the Snake River valley's fertile soils, ample sunshine, and water—all extensively advertised by the railroads—lured fresh settlers to the area. Some newcomers moved into the lower valley, but many others established farms and irrigation systems in the upper valley above American Falls. The population rose accordingly: by 1890, the year Idaho achieved statehood, the total stood at roughly 88,000; by 1900, some 161,000 people made their homes here.

The stream of migrants peaked during the first two decades of the twentieth century, as thousands of people claimed farms under the 1894 Carey Act and the 1902 National Reclamation Act. Most of them settled along the central Snake, midway between the upper and lower valleys. Largely because of this intensive round of irrigation development, Idaho's population more than doubled between 1900 and 1910, rising to about 325,000. The rate slowed in the following decade, but growth continued. By 1920, some 431,000 people inhabited the state.

Although they shared an interest in irrigation, social and cultural diversity characterized the people who sought to fashion the Snake River valley into an agricultural landscape. Real estate salesmen, implement dealers, bankers, lawyers, railroad managers, investors, journalists, and public officials—these individuals promoted, financed, and administered irrigation systems and farms. Engineers designed dams and canals, and construction companies and their workers built them. Most important were the people who actually settled and farmed the land. Quakers, Mormons, and Mennonites became irrigators, as did Japanese, Germans from Russia, and Indians on the Fort Hall Reservation. Farmers from the Midwest, the Great Plains, the South, and even other irrigated areas in the West settled in the region. Probably the greatest single migration to the Snake River valley came from Utah and the Church of Jesus Christ of Latter-day Saints. The Mormons established their first agricultural settlement in the upper valley in 1879; by 1914 about 36,000 church members, including recent arrivals from Europe, made their homes here.[7]

No single factor drew these folk to the irrigated landscape, but in almost every case the environment had something to do with their relocation. The belief that they would find cheaper land and a more dependable water supply attracted some; the prospects for land speculation lured others. A few came in the hope that the Snake River valley's dry, sunny climate would restore lost health. A substantial number recognized that irrigation offered one of the last opportunities in industrializing America to acquire a farm. To those who would leave urban homes and industrial jobs and migrate to places like the Snake River valley, the agrarian dream still seemed possible. By turning land and water into an agricultural landscape, this new generation of pioneers sought to recapture a vanishing way of life.[8]

Idaho irrigators of whatever sort faced a difficult task. They would have to dam streams and divert them into extensive networks of canals and ditches, and they would have to strip the land of sagebrush, then painstakingly smooth, grade, and plant it. They would have to grow crops and raise livestock in an environment that had been, but a short time before, a desert. Yet, on the whole, they had great confidence in their power to manipulate earth, water, plants, and animals as they wished. Thus their guiding vision, so well expressed by D. W. Ross, the *Twin Falls News,* the *Idaho Republican,* and W. F. G. Thacher: through irrigation they would make a landscape in which human ingenuity prevailed over the vagaries of nature.

To achieve their objectives, farmers and engineers sought to intercept the unceasing physical process that kept water moving through the Snake River valley: the hydrologic cycle. The irrigators had to shift the cycle out of its regular course and make it serve agricultural purposes. Farmers and engineers often referred to this work as "conquest"; they had to defeat nature, "tame" the Snake, and make it do their bidding. Yet conquest was not the only way that they described their manipulation of the hydrologic cycle. They also believed that they were working with nature. To move water, irrigators had to site dams and canals in accordance both with existing landforms and watercourses and with a powerful, elemental force: gravity. Partly this siting was mere expedience, a practical, efficient, commonsense use of nature. But by merging their systems with the land and the hydrologic cycle, the farmers and engineers believed that they were developing and perfecting the earth's raw potential. Indeed, the earth itself at times seemed like God's unfinished construction site; its topographical features already carried the outlines of a future landscape that the irrigators would complete. As a manifestation of providential design, the earth seemed to invite the dams and canals that the irrigators would build.[9]

Along the Snake and its tributaries, the irrigators found "natural" sites for erecting dams. Early farmers tended to settle on bottomland, places at which they could most easily and efficiently turn water out of streams and into ditches that carried water to the fields. During the 1880s, Mormon pioneers converged on the Snake River forks country, around where the Teton, Henrys Fork, and Snake (South Fork) flowed together. Topographic maps from 1876 and 1879 labeled portions of this low-lying area "beaver swamp." The Mormons congregated in the forks country for much the same reason as the beaver: it was easy to build dams there. Meander bends or low-lying banks allowed them to divert water with relatively minimal effort. Each year the farmers built small rubble, brush, and canvas dams that channeled water into their ditches. Some farmers used natural river riffles to help guide the water. Construction of small dams was usually a seasonal routine. High water in spring wiped out the structures, making new construction necessary before irrigation could begin.[10]

Large irrigation projects, sponsored by corporations and government agencies, faced greater technical and environmental challenges in building dams than did the small irrigators. But like the folk builders, professional engineers situated their large earth and concrete structures in places that seemed natural for catching and diverting water. Canyons, valleys, and lakes provided efficient and thus relatively inexpensive places for dam construction. Paul Bickel and his staff, engineers of irrigation works around the town of Twin Falls, designed dams and reservoirs that incorporated natural topography. On the south central Snake, the engineers incorporated two volcanic rock islands into Milner Dam, a rockfill and concrete structure. The *Twin Falls News* noted that the earthen dam that formed Dry Creek Reservoir, which fed water into the Main Canal on the south side of the Snake, appeared to be well suited to the surrounding land. "It would seem," remarked the paper, "that bountiful nature had provided the reservoir in the most convenient location for the convenience of man."[11]

While working for the U.S. Geological Survey (USGS), engineer Arthur Foote identified landforms and water bodies suitable for dam and reservoir sites. On the upper reaches of the Snake River in Wyoming, Foote located a dam and reservoir site at Jackson Lake, just below the Teton mountains. The U.S. Reclamation Service completed Jackson Dam in

1907, raising the level of Jackson Lake and creating a reservoir for irrigators downstream in Idaho. The lake basin alone was outstanding, but scientist and USGS director John Wesley Powell saw another natural advantage in the site's relatively high elevation and cool climate, which lessened evaporation. Subsequent observers agreed with Foote and Powell. In 1911, engineer D. G. Martin stated that Jackson Lake "is one of the most natural reservoir sites that I know of in the entire Northwest."[12]

Irrigators turned geological formations into natural dam and reservoir sites; similarly, they transformed rivers into natural canals. Government officials and engineers used the Snake to convey water from Jackson Lake to downstream irrigation projects. Engineers built canals between rivers ("interbasin transfers"), forming vast integrated networks of man-made and natural water channels. The Milner-Gooding Canal (1931), a U.S. Bureau of Reclamation project, took water from the Snake at Milner Dam and conveyed it nearly eighty miles to the vicinity of Shoshone. There the canal connected first with the Little Wood River, then the Big Wood River, and finally the irrigation system of the Big Wood Canal Company. Water shunted into the Wood rivers flowed to downstream headgates and into canals that went to irrigated farmland.[13]

While irrigators used rivers as canals, they found it expedient and economical, even necessary, to build actual canals and ditches in accordance with the land's natural features. The artificial waterways had to follow the earth's contours, its curves and elevations. And the conduits had to incorporate one of nature's most powerful forces, gravity, which supplied the motive power for the water. Indeed, landforms, water, and gravity were the basic influences on canal system design. In one way or another, every canal system was an equation of these three factors.[14]

In general, Idaho irrigators laid out their canal lines along the overall east-to-west slope of the Snake River valley. Water flowed along this gradient under the force of gravity. But within the valley's general east-to-west trend lay an uneven surface of hills, plateaus, ridges, canyons, and bottoms. Consequently, the irrigators had to situate canals to accommodate these varied landforms as well as gravity. In addition, the canals had to achieve a precise relation between gravity, water, and soil; if builders inclined the canals too steeply, water gained too much momentum and eroded earthen canal banks. The canals thus carried water

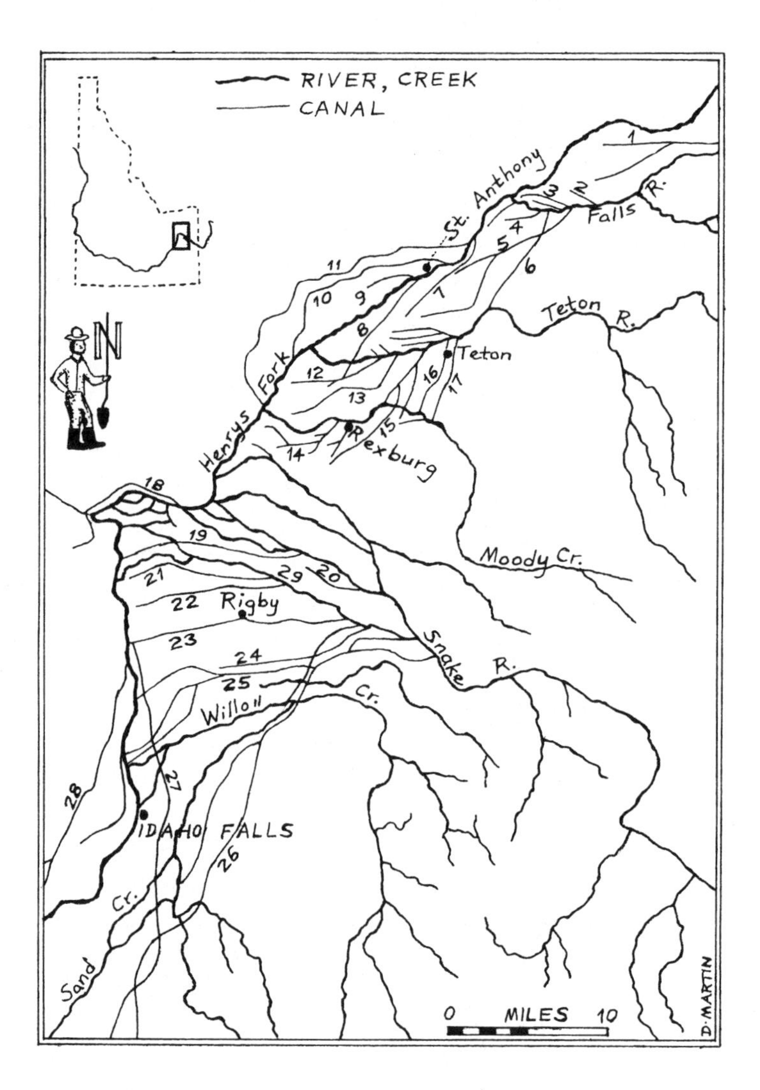

Map 2 Canal network of Snake River forks country, about 1900. Canals and natural waterways interconnected, creating a unique human and natural ecology. Note that irrigators named the canals after people, places, or natural features—primarily the streams from which the canals diverted water. (1) Springfield; (2) Terrill; (3) Anderson; (4) Kerr; (5) Fall River; (6) Fall River; (7) Birch; (8) Farmers Consolidated; (9) Mason; (10) Egin; (11) St. Anthony; (12) Salem; (13) Boqua; (14) Westfield; (15) Rexburg; (16) Teton; (17) Sidney; (18) Market Lake; (19) Menan; (20) La Belle; (21) Lewisville; (22) Rigby; (23) Burgess; (24) Farmers Friend; (25) Eagle Rock and Willow Creek; (26) Cedar Hollow and Foothills; (27) Idaho; (28) Great Western; (29) Old Channel/Dry Bed of Snake R. *(Map by Dale Martin)*

down and conformed to the land. They registered the force of gravity and negotiated the earth's surface.[15]

Pioneer irrigators of the nineteenth century merged their canals with the land in remarkable ways. In the Boise Valley, the Andrews Canal (1864) integrated canals and natural sloughs in a single system. In the forks country of the upper Snake River valley, the small folk dams diverted water into canals and ditches that settlers conveniently routed through low-lying swales, meander bends, and creeks. The Farmers' Friend Ditch (1884) diverted water from the Snake and moved it part way through a creek that settlers had enlarged. The Farmers' Progress canal system (1887), which delivered water to irrigators through the branches of Willow Creek, prompted the *Idaho Register* to observe: "Thus has nature provided a large scope of country with natural canal channels."[16]

What was true of the small, folk irrigation systems typified the largest and most sophisticated networks as well. On these huge projects, built on the arid lands above the stream bottoms, engineers wove their great canal systems into the earth. The Twin Falls Land and Water Company's irrigation project, first opened in 1905, was an elaborate synthesis of waterway, landform, and gravity. Designed by chief engineer Paul Bickel, his assistants, and consultants, it constituted the single greatest piece of engineered landscape in the Snake River valley. The project encompassed about 350 square miles on the Snake's south side. Here the land generally sloped east to west but also south to north, toward the river. Bickel and his associates blended their earthen hydraulic works into this topography. The system began at Milner Dam on the Snake River. At the dam, water flowed into the Main Canal, a massive conduit that measured 124 feet across at the top, 80 feet at the bottom, and could carry a flow of water up to 10 feet deep. Excavated into the ground, the Main Canal ran due west from the dam, roughly parallel to the river. About thirty-five miles from Milner Dam the Main Canal forked into two smaller but still substantial conduits. The High Line Canal, which formed the upper edge of the project, continued due west along elevation contours until it ended at the canyon of Salmon Falls Creek, about eighty miles from Milner Dam. The Low Line Canal carried water into the heart of the project. On the downhill side of all three principal canals (Main, High Line, Low Line), water passed through mechanical gates into smaller conduits, laterals, that conveyed water to the farms. Many of the laterals consisted of ditches, but Bickel and his associates also incorporated some 400 miles of natural creeks and draws, or coulees, as they called them, into the system. In its entirety, the vast, branching network of canals, ditches, creeks, and coulees eventually totaled about a thousand miles in length and embraced approximately 200,000 acres.[17]

Much like irrigation systems as a whole, individual farms joined technology, land, water, and gravity. Farmers built ditches on high ground and then distributed water to lower-lying fields. On farms encompassing uneven terrain, irrigators built their ditches along small ridges, a practice that gave their fields a patchwork appearance. When actually spreading the water on crops or meadows, farmers either channeled the water through furrows or flooded the land. Most irrigators graded their fields with horse-drawn scrapers, smoothing and sloping the ground so that the water would flow evenly and without so much velocity that it would wash away crops.[18] Even while wielding a shovel a farmer was manipulating the trio of water, land, and gravity. As an Idahoan named Rossi explained in 1887,

> A man lets on from the main ditch a stream of water . . . [then] follows the water up, builds little dikes in places where it is liable to cut the soil, or where he finds depressions in the land, leads the water to the highest places in that portion of the field, and generally assists the water in covering the whole piece of ground. At night he turns off about half the stream and lets the balance do its own work.[19]

The irrigators' integration of hydraulic technology with land and water was almost an equal human-nature partnership. But equality was not quite what the irrigators intended. In Rossi's description, the farmer was still the master who led, guided, and cut off the stream. Like a draft animal, water worked for the farmer. Indeed, irrigators often described their relationship to water in just such terms. They often spoke of the Snake River as a creature that must be "tamed," "harnessed," and "made to do duty" for man's benefit. Water was, in their view, a useful but obstreperous form of nature that awaited their steadying and constructive hand. Many Snake River valley inhabitants went even further in depicting human dominion over water, for the highest compliment one could pay a farmer was to say that he could make the fluid defy gravity. The master irrigator, as the phrase went, could "make water flow uphill." Thus irrigators integrated human and natural systems, but in their view, nature was the junior partner. Man was still in control.[20]

This assumption of human control over a pliant and even willing nature, basic to the outlook of virtually all farmers and engineers, was consistent with another popular belief: that irrigators were master technicians whose work realized the inherent potential or purpose of the land. The *Twin Falls News* expressed such a view when it lauded the vision of developer Ira Perrine, who first advocated reclamation projects around Twin Falls. Perrine, stated the journal in 1904, "saw before all others" that irrigation was "the one thing needed to complete Nature." The paper further extolled the project's water source, the Snake River, and proclaimed that the stream had been "stolen from Nature to strengthen Nature." For other observers, irrigators were not merely men who completed the land but the agents of God's great plan for the earth. E. B. Darlington, chief engineer for the Twin Falls North Side Land and Water Company, wrote in 1920 that irrigation engineers were "understudies of the Creator," men charged with the task of making the earth a better place. Darlington's statement reflected the widespread notion that dams, built according to the "laws of nature," were part of a divine scheme for earthly improvement. God, a sort of Chief Engineer, had drawn up the blueprints and built the framework; now his understudies, the irrigation engineers, would finish the job.[21]

By the time that Darlington made this statement, farmers, engineers, developers, businessmen, and their allies had good reason to believe that they had attained their environmental and theological goal. They had achieved great success in tapping the hydrologic cycle and using the water to transform the Snake River valley into lush, productive farmland. By 1920, numerous dams in the Snake River basin captured and held water for dozens of irrigation projects. About 13,000 miles of ditches and canals interlaced the region and carried water to approximately 18,000 farms that covered roughly 2 million acres along the Snake and other streams. The irrigators had apparently fulfilled God's wish and their own.[22]

Or had they? Had they really created an environment in which farmers no longer needed to worry about drought, a landscape that they had assembled to their liking and in which they operated with laboratory precision? The answer is no. Irrigators indeed had left their mark on the land, indeed had created productive farms. But they had not created a perfect landscape. In fact, it was so imperfect that it often left them frustrated and doubtful of their own powers. For human design alone did not determine how their irrigation systems functioned and developed; uncontrollable natural processes did, too. If dams and canals appropriated the hydrologic cycle, the hydrologic cycle in turn preempted, altered, and compromised irrigation systems. Floods, drought, groundwater, and an array of other earth forces denied farmers the control that their hydraulic technology was supposed to give them. Instead of shaping the land to their liking, the irrigators often found themselves reacting to natural processes that they could not master and that they dimly understood. They had stolen from nature only to find that nature stole back.

Contrary to D. W. Ross's prediction, irrigators never came to laugh at the cloudless skies. Precipitation, not dams or canals, still determined the total amount of water available. Thus, too many days of clear weather more likely evoked alarm than defiant glee, as worried farmers watched for some sign that nature would give them the moisture they desperately needed. When precipitation finally arrived, fear turned to jubilation. In February 1906, the *Gem State Rural* reported "general rejoicing in all the arid portions of Idaho" because "a heavy fall of snow in the mountains" portended "an unusually large supply of irrigation water this season." The intensity of the farmers' happiness doubtless corresponded to the degree of their anguish in those years when not enough snow fell on the nearby ranges. Prolonged drought occurred as early as the 1870s and then periodically thereafter: the late 1880s, the early 1900s, from 1919 through the 1930s, and so on. Indeed, what probably inspired the rejoicing of February 1906 was the prospect not just of ample water but of an end to prolonged drought. Nature could give and it could withhold.[23]

Water shortages only intensified as increasing numbers of farmers brought additional land "under the ditch." In 1890, the irrigated landscape in the Snake River valley encompassed about 150,000 acres; by 1920, farmers diverted water to about 2 million acres. By the early 1900s, decreasing precipitation and increasing irrigation sucked streams dry;

even the Snake sometimes ran out of water downstream from the town of Blackfoot. Again, dams furnished no guarantee against such shortage. On the upper Snake River, even the addition of two huge dams and reservoirs, American Falls (1927) and Palisades (1958), did not totally forestall the combined effects of drought and increasing water use.[24]

The technical difficulties of dam and reservoir operation exacerbated the problem of water scarcity. Dams and reservoirs were not impervious containers that held water like tanks or bowls; natural groundwater processes acted on them. Portions of impounded water seeped into surrounding alluvium and rock formations. When irrigators drained reservoirs and the hydrostatic pressure decreased, water flowed back in. Some reservoirs did not even have the benefit of this "bank storage"; they completely lost large quantities of water. A reservoir (1910) on Salmon Falls Creek leaked water into surrounding lava rock. By 1935 the reservoir had yet to fill. On the Twin Falls North Side Land and Water Company project, leakage from two offstream storage facilities, Wilson Lake and Jerome Reservoir (both c. 1909), eventually forced irrigators to abandon them for storage purposes. Water from Jerome Reservoir seeped into the volcanic rock at a tremendous rate, about 2,000 acre-feet per day, eventually filtering back into the Snake River canyon. (An acre-foot equals an area of one acre covered to a depth of one foot.)[25]

Erratic river hydrology compounded the difficulty farmers had controlling water. Rivers were dynamic, not regular; the hydrologic cycle merely described a broad pattern of water movement within which unpredictable variation occurred. High water in spring wiped out the small folk dams, and rivers under heavy flow changed course. When the upper Snake shifted in its bed, it left the heads of canals high and dry. The Great Feeder Canal symbolized the struggle of irrigators to wrest water from a shifting upper Snake. In 1894, extreme high water moved the South Fork of the Snake out of an old channel (immediately dubbed the Dry Bed) from which many Mormon canals diverted water. In response, the farmers in 1895 constructed the Great Feeder Canal to restore water to the Dry Bed. In subsequent years, the upper Snake continued to shift. In 1930, the Idaho commissioner of reclamation concluded that, because the river moved "at will" through its channel, "herding" the water with levees and riprapping would be necessary.

When running high, the river was more like a wild stallion than a harnessed, tame horse.[26]

The dynamic hydrology of creeks and rivers posed major problems for the irrigators' efforts to divide the streams and turn water into private property. To apportion streams, Idaho in 1881 officially adopted the law of prior appropriation. In theory, priority was a realistic attempt to merge the cultural convention of private ownership with streams that ran low at the end of the summer and during drought. Prior appropriation, in short, attempted to address the dynamism of nature. According to the law, irrigators diverted water under the principle of "first in time is first in right." When a creek or river ran low, the earliest claimants—those holding prior rights—would take water before those farmers holding later rights. In practice, prior appropriation failed to address nature's complexity fully. A creek or river not only decreased in volume but also fluctuated along its length; groundwater inflow, precipitation, and loss through seepage influenced the quantity that a stream carried at any given point. Such natural irregularity overwhelmed prior appropriation. Irrigators scattered along a creek or river simply could not match the abstract order of priority with fluctuating, unknown quantities of water. The problem became even more complicated when irrigators had to differentiate between natural flow and reservoir water, or stored water, that they released into a river for downstream diversions.[27] A 1902 federal government report succinctly explained the problem of water distribution:

> If [a] stream channel were like an iron pipe or conduit [in which water] must pass until discharged at determined points, . . . water supply would be very simple. . . . whatever water came into the pipe must come out at some point, or, in other words, . . . the quantity to be dealt with would be constant. This, however, is not the case in nature.

Irrigators certainly desired the precise technical control of water. They knew, however, that this was impossible "in nature."[28]

Hydrological dynamism and the irregularities of water allocation upset the dreams of farmers and engineers and compelled them to modify their water distribution methods, including prior appropriation. To overcome the problem of scarce and elusive water, irrigators developed cooperative water allocation practices. A lawsuit among Boise River irrigators, for example, resulted in a ruling, the "Stewart Decree" (1906), that instituted a "sliding scale" method of water distribution. As stream flow decreased, a watermaster gradually reduced each diversion, beginning with the latest water right and ending with the earliest. The sliding scale thus respected priority but required all irrigators to share the shortage. During the early 1900s, irrigators on the upper Snake River attempted extralegal, cooperative methods. In 1923 they created the Snake River Committee of Nine, which established procedures for the differentiation and distribution of stored water and natural flow. Even within canal systems, irrigators cooperated. Farmers on the Danskin Ditch, a small canal system in the upper Snake River valley, shared water with each other during 1919, a year of severe drought.[29]

Irrigators experienced difficulty controlling and regulating natural waterways. But many of the same uncontrollable geomorphic and hydrological processes that shaped streams also affected the waterways that the farmers and engineers had constructed. Idaho irrigators built their canals out of earth, either excavating them from the ground or building them up from soil and rock. As the earthen material alternately froze and thawed, canal banks gradually slumped. Runoff from rainstorms on frozen or steep ground deluged canals and ditches, eroded their surfaces, and filled them with silt. Even the creeks and rivers that fed irrigation systems carried silt into them. And irrigators observed that the gravity that moved their water could destroy their canals. In 1918, D. W. Cole, a Reclamation Service engineer on the Boise project, wrote of the irrigators' "audacity in snatching a river from its normal course and inducing it to flow" to farmlands. Even after the canals were built, Cole stated, "the law of gravity and the persistent disposition of water to run down hill are unceasingly at work for your undoing." Irrigators devoted a great deal of their resources to maintaining the physical shape of canals and ditches in the face of powerful natural forces. Irrigation companies and districts retained staffs of workers whose responsibilities included not only shunting water through canals and ditches, but also dredging and reshaping the structures. Here again, humans had to adjust to environmental conditions that they could not completely control.[30]

The process of seepage drained canal systems much as it took water from reservoirs. Irrigators sought to intercept and control the hydrologic cycle, but the same natural process captured water from the canals. Some of the water evaporated, but tremendous volumes—as much as 30 to 60 percent—infiltrated the ground, especially where porous soils underlay canals, ditches, and fields. In 1912, primary canals on the north side of the Minidoka irrigation project lost 96,000 acre-feet of water out of a total 314,000 diverted. Engineers and farmers sometimes puddled clay into the canals in an attempt to stop the seepage. They might have lined the canals with concrete to prevent water loss and retain bank shape, but except in a few places they did not. The cost of construction stopped them, but more important, climate discouraged the practice. Idaho's frosty winters caused concrete to flake and crumble, much more than in the warmer irrigated landscapes of Arizona or California. Maintaining concrete canals under such conditions would only have added to heavy installation expenses.[31]

Seepage from canals as well as reservoirs was a severe problem, and engineers had to admit their inability to comprehend it fully and to predict it. Arthur Powell Davis, chief engineer for the U.S. Reclamation Service, came to this conclusion in 1918. "Some of the difficulties encountered," he said, "are of such nature that no method of foreseeing them appears to be possible." The natural world that technocrats such as Davis sought to control and perfect always proved at least somewhat recalcitrant and elusive. When addressing the problem of seepage, irrigation engineers could not present themselves as God's omnipotent understudies.[32]

Natural processes and conditions that affected canals and reservoirs made irrigation less than perfect on the farm itself. Not all irrigators, contrary to Rossi's 1887 description, were skilled at preparing the land and then coaxing and leading the water through ditches, across pastures, and down furrows to crops. This was especially true during the early days of irrigation, when newcomers from more humid climes struggled to learn the technique. Often they found that, instead of mastering land and water, land and water mastered them. One farmer near Twin Falls failed to grade his land properly and thus "discovered that no amount of damming either verbally or with a shovel would make the water flow uphill." In 1910, Patrick McCaffrey, the agricultural editor of the *Hagerman Valley Sun,* observed the clumsy efforts of neophyte irrigators on a reclamation project north of Twin Falls. These people had difficulty working with gravity, much less defying it. His description offers a counterpoint to Rossi's master irrigator:

In traveling over the great North Side tract, one notices with pity the awful attempts of most eastern settlers to perform the simple trick of training water to run down hill, known in this tailormade country as irrigating. You will see them up to their knees in mud and struggling like a toiltrained cross between a beaver and a galley slave, and all of this toil worse than wasted, for instead of improving the land—as the agency of sanely applied irrigation should, they are trying their best to run it into a brick pavement.[33]

But of all the problems inherent in farm irrigation, none was greater than seepage. Even the better irrigators often applied too much water to the land. Overapplication was partly a consequence of the "continuous flow" method of distribution. Under continuous flow, an irrigator received a steady stream of water—"an inch to the acre" (the inch was a measure of flowing water). Water constantly trickled through laterals and farm ditches, even when individual irrigators did not actually use it for crops. As the water passed through the ditches, much of it seeped into the ground. The other cause of seepage was the farmers' tendency to spread too much water on their fields. Only a percentage actually went into plant growth; much of it evaporated or infiltrated the earth. Engineers and government officials found it convenient to attribute this water loss to the farmers' incompetence and greed. But to be fair, gauging just how much water a crop needed was extremely difficult if not impossible and depended on an array of variables: soil type, temperature, precipitation, plant variety. Here again, the problem was one of a complex natural world resisting the irrigators' efforts to control and regiment it. The typical response to this complexity was simply to soak the land in the belief that the plants would get the water they needed. It was a logical course of action, although costly. On the north side of the Minidoka project in 1912, laterals and field irrigation lost an additional 144,000 acre-feet of the 314,000 diverted from the Snake River.[34]

Irrigators lost water through seepage, but much of that moisture returned to them in an unwanted form: as rising groundwater that saturated and ruined farm soils. When water seeped out of canals, ditches, and fields, it eventually encountered impervious layers of volcanic rock or clay subsoils that restricted its downward movement. In addition, seepage from irrigation on high ground percolated along gradients to lower land, especially river bottoms. After prolonged irrigation, groundwater rose and created springs, filled depressions, formed shallow pools, and in general made the soil soggy. Crops grew poorly or not at all in "seeped" land, as the farmers and engineers called it.[35]

A related and equally serious problem developed in poorly drained bottoms. Here, seepage water concentrated salts in the soil, making it increasingly toxic to cultivated plants. The salinity problem began deep in the ground, where subsurface water dissolved salts that occurred naturally in arid soil. Then, as the water rose and evaporated, the salts precipitated and formed patches of thin, poisonous crust on the earth. One can hardly overlook the irony that waterlogged, salty land posed: the water that the irrigators coveted and purported to control had become a serious nuisance.[36]

Irrigated areas that covered extensive areas of low ground developed the worst cases of seepage. Such was the situation on the north side of the Minidoka project, adjacent to the south central reach of the Snake River. Built in 1907–1909 under the direction of the U.S. Reclamation Service, this great irrigation system encompassed roughly 70,000 acres. Much of the project consisted of relatively flat land, its sandy soils deposited over the ages by wind and the action of the Snake. This topography had little of the slope necessary to make the water flow properly. Indeed, over much of the project, the Reclamation Service built canals up from the ground to give the water the elevation it needed to reach distant farm fields. Because of little slope, water moved slowly, and this languid pace prolonged the water's contact with the sandy, porous soil. High rates of seepage developed. And the losses grew still higher when irrigators applied water to their fields. Because the sandy soil would not hold moisture, farmers had to apply irrigation water in copious amounts. But the loss of water alone was not the worst of the farmers' difficulties; indeed, government soil analysts had anticipated that the excess water would drain away from the project. They were wrong. Seepage water percolated downward until it reached a layer of clay; then, as seepage continued, the groundwater level began to rise. Within a short time, by fall 1909, approximately 10,000 acres were so soggy that farming suffered.[37]

What was it like to live on waterlogged land, on an irrigated farm so wet that crops had difficulty growing? During the early 1900s, J. P. Butler and his wife had started an 80-acre farm in the lower Boise Valley, in an irrigated area of about 6,800 acres known as the Arena Basin. Butler worked hard to prepare his rough land for irrigation, and at first the farm went well. The area's extremely sandy soil, however, soon led to problems. The Reclamation Service estimated that each acre of Arena Basin farmland required eight to ten acre-feet of water, perhaps four or five times the amount applied to most Idaho fields. Furthermore, as much as half the water that flowed through canals in the locale seeped into the porous ground. By 1917, many farmers in the Arena Basin, including Butler, found some of their fields so swampy that the land could not be worked. And conditions only grew worse. Water began to flow out of the Butler well. It seeped from the ground and formed a stream that ran between the house and barn, and the barnyard became so muddy that the Butlers could no longer keep horses there. Water then seeped into the cellar. In 1920, to escape the rising moisture, the Butlers moved their house and other buildings to higher ground. And still, slowly, insidiously, the water came on, until by 1923, all but eighteen acres of the Butler farm were waterlogged. "There is no one who realizes what it means to live here and watch the water creep up on you until you have to leave it," Mrs. Butler said, "unless some one has experienced it." Ironically, water had almost completely ruined the land that the Butlers had originally sought, through irrigation, to reclaim.[38]

Waterlogged, salty land eventually provoked farmers and engineers to take action. And so, in response to uncontrolled environmental conditions, a new phase in the development of the irrigated landscape began: the construction of drainage systems. These waterways looked much like irrigation canals, except that they were dug below the level of fields and canals. Some were only a few feet deep; others could be ten feet across and ten feet deep or more. The drains intercepted and collected the seeping water, then carried it back to creeks and rivers. By 1920, Idaho irrigators had excavated roughly 290 miles of drains, covering a total of about 65,000 acres. Ten years later, they had excavated 651 miles of ditches that served approximately 375,000 acres. Ironically, the land that engineers and farmers had first sought to reclaim from aridity they now had to reclaim from water. It is a little-known fact that many irrigated areas, not just in Idaho but in other parts of the American West, underwent two forms of reclamation.[39]

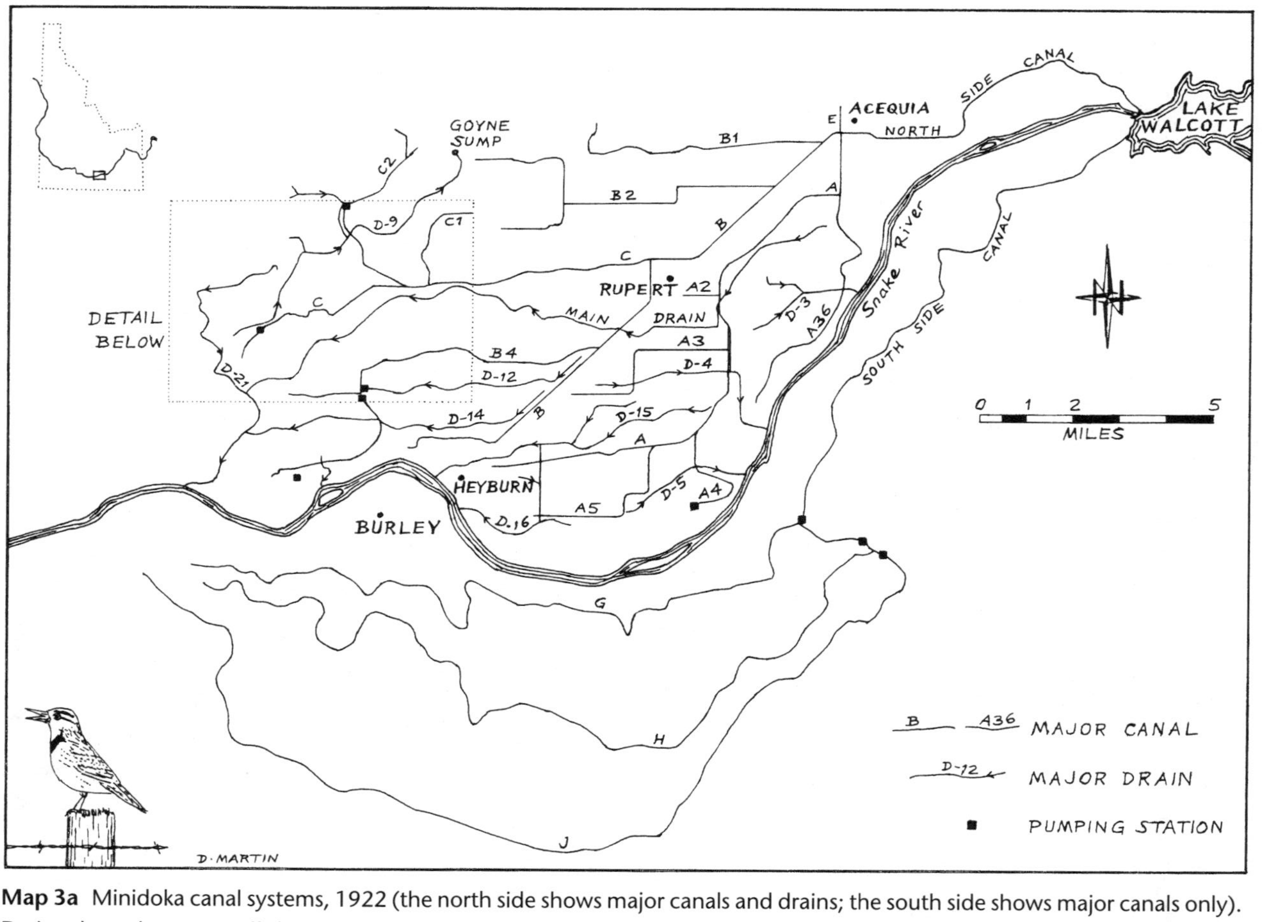

Map 3a Minidoka canal systems, 1922 (the north side shows major canals and drains; the south side shows major canals only). Drains drew the water off the project and conveyed it back to the Snake; however, project engineers also conveyed water through the D-9 drain to Goyne Sump, from which the water seeped through underground formations and back to the river. Note the names of the conduits, an industrial numbering system reminiscent of tracks in a railroad yard . *(Map by Dale Martin)*

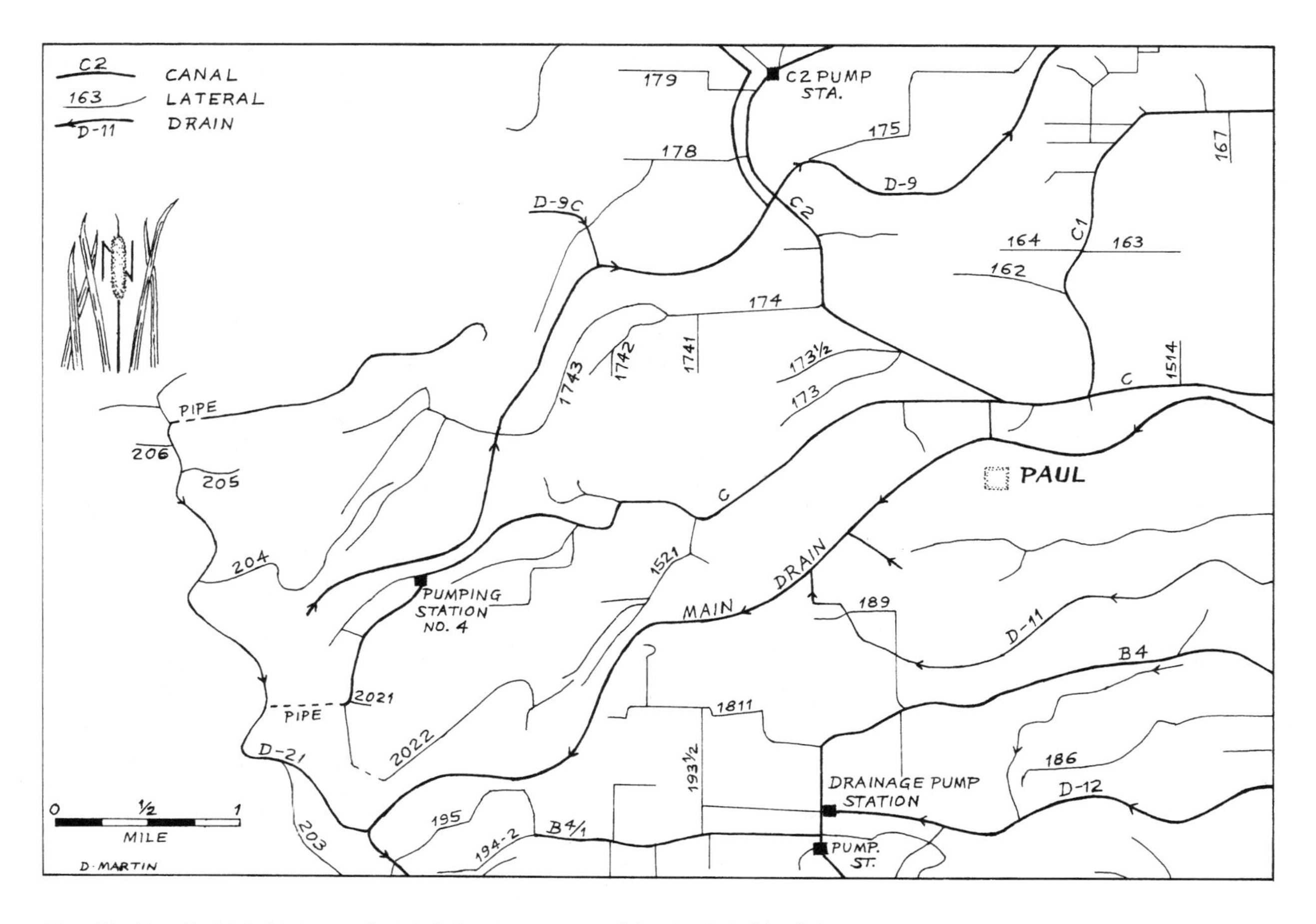

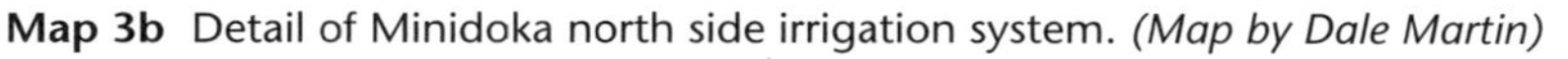
Map 3b Detail of Minidoka north side irrigation system. *(Map by Dale Martin)*

In designing and building drainage systems, engineers often returned to a technique that they had originally employed in the construction of dams and canals: they incorporated natural landforms and waterways. On the Twin Falls Land and Water Company's project, the coulees and creeks that bisected the tract gathered unwanted water. After 1912, project managers gradually installed wells, drain tiles, and tunnels to collect seepage and convey it to the coulees and creeks. On the north side of the Minidoka project, the Reclamation Service built drainage ditches that connected borrow pits used for canal construction with natural depressions and swales in the land. As early as 1910, engineers channeled water into Goyne Sump, a small, shallow, intermittent lake on the northwest edge of the project. Workers then sank a shaft, eventually six feet in diameter and ninety feet deep, through the sump bottom and down into the underlying volcanic rock. By the late 1930s, twenty-two cubic feet per second of drainage flow poured into the shaft, percolated through the "basalt and red clinkery lava," and eventually seeped into the Snake River canyon. In their optimistic moments, engineers and farmers liked to believe that the integration of hydraulic works with geological formations represented the fulfillment of a divine plan. But in the case of drainage systems, the irrigators could not be so certain. Now they were not so much following a divine blueprint as responding to the contingent behavior of a natural world that they could not fully control.[40]

Seepage had compelled irrigators to build drainage systems. And yet seepage was not just an environmental problem, and drains were not merely a physical response to it. Seepage was also a social problem: it affected important social relationships. And because seepage was a social problem, the drains that alleviated it were also more than just ditches: they were expressions of social action as well. Thus seepage led not only to alterations in the irrigated landscape—the excavation of drains—but also to changes in the irrigators' society.

Seepage became a social problem that contributed to social change when it quite literally undermined an established human institution: private property. As water seeped from canals and fields, as it flowed through the ground, it passed underneath property boundaries. Often, water from canals and individual farms on high ground infiltrated downward and waterlogged neighboring low land. Farmers who owned the waterlogged land could hardly be blamed for this seepage, nor did they have the resources to construct drains. On most irrigation projects, government officials, project managers, and farmers themselves recognized that seepage and drainage were community problems. In essence, seepage turned a landscape of private farms into a sort of hydrological commons. Usually, all the irrigators on a project or within the same canal system paid for drainage, even the farmers whose fields were not waterlogged or salty. Under Idaho law, irrigators also could petition to form drainage districts, which then assessed farmers to pay for construction.[41]

Farmers on high land, however, did not always accept their responsibility within the hydrological commons. During the 1910s and 1920s, some of them argued that they should not have to pay for drainage because the seeped lands were not theirs and they received no benefit from drain construction. In 1917, about 150 farmers in the Nampa-Meridian Irrigation District, a project in the Boise Valley, objected to a drainage assessment on their high lands. The Nampa-Meridian board of directors then petitioned for a judicial ruling on the legality of the levy. In 1918, the court approved the blanket assessment and also confirmed the district's power to apply the same per-acre charge to both seeped land and high ground. Law, which in the United States has often served as a bulwark against challenges to capitalist conventions and institutions, in this case helped to qualify the private property right. The irrigated landscape was not the sum total of the private lands that constituted it; seepage turned it into a hydrological whole, and irrigators at least tacitly recognized this fact.[42]

Drainage was the most tangible response to seepage, one that gave new form to the irrigated landscape and its human community. There was another response as well, although one that ultimately left no mark on the land and had no influence on the irrigators themselves. To further minimize seepage, engineers and government agents sought to devise a system for reducing the amount of water that farmers used. Indeed, the technocrats reasoned that more efficient irrigation not only would reduce seepage, it would also defuse conflicts over scarce water and leave more of it for new settlers to establish farms. But to get irrigators to use less water, engineers and public officials first had to determine exactly how much water was necessary to irrigate particular crops. To establish criteria, called the "duty of water," the U.S. Department of Agriculture conducted numerous tests on Idaho farms and irrigation projects during the 1890s and early 1900s. With a scientifically determined duty of water, engineers and other government officials could then tell the farmers exactly how much water they should apply to their crops.[43]

But the technocrats failed to achieve their goal. Farmers, afraid that the standards would rob them of their water rights, railed against the experiments. More important, nature frustrated the investigations. As Reclamation Service official I. D. O'Donnell explained in 1916, various environmental factors constantly fluctuated and made the duty of water an imprecise concept. He stated that under "fixed and artificial conditions, such as irrigating in tanks indoors," an exact duty could be established. But, said O'Donnell, any "conclusions" derived from laboratory tests "will not hold good in the big outdoors." There, soil, topography, heat, wind, "and more than 57 varieties of other natural controlling factors" affected the quantity of water needed to raise a crop. "Nature," he also pointed out, created "a continuous change of style in the seasons," so that basic climatic elements—including precipitation—were different each year. Nature, in short, was too variable for the duty of water to have a precise meaning. For an official in a government agency bent on the control of the natural world, O'Donnell provided a frank discussion of its intractability.[44]

Seepage, drainage, and the government's failure to establish a precise duty of water demonstrated that nature exerted a powerful influence on the irrigated landscape and its institutions. The irrigators had intercepted the natural movement of water, but they had not mastered the land and the hydrologic cycle. Fields and crops were not final destinations for streams; rather, they were but minor diversions in a much grander, uncontrollable passage of water over the earth. In attempting to co-opt the hydrologic cycle, irrigators and their hydraulic works had been co-opted by it.

It would be a mistake, though, to conclude that seepage always worked a hardship on farmers. The irrigated landscape was not that simple. In many cases, seepage benefited irrigators. Some farmers collected the water in wells and ditches and then channeled or pumped it back into canals for reuse. Seepage filled creeks and ravines that had previously run dry in the summer; irrigators then diverted this water. Irrigation water that percolated through the ground eventually made its way back into a few conduits. Some of the water, too, made its way back into creeks and rivers as "return flow." (Indeed, return flow actually kept some streams flowing more regularly.) Irrigation depleted stretches of rivers such as the Snake and the Boise; return flow filled them again downstream.[45]

Some irrigation projects came to depend on return flow (from ground and surface) to fill natural flow priority rights; in 1916, for instance, the Reclamation Service reported that return flow could supply all of the priority rights downstream from Star in the Boise Valley. Return flow even constituted an important element of legal cases concerning water rights. An example of such a case was *United States v. Boise Fruit Tracts Co. et al.* (1916). Boise Fruit Tracts sought to transfer water rights from the lower end of the Boise River to the head of the Farmers' Union Canal upstream. The federal government objected, contending that return flow supplied the water rights in question and that a transfer of these rights would jeopardize the water supply of appropriators below the Farmers' Union Canal. The court granted a permanent injunction preventing the transfer. In the Boise Valley, return and natural flow were virtually indistinguishable; irrigation, seepage, and river merged.[46]

Within irrigation projects, seepage and groundwater proved useful to engineers and farmers in ways that they probably did not realize. Deer Flat Reservoir in the Boise Valley, built by the Reclamation Service and placed in operation in 1909, provided an example of the inadvertent benefits that seepage furnished irrigators. To create the off-stream reservoir, the Reclamation Service erected embankments between low hills. No doubt the site seemed ideal, but the reservoir at first caused people to wonder about the skill of the government engineers. Deer Flat lost tremendous amounts of water to seepage in the first few years of its operation. The worst year was 1911, when about 135,000 acre-feet drained away. Then, by 1917, seepage had lessened dramatically, to roughly 21,000 acre-feet. Seepage had created a large groundwater body that could no longer absorb much from the overlying reservoir. Ironically, the reservoir water rested on the groundwater; the subsurface water body as much as the embankments kept water in Deer Flat Reservoir.[47]

Irrigators on the Egin Bench in the upper Snake River valley actually made a practice of manipulating groundwater levels to irrigate their crops. During the 1880s and 1890s, settlers moved onto benchland on the west side of Henrys Fork of the Snake River in the vicinity of St. Anthony. This area was about three miles wide and fourteen miles long; farmers eventually cultivated about 28,000 acres. The first irrigation water applied to the sandy loam covering the bench quickly drained away; the soil would not hold water like other soils. The farmers, though, found that this situation worked to their advantage. Water percolated down until it reached a layer of volcanic rock; with continued irrigation, the groundwater level rose until it reached the root zone of crops. Farmers realized that they could actually manage the groundwater level, keeping it within six to eighteen inches of the surface. They developed a system, called subirrigation, in which they did not apply water directly to their fields as irrigators commonly did. Instead they ran water through canals, into borrow pits along roads, into swales, and into special ponding areas; from these, water seeped down, causing the groundwater to rise. Irrigators channeled water through their own ditches to adjust the groundwater level to meet the exact conditions of their individual farms. The Egin Bench farmers used a tremendous amount of water, but subirrigation worked reasonably well for them. They still practice it today.[48]

Egin Bench subirrigation, furthermore, had unforeseen consequences for nearby groundwater formations. Water from Egin Bench eventually seeped west about twenty miles until it reached Mud Lake, an intermittent body of water. Around 1900, Mud Lake began to fill with water from Egin Bench; by 1927, it was a permanent lake covering 15,000 acres to an average depth of seven feet. Mud Lake then attracted irrigation companies, which started new projects with its water.[49]

Seeping irrigation water had one other inadvertent benefit: it infiltrated deep underground aquifers that irrigators eventually tapped. Engineers had recognized during the early 1900s that seepage augmented the aquifers and that wells sunk into them might furnish a substantial amount of water. They even described the aquifers in terms appropriate to their belief that the earth's features manifested an ultimate utilitarian purpose. In 1900, D. W. Ross stated that percolating water had "converted" the ground under the Boise Valley "into a large reservoir, absorbing like a sponge a large percentage of the water diverted through the canals. . . . This reservoir . . . is being filled without expense; it will be indestructible and will never fail to discharge its store at the season most needed."[50]

Idaho's greatest aquifer lay under the Snake River plain. Nearly two hundred miles long, twenty-five to fifty miles wide, lying on the north side of the Snake, the aquifer extended from the Ashton vicinity on the northeast to the Hagerman Valley on the southwest. Geologically, it consisted of loose volcanic and sedimentary rock layered between porous lava flows and permeated with water that trickled down from the surface. Rain, snow, and streams such as Henrys Fork, the Big and Little Lost rivers, and the Snake contributed water to the aquifer, which flowed southwest until it gushed from the dark volcanic wall of the Snake between Milner Dam and King Hill. Here, Thousand Springs and other founts fascinated and pleased the many travelers and tourists who stopped to look. But, ironically, the gawkers delighted in a splendid "natural wonder" that irrigation systems had helped to create. As hydrologists eventually discovered, the water that sprang from the cracks came not only from creeks, rivers, and precipitation; much of it had passed through irrigation systems in the upper Snake River valley, at places such as Egin Bench. The technicians took measurements, and over the years they noticed in the aquifer discharge an increase that paralleled the expansion of irrigated agriculture. In 1902, the Snake Plain aquifer discharged about 3,800 cubic feet per second; by 1917 the figure had risen to over 5,000 cfs; by 1956, approximately 6,000 cfs poured from the springs. Irrigation, the scientists and engineers estimated, had expanded the aquifer's flow approximately 60 percent, increasing its discharge by about 1.8 million acre-feet per year.[51]

Bloated with irrigation seepage, the Snake Plain aquifer then served as a water source for a new round of irrigated agriculture. In the late 1940s, high crop prices and electric pumps invited irrigators to sink wells into the aquifer, which by then contained 60 to 150 million acre-feet and which seepage had boosted within 70 feet of the surface in some places. Over the next three decades, farmers and agricultural corporations used the aquifer to bring about 1 million more acres into production. Another phase in the evolution of the irrigated landscape had begun. The newly developed areas, though, neither looked nor functioned like the older parts of Idaho's irrigated regions. The curving canals and ditches, the graded and patchwork fields—these were largely absent. Now the irrigators applied water with mechanical sprinklers, including huge, wheeled contraptions that, propelled by electric motors, circled endlessly around pivots. The sprinklers were essentially vehicles that rolled over the earth's irregular surface, showering huge expanses of land as they went.[52]

Of course, there was no assurance that the sprinklers would roll on forever. There was no guarantee that this part of the irrigated landscape would stay the same, or even that it would keep functioning at all. The farmers' deep wells and powerful pumps gradually drew more water from the underground reserve than seeped into it. As early as the 1950s, aquifer levels began to drop, making the water increasingly difficult and expensive to extract. It was a trend that would continue into the future, and it hinted at the potential of farm abandonment and the possibility that the sprinkled lands might one day cease to exist.[53]

Cessation of the sprinklers would demonstrate once again a basic characteristic of Idaho's irrigated landscape: this was a dynamic environment in which engineers and farmers seldom if ever achieved the control over nature that they desired. Formed in part by the continual interaction of hydraulic technology and hydrological processes, it was always in a state of change, always undergoing definition and redefinition. In a few instances the changes in the land benefited the irrigators—groundwater might actually prove to be a boon, not a bane. But most often this was not the case. Mostly the disturbances posed new challenges that engineers and farmers continually strove to overcome. The irrigated landscape, it turned out, was not about controlling rivers as if they were pipelines, not about arranging the land as if soil and water were building materials at a construction site, not about operating farms as if they were laboratories. This was a landscape of work, of constantly trying to impose a semblance of order on the land so that agricultural production could continue.

Confronted with this reality, irrigators had to find more appropriate images and metaphors to explain the land in which they made their homes. So it was with the engineer E. B. Darlington. When he reflected on the meaning of the irrigated landscape, he sometimes emphasized the ideal of control; it was he, after all, who had said that engineers were understudies of the Creator. But at other moments Darlington turned to less grand, more serviceable metaphors to explain the work to which he had devoted his life. After he had joined the Minidoka project as its superintendent, after he had labored on its drainage systems and developed methods of reusing seepage, he resorted to an alternate image. "Reclamation," he wrote in 1928, "is something like housekeeping. It is never finished."[54] Irrigators, Darlington had to acknowledge, did not master and complete the earth and run it with mechanistic precision. "Housekeeping" was his attempt to explain accurately a complex world in which humans constantly altered nature, "reclaimed" it, and then responded again and again to its unpredictable, uncontrollable changes. The process was perpetual and reciprocal, and it was consistent with one of the great natural forces shaping the irrigated landscape. In keeping with the unceasing round of the hydrologic cycle, the work of the irrigator was never done.

2/ **Habitat**

The Irrigated Landscape and Its Biota

With dams and canals, irrigators had begun the transformation of the Snake River valley into an irrigated landscape. They moved the process another step forward when they stripped the earth of its preexisting flora and fauna and supplanted them with crops and livestock. Sagebrush, greasewood, rabbitbrush, bunchgrass—these would give way to expanses of alfalfa, potatoes, beets, wheat, and other cultivated plants. And where jackrabbits, coyotes, and deer once roamed, the farmers would seed pastures and fill them with cattle, sheep, pigs, and horses. A new world would replace an old.

To the irrigators, this biological transformation seemed to fulfill the Creator's ultimate plan: that they reclaim the desert waste and live again in an Edenlike garden. It was a grand myth, and they expressed it in vivid, dramatic ways. On 2 March 1905, the Twin Falls Land and Water Company inaugurated its 200,000-acre irrigation project with a ceremony at Milner Dam on the south central Snake River. Surrounded by onlookers, company officials raised the gates to the Twin Falls Main Canal. "As the water flowed into the big canal to be carried out to transform and redeem a desert," the publicist J. B. Warrington later recalled, "cheer after cheer was given and bottles of wine were broken over the gates." In 1920, Irene Welch Grissom, operator of a Snake River valley irrigated farm and Idaho's official poet laureate, told the story in a quieter but no less profound way, in "The Desert Reclaimed": "Where sagebrush once grew rank and dry / And lean gray wolves howled to the night / Vast tracts of fertile farm lands lie / Green clad in summer's golden light." In these and other forms, in stories, engineering reports, memoirs, and in the plain speech of everyday life, the irrigators celebrated the demise of a gray howling desert and the creation of a green, bountiful landscape.[1]

But there were problems with these scenarios. Just as the irrigators' metaphorical formulations could not contain the untidy reality of hydrology, so the garden myth could not be reconciled with the messy facts of biology. Just as complete control of water and earth eluded irrigators, so did complete control of living things. God bade them subdue this land and make it fruitful, and in doing so he gave them the chance to restore some semblance of humanity's original Edenic condition. But, as Idaho irrigators would learn, the garden myth could not explain the realities of the Snake River valley as they experienced it. They could not arrest, and so escape, the course of secular time. They could not control flowering, crawling, flying, wandering, always growing and moving organisms; they could not stop biology and history. This land was a place of profound ecological change that the irrigators could not halt.

The Snake River valley had been undergoing change for thousands of years before the irrigators appeared on the scene. Centuries of aridity that followed the Ice Age eliminated some plants and animals while encouraging better-adapted species. Forests retreated; mammoths, giant bison, horses, large cats, sloths, and other megafauna died out completely. In turn, dryland plants such as sagebrush and grasses advanced, along with a smaller, tougher complement of animals: pronghorn antelope, a new bison, sage grouse, deer, jackrabbits.[2]

Humans first entered the Snake River valley as many as 15,000 years ago, and they made still more changes to the region's flora and fauna. How much effect they had on the biota is not entirely clear, but one thing is certain: irrigation farmers and engineers did not come to a land unaltered by people. Like environments elsewhere on earth, the post-Pleistocene Snake River valley was, to borrow archaeologist Steven Simms's words, a "human wilderness." Although scholars debate the matter, the first peoples' hunting practices probably contributed to megafaunal extinction. No doubt, too, these folk altered the land through their use of fire. Subsequent inhabitants of the area continued to shape valley life. They hunted and burned, and by 1700 they adopted an exotic species, the horse, that may have competed with established grazers.[3]

In the nineteenth century, new peoples migrated into the Snake River valley and accelerated the human modification of the natural environment. Between 1810 and the 1830s, fur trappers extracted beaver pelts by the thousands. To drive away its American competitors, the British Hudson's Bay Company attempted with some success to turn the Snake River basin into a "fur desert." From the 1830s to the 1860s, the horses, mules, and oxen of overland migrants consumed grasses that flourished in the shadow of the sagebrush and in the moist stream bottoms. In 1863, cowboys herded about 300 head of cattle into Idaho, and such drives continued into the 1870s. The cattle grazed so heavily that by the winter of 1889–1890 insufficient forage remained, and thousands of starving animals perished in the cold.[4]

When irrigation farmers arrived in the Snake River valley with their crops and livestock, they added another level of change to an already changing land. They did not so much conquer the desert and create a mythic garden as transform the land into a strange new habitat, a hybrid mixture of artifice and nature. In Idaho's irrigated landscape, agricultural plants and animals blended with flora and fauna that farmers generally considered wild. The result was a landscape that often resisted the irrigators' efforts to control it. A new world indeed replaced an old, but it was not one that the human inhabitants of the irrigated landscape had fully anticipated.

The development of this new world and its biota can be traced by following the flow of water across the landscape, from rivers, through canals, across fields, into drains, and back again. On rivers, dams and reservoirs did not merely catch and store water; rather, they produced new environments in which human actions became tangled with uncontrolled flora and fauna. This ecological transformation began with destruction: when engineers and farmers intercepted the hydrologic cycle, when they built dams and canals, they ruined the habitat of fish and other wildlife. But destroyed habitat for some species became, paradoxically, enhanced habitat for others. This fact runs against the way that modern Americans tend to see land and water; our impulse is to see human-altered environments as "biological deserts." But in fact, when humans disturb the environment, they almost invariably make it better for certain flora and fauna; destruction and creation go hand in hand.[5]

Dams created reservoirs that inundated the riparian habitat of terrestrial animals. But those same reservoirs provided a new habitat that attracted many kinds of birds. During the early twentieth century, thousands of birds congregated around Idaho reservoirs—American Falls, Lake Walcott, Lake Lowell, and others—and here they found resting and feeding sites and, on occasion, places to nest. The Minidoka project's Lake Walcott, for example, which began to fill with water in 1909, soon attracted an array of avian species, including white pelicans, great blue herons, cormorants, ducks, and numerous other birds that frequented lakes and wetlands. Off-stream reservoirs, fed by canals, similarly drew thousands of birds. Lake Lowell, in the Boise Valley, first filled in 1910; according to T. S. Palmer of the U.S. Department of Agriculture's Bureau of Biological Survey, "The following autumn blue herons, ducks, geese, swans, and other birds appeared in considerable numbers." In the spring of 1911, the Biological Survey estimated that ducks, coots, and grebes had built between 150 and 200 nests along the shore.[6]

A similar pattern of devastation and creation occurred in the habitat of fish. Irrigation turned rivers into deathtraps for trout, salmon, and other fish that needed cold, clear water. Dams hindered migration and spawning runs, and they slowed water and raised temperatures to intolerable levels. Warm water, suffused with nitrogen and phosphates from agricultural runoff, promoted the growth of plants such as algae; in turn, the algae consumed oxygen that the fish needed to survive. Irrigation diversions, especially during drought, exacerbated the harsh conditions, as did farm pesticides that eventually filtered into streams. Irrigation systems also trapped fish. A trout that swam into a canal might never get out; death in a farmer's field might be its fate. Yet while trout and salmon died, other fish thrived. In the new world of Idaho rivers, in the warm, deoxygenated, muddy water that stifled salmon and other cold-water species, fish such as Utah chub and suckers proliferated.[7]

Irrigators gave little if any thought to habitat destruction, which they probably saw as a regrettable but necessary consequence of their quest to tame the Snake and other streams. Of greater interest to them were the animals that flourished in the new environments of rivers and reservoirs. Government officials focused their attention on the birds. In 1909, an executive order established the Minidoka Bird Reservation at Lake Walcott and the Deer Flat Bird Reservation at Lake Lowell. The Bureau of Biological Survey managed the sites, which included not only the reservoirs but adjacent land along the shorelines. Of the two refuges,

Minidoka was more satisfactory. Lake Walcott's water level fluctuated less than Lake Lowell's; thus, Lake Walcott provided better habitat for nesting and foraging. By the early 1940s, the U.S. Fish and Wildlife Service had enhanced the refuge by installing special cut-off dykes that prevented water from flowing out of shallow arms and inlets when the reservoir level dropped. The bird habitat at Lakes Walcott and Lowell clearly demonstrated the hybrid nature of the irrigated landscape: these were wildlife refuges, "natural" areas, on man-made bodies of water.[8]

The mixed-up quality of this new world extended into the water itself. While government officials were busy monitoring birds and enhancing their habitat, irrigators were stocking the streams and reservoirs with their favorite game fish. Anglers with a liking for a particular species simply transported the fish to the water and released them. In 1917, for example, L. N. Huggins "planted" bass and perch, two fish that liked reservoir habitat, in Lake Walcott. Irrigators perhaps did not directly associate their reservoirs with myth, but fish stocking showed that the two were at least compatible. The dam had tamed the river; now the reservoir, like a fresh pasture or newly plowed field, could be planted and turned into the aquatic equivalent of the terrestrial garden.[9]

The connection between reservoirs and the garden myth became more clear when irrigators planted trees and created parks along the shorelines. Groves of cottonwood, willows, poplars, and other trees provided shady picnic sites where farm families sought respite from work and the hot summer sun. On the shore of Lake Walcott, near the Minidoka Dam powerhouse, the Reclamation Service set up an experimental farm that included trees and grass. Local settlers discovered that the site was ideal for summer recreation, and they began to congregate there on weekends and holidays. In 1917, Barry Dibble, the project manager, noted with pride that 400 to 500 people might come to the park on one of these days. Irrigators may not have overtly likened the sylvan glades to the biblical garden, but when they took their repose in the cool shade they certainly acted out a pastoral idyll that harkened back to the lost Eden.[10]

Yet all was not well in and around the reservoirs that the irrigators had fashioned. Certain species eventually turned unruly and upset the desired environmental order. Farmers cursed the birds, which, while flocking to the reservoirs, swooped down and fed on nearby grain fields.

Even species that the farmers deliberately planted ran out of control. No aquatic organism better illustrates this than the carp. During the 1880s and 1890s, Idahoans brought these mud-colored fish to the irrigated landscape with the intention of raising them in ponds for food. In a sense, the carp was the aquatic equivalent of the hog—its virtue was its ability to grow a prodigious amount of flesh in a short time. The farmer's carp pond, moreover, was the aquatic parallel of the hog pen. But within a short time, the fish did what some hogs are prone to do—run wild. Both by accident and by intentional introduction, carp made their way into Idaho rivers and there flourished. Not only were they ideally suited to murky, warm water, but they helped to create their own habitat. Voracious feeders, they edged out other fish, and by rooting in the mud for plants, they stirred up sediments that made the water intolerable for species needing clear water. Idahoans never developed a taste for the carp's bony, bland white flesh, and they came to detest the animal because it competed with more desirable game fish. But the people of the Snake River valley eventually discovered that they could not eliminate carp from their rivers and reservoirs. Manipulated and introduced by the farmers, carp became the dominant fish in a hybrid environment over which humans exercised incomplete control.[11]

This new environment, this new biotic world, grew and ramified as water flowed from the rivers into the irrigators' canals. These conduits did not just carry water to the farms; open to all of the biological processes at work in the Snake River valley, they became habitat. An astonishing array of plants spread through them. With the plants came an equally astonishing complement of animals. Canals became a unique aquatic and riparian environment, home to a biota that often baffled the irrigators and caused them a great deal of trouble.

When water passed from streams, through headgates, and into the canals, it carried uprooted aquatic plants and their seeds. As the current slowed, roots, stems, and seeds settled in the warm mud at the bottom, and they began to grow. And as they grew and matured they shed seeds and parts again and again until, within a few years, the plants spread through all reaches of canals and smaller ditches. Farmers and engineers apparently knew little about these flora, and they categorized them generally as moss, although little if any actual moss grew in the canals. What was "moss," then? Certainly it included algae—long, limp, green filaments of it waving languorously in the current. At least one species of pondweed, likely more, flourished here as well. These plants put out stems, several feet in length, with long, narrow, pointed leaves; and just under the surface of the mud, each plant sent out rhizomes from which sprouted still more stems. Elodea, or waterweed, a plant with whiplike stems and tiny leaves, probably grew in thick patches in the canals. Bulrushes and cattails, rooted in the mud, rose in dense stands above the waterline.[12]

As aquatic vegetation filled the canals and ditches with lush greenery, terrestrial plants spread across their banks. The irrigators were partly responsible: the soil of new canal banks was soft, uncompacted, and easily eroded; canal managers tried to stabilize it by planting grass and willows. Here was an intentional conjunction of nature and artifice: sod and tough tree roots would hold the hydraulic technology together. Yet grass and willows were not the only plants that grew here. An unplanned flora took root as well. Carried by wind, irrigation water, and the fur and digestive systems of wild animals, seeds of numerous kinds of forbs sprouted in the exposed soil. Cultivated plants, such as alfalfa or sweet clover, migrated from adjacent pastures and fields. And so, too, did plants for which farmers had no use: weeds. The banks of newly constructed ditches and canals provided ideal ground for the growth of weed seeds. Weeds of all sorts probably began to sprout along canals and ditches shortly after irrigators excavated and built the conduits. But these plants first drew widespread attention in 1894, when Robert Milliken, secretary of the Idaho State Horticultural Society, identified rank growths of Russian thistle—one of the classic tumbleweeds—on the banks of the Phyllis Canal near Nampa.[13]

Overspread with vegetation, canals and ditches provided ideal cover and food sources for a host of animal life. Birds flew over them and lighted on the banks, and they gorged on seeds and the many kinds of insects that lived on the ground and in the water. Mice, which also consumed seeds and insects, constructed miniature runways through the grass. Ground squirrels and pocket gophers tunneled into the canal banks, where they consumed roots. Yellow-bellied marmots, "rock chucks," ate grass and forbs and made their homes where canal construction had produced piles of rock. Mice, squirrels, gophers, and marmots in turn attracted mammalian predators. Coyotes and foxes roamed the canals in search of the rodents. Skunks added mice to their diet of insects and carrion. Badgers burrowed into banks in search of their underground prey.[14]

An aquatic fauna accompanied the terrestrial animals. Fish made their way into canals; irrigators sometimes angled for them from the banks. Shellfish moved in as well. The bed of the Snake River was lined with freshwater clams, reported the *Twin Falls News* in 1906, "and they have found their way into the irrigation ditches of the Twin Falls tract. School-boys are in the habit of digging huge clams from the little ditch which passes by the Bickel school. In all the ditches where water has run for any length of time clams may be found embedded in the banks and bottoms."[15]

Canals were coming to resemble nothing so much as the rivers from which they drew water; no wonder, then, that they attracted beavers and muskrats almost immediately. These animals recognized no distinctions between creeks, rivers, sloughs, and irrigation systems, especially when their favorite plant foods began to grow in the canals and ditches. Muskrats thrived on the fleshy, starchy roots of cattails that grew in irrigation systems, and they built their houses in the canal banks. Al Peters, a former manager of the Twin Falls Canal Company, recalled how an attempt to stop erosion inadvertently enhanced muskrat habitat. Until 1947, the company made a practice of laying sagebrush on the inside of canals, holding it in place with wire fence material and stakes. "The brush was a haven for muskrats," Peters said. "You could see 30 to 35 rats swimming in the canals in less than a mile." Beavers, too, liked canals and ditches; willow trees and other vegetation provided the animals with nourishment as well as building material for their characteristic lodges and dams.[16]

Hydraulic technology, the canals, had filled with a rich flora and fauna. Perhaps the irrigators found this new biota compatible with their myths. Perhaps on a summer evening they gazed in awe and admiration as the setting sun swathed their fields and canals in its slanting rays, and breezes swayed the willows and grasses on the banks, and swallows dived over the tranquil water. Perhaps at such moments they thought: the dream is realized. Perhaps, too, their children, when they fished and swam in the canals and tramped along the banks, lived out their own form of the Edenic fantasy. But most often the irrigators did not experience these epiphanies. The hybrid biota, which otherwise might appear so beautiful as part of a pastoral setting, disrupted water delivery and quite literally undermined the physical integrity of the canals and ditches. From the moment that they had completed the conduits, the irrigators had begun to lose biological control over them.[17]

As the plants spread, they interfered with the flow of water, a severe problem in an arid land where farmers put a premium on irrigation. The sheer mass of aquatic vegetation drastically reduced canal and ditch capacity, blocking the flow of water to the farms. The plants trapped silt, further diminishing the amount that a system could carry. In 1911, irrigation expert W. M. Wayman reported that "moss . . . will obstruct the flow to such an extent that not more than half the ordinary capacity of the ditch can be delivered." The aquatic plants contributed to other problems as well. They slowed water movement, increasing ground absorption and thus seepage. They interfered with the measurement of water in flumes and weirs, crucial operations that determined how much water irrigators would receive. Finally, they transpired water in large quantities; moisture intended for crops went into the growth of an aquatic flora that the farmers in all their mythic dreams never imagined.[18]

And these species were only half the problem. Terrestrial plants fouled the conduits, too. Tumbleweeds clogged canals and ditches, and grass growing on the banks flopped over into the water, further inhibiting the flow. Indeed, terrestrial plants could so choke a ditch that they caused it to overflow. Some plants sank their roots into the banks and absorbed water that might have gone to crops. Perhaps the most troublesome water-using "weed" was one that irrigators had deliberately introduced to the canals: willows. Water-loving phreatophytes, willows spread quickly in the moist riparian soil of the canals. Their dense, tough roots held the canal banks in place, but they exacted a price in the tremendous quantities of water that they took up. And whenever the irrigators cut the trees down, the roots immediately sent up new shoots.[19]

But it was the mammals that posed the greatest threat to the canals. Muskrat houses and the lodges and dams of beaver interfered with the flow of water. Even more devastating were the burrowing habits of these and other species. Seeping water penetrated mammal burrows and saturated the surrounding soil; canal banks gradually weakened and collapsed. Difficult and costly to repair, canal breaks could spell disaster for farmers. Breaks flooded fields and could leave farmers without water for days. In July 1923, ground squirrels caused a serious washout on the main conduit of the Aberdeen-Springfield Canal Company system. On one section of canal consisting of earthen fill, the "squirrel holes had been giving some trouble and had been the occasion for anxiety." When the bank finally collapsed, water surged out and cut roads, washed away railroad grades, and flooded several farms. Workers had difficulty closing the headgates at the Snake River, and water continued to flow from the breach for twenty-four hours. The destruction halted water delivery for about a week, until the canal company rebuilt the structure. The break also interrupted railroad service on the Oregon Short Line's Aberdeen branch for six days. When laborers repaired the railroad bed, they missed a section of embankment that the flood had weakened. As the first train passed over, the bed crumbled, and the locomotive tumbled into the water. Collapsing technology—canal and railroad—underscored the vulnerability of human systems in the irrigated landscape. Ironically, small burrowing mammals endangered the very waterways and machines that for many people symbolized man's ability to dominate the natural environment.[20]

While the hybrid biota grew tangled and troublesome on the canals, a related biological revolution worked itself out on the land that was to be the final destination of all that water: the farms. Settlers cleared the sagebrush and other desert plants in great swaths, with horse-drawn cutters, pieces of steel rail, and heavy, toothed rakes. Then they stacked the vegetation and burned it. Next came grading, plowing, ditch digging, and planting; and at last it was time for irrigation. Farmers diverted water from main canals into their individual ditches and then, by shoveling openings in the banks or by raising small headgates, allowed it to flow across the fields. The results were indeed profound. Where desert plants had stood but a few months or even weeks before, green pastures and row after row of sprouts flourished. Yet this change represented more than simply the conquest and reclamation of the desert. Rather, farming involved a relentless struggle between what the irrigators sought to grow and what nature would allow. Shifting environmental conditions, including climate, soil, and availability of water, always limited the crops that farmers could raise. And when the farmers cleared land, dug ditches, planted seeds, and opened headgates, they unwittingly created more habitat for organisms that often disrupted the very agriculture they were trying to establish. The end product of this human-nature interplay was less an ideal garden than something far more complicated: a hybrid landscape in which environmental control was, at best, temporary.[21]

Climate and soil were basic environmental factors that always shaped the farmers' crop choices. Although nominally arid, the Snake River valley was not uniformly hot and dry. Seasonal temperatures and the length of growing season varied considerably. Farmers in the upper valley, around the town of Ashton (elevation 5,240 feet), had fewer than ninety frost-free days in which to grow their crops. By contrast, irrigators in the Boise Valley (elevation about 2,700 feet) enjoyed a growing season of roughly 180 days. Soils varied also, but generally shared a common characteristic: they contained relatively little humus and nitrogen. Hoping to find the crops that grew best under these conditions, irrigation farmers experimented: they tried cotton, tobacco, cranberries and strawberries, figs, peas and beans, hay, grain, potatoes, lettuce and onions, and several kinds of tree fruit; some irrigators even considered raising rice and cactus.[22]

The farmers of the Snake River valley might have interpreted this quest as an indication of nature's power, as a sign that they could grow crops only on nature's terms. But this was not always their attitude. Rather, they often viewed the search for the best crops as part of their overall effort to realize the inherent, God-given potential of the land. Indeed, much as the irrigators found "natural" sites to build dams and excavate canals, they looked for places that were "naturally" suited to the production of certain kinds of crops. Even preexisting vegetation indicated the best places to raise those plants; settlers believed that stands of big sagebrush signaled soil fertility.[23]

By the early twentieth century, irrigation farmers had made their choices, and they began to concentrate on a few crops. In part, irrigators in southern Idaho emphasized hay, and especially alfalfa. This was a popular crop throughout the irrigated landscape; it was, as one farmer called it, the "backbone of the country." One reason farmers liked the plant was that it improved the soil for other crops. It collected nitrogen from the air and stored it; when plowed, it passed the nitrogen as well as organic matter—humus—into the soil. The popularity of alfalfa also stemmed from a mutually supportive relationship that developed between irrigated hay production and livestock raising. During the 1880s and 1890s, cattle and sheep overgrazed Idaho and Great Basin rangelands. With little forage left to support their economy, stock growers began purchasing nutritious, nitrogen-rich alfalfa hay from Snake River valley farmers. Thus by the early 1900s, alfalfa fields were everywhere in the irrigated landscape, acre after acre of the dark green plant, its long stems sprouting clover-like oval leaves and distinctive purple blossoms.[24]

Wheat and other grains constituted a second major crop in Idaho's irrigated landscape. Wheat grew relatively well in the newly cultivated soils that lacked humus and nitrogen, and it required far less water than other crops, particularly during the late season when water was scarcest. These characteristics made the crop especially appealing to the first settlers in Idaho's irrigated lands. Their soils were as yet unimproved by alfalfa culture, and the supply of water from small or new irrigation systems could be especially uncertain. But even in later years, farmers tended to raise more grain during droughts. Sometimes this tactic had the added benefit of leaving wheat farmers with an actual surplus of water. In the drought year of 1919, for example, farmers with matured wheat shared their excess water with neighbors whose crops were shriveling in the heat.[25]

Alfalfa and wheat were basic crops, dependable but usually not highly profitable. To offset the generally low value of hay and grains, irrigation farmers in the Snake River valley turned to more lucrative crops. Sugar beets and potatoes were two of the most important. Sugar beet culture in Idaho rapidly expanded after 1900 under the direction of the Utah-Idaho Sugar Company and other corporations. Southern Idaho's soils and climate—its high altitude, warm days, and cool nights—also favored potato cultivation. The Russet Burbank, which resisted disease and flourished in the Snake River valley environment, was widespread by the 1930s. Sugar beets and potatoes required intense labor and irrigation, so farmers usually grew them in combination with hay or wheat, which needed less care.[26]

Although these predominant crops did well in southern Idaho, farmers still had to tailor cultivation to conform to local environmental conditions. Irrigators in the upper Snake River valley, for example, once grew sugar beets, but the cooler climate and other environmental factors prevented them from producing high yields. In place of beets, the farmers took up potatoes, including the production of seed potatoes. The altitude, brief growing season, and cool weather yielded vigorous potato seed and arrested the growth of diseases. By the 1920s, the upper valley was developing into an important source of seed potatoes for the rest of Idaho and other potato-growing regions in the American West.[27]

In addition to their leading crops, farmers discovered that the Snake River valley environment favored certain forms of specialized production, including seeds. Farmers would pay good prices for healthy crop seed that germinated well; growers who could raise this seed received ample monetary reward for their efforts. When prices for wheat and other crops collapsed after World War I, seed production took on special importance for Idaho irrigators, who became specialists in the production of alfalfa, clover, bean, pea, sugar beet, potato, and other vegetable seed. In part they succeeded because of environmental conditions. Farmers on the Aberdeen-Springfield irrigation project, for example, raised a substantial portion of the nation's Grimm alfalfa seed. A particularly winter-hardy variety, Grimm was in much demand from farmers on the northern Great Plains. Henry K. Wiley introduced the seed to the Aberdeen-Springfield area in 1907 after learning about it from Professor Paul K. Blinn of the Colorado Agricultural Experiment Station. Wiley and other farmers found that Grimm alfalfa survived Idaho's cold winters and flourished on their sloping, well-drained land. And in periods of water shortage, the crop provided an added bonus: seed maturation required less water than plant growth for forage. By 1925, Wiley and his associates were producing 25 percent of all the Grimm seed in the United States.[28]

Idaho's irrigation farmers also secured a place as seed producers because agriculture was new to the Snake River valley. By the twentieth century, various weeds, diseases, and insects plagued fields in warmer, more humid, and older agricultural lands in the nation. Idaho's relative isolation from these areas and its "clean" fields provided an environmental haven for seed production. Pea seed companies, for example, moved westward to flee the pea weevil, leaving infested fields behind. In Idaho, the upper Snake River valley provided safe ground for seed production. Long, hard winters here made farm fields inhospitable for the pea weevil. Rogers Brothers introduced seed pea production to the area in 1906; by the 1920s, numerous midwestern and eastern seed companies had established themselves in and around Idaho Falls, St.

Anthony, Rexburg, and Ashton, where farmers cultivated thousands of acres of peas. By 1922, Fremont County alone produced one-third of all the seed peas in the United States.[29]

Over the short term, farmers achieved remarkable success in matching agricultural production to environmental conditions in the Snake River valley. But the plants and cropping systems that brought them wealth simultaneously produced changes in the land that in turn altered the environmental basis of irrigated agriculture. While irrigation fostered crops, it inadvertently created a vast habitat for weeds, the plants that humans found undesirable or sought to exclude from cultivated ground. The settlers who at first saw their land as new and clean, as the realization of the garden, soon found it infested with weeds and other agricultural pests.[30]

The biological characteristics of weeds suited the plants to most agricultural environments, including the irrigated landscape. Many weeds had evolved in conjunction with Eurasian modes of farming. Plowing and grazing created opportunities for them, and they thrived alongside the cultivars. Indeed, some weeds botanically resembled crops, and ancient farmers once actually raised them. It was these croplike characteristics that often ensured their survival in cultivated fields. Farmers, for example, inadvertently harvested weed seeds along with the crop seeds that they resembled. Weeds also tended to be vigorous plants that flourished under conditions that weakened many crops. As crops withered in dry, waterlogged, salty, or nutrient-poor soil, weeds took over. Weeds also reproduced prolifically, and they took advantage of every opportunity to spread their seeds, through wind, water, animal fur and excrement. And perennial species produced tough roots that resprouted after being chopped into pieces. Not all weeds originated in Europe or Asia, but even the American species shared with the Old World plants the biological characteristics that adapted them to agriculture.[31]

Eurasian weeds entered the Snake River valley through several routes. The grain stocks that the first irrigators brought to Idaho probably contained weed seeds. Subsequent importations of crop seed, especially from places such as central Asia, often carried them. Beginning in the 1880s, railroads opened markets for farmers and also provided a means for weeds to enter the Snake River valley. In the 1890s, University of Idaho botanist L. F. Henderson observed that new weeds often appeared along railroad lines, at Nampa, Mountain Home, Boise, and other towns. He surmised that cattle-car offal, baled hay at construction sites, and sweepings from freight cars contributed to the spread of weed seeds.[32]

Once weeds entered the irrigated landscape, a combination of farming practices and changes in the environment assisted their proliferation. When farmers cleared large acreages, they removed the sagebrush that had checked the movement of the tumbleweeds: Russian thistle, tumbling mustard, and tumbling pigweed. The farmers had created, in effect, a tumbleweed racetrack. When the wind broke off the mature, dried plants at ground level and blew them over the bare soil, they scattered their seeds by the millions. The weed problem only became worse in land altered by seepage, waterlogging, and the buildup of salts in the soil. Species such as seepweed and foxtail barley almost immediately spread into wet, salty patches of soil, even in the midst of alfalfa or some other crop. Intensive grazing on waterlogged or salty pasture also encouraged the expansion of foxtail barley. Livestock could eat foxtail when it was young and fresh, but the sharp, spiked head of the mature grass produced sores in the eyes, mouth, and stomach. Consequently, the animals tended to avoid foxtail in favor of more palatable grasses. As they overgrazed the more attractive species, they created more openings for foxtail to expand.[33]

The market also helped to spread weeds. In 1875, D. W. Fouch, a farmer in the Boise Valley, reported declining crop yields. He attributed the situation to three factors: continuous cropping with neither rotation nor fertilizer; weeds such as cockle and wild oats overrunning fields; and lack of water due to the Boise River's failure to rise in May and June and then a subsequent period "of great heat." Responding to the intense demand for food from Idaho mining centers, farmers such as Fouch produced crop after crop without trying to restore the soil. Poor soil and then drought weakened wheat and other crops, and the resulting patchy growth provided openings for weeds that thrived in poor, dry, disturbed soil. In turn, without the rotation or tillage that would control them, the weeds proliferated through the fields. The interplay of the market and nature that encouraged weeds in the 1860s and 1870s continued into the twentieth century. During the 1910s, economic conditions helped to propel an expansion of weeds that threatened Idaho's nascent seed industry. In attempting to cut costs and make as much profit as possible, seed farmers adopted practices that favored weed spread. Rather than buy seed that was certified pure, they used their own seed year after year, the proportion of weed seeds in their stocks steadily mounting. In some instances they screened their seed to remove inferior seed and weed seed; to save money they planted the screenings. Other farmers purchased untested seed because it was less expensive. As farmers planted more and more weeds they increasingly endangered the productivity of their fields.[34]

But the most insidious means by which weeds spread into fields was through the one thing that Idaho farmers most valued: water. The very substance that purported to give humans the power to conquer the desert and turn it into a garden served as an avenue for the spread of weeds. Any weeds growing close to water—along rivers, next to canals and ditches, and on the farms themselves—cast their seeds into the current, which took them through the irrigation systems and onto the open soil of fields. The weeds dropped their seeds in mind-boggling quantities; researchers at the Colorado Agricultural Experiment Station calculated that millions could pass through a twelve-foot ditch in twenty-four hours.[35]

And then there were the crops that, ironically, became weeds. Alfalfa and sweet clover, for example, moved outside hayfields and pastures and grew along fencerows, irrigation canals, and drainage ditches. From these areas the plants moved into cultivated fields where the farmers did not want them. Barnyard grass provided outstanding hay and forage, but was "a great curse in many of the irrigated fields and orchards" in the Snake River valley. As Alexander McPherson, an agricultural expert for the Twin Falls Land and Water Company, put it in 1906: "A weed may be a weed in one locality and a plant in another, or it may be a plant which when cultivated is of use to man but becomes a pest when allowed to run wild." There could be no better illustration of the hybrid nature of the irrigated landscape than crops that became weeds.[36]

Weeds disrupted the environmental control that farmers sought to establish and maintain in their fields; so, too, did an array of animals. Among other fauna, insects found outstanding habitat here. Certain insects were specially adapted to Eurasian crops. Some species, native as well as introduced, flourished in both crops and adjacent uncultivated areas like the weedy peripheries of ditches, canals, fencerows, railroad tracks, highways, and stream bottoms. Periodically insects migrated from adjacent vegetation into farm fields, often serving as disease vectors in the process.

Perhaps the alfalfa weevil best exemplified those foreign species that came to inhabit particular crops. The initial isolation of the irrigated landscape gave Idahoans a brief holiday from the weevil. By the early 1900s, farmers had been raising alfalfa in the Snake River valley for decades, apparently free from any insect infestation other than native grasshoppers and crickets. But as modern transportation and economic networks linked Idaho more directly to the wider world, opportunities arose for nonnative insects to migrate into the region. Around the turn of the century, a shipment of furniture packed in alfalfa arrived in Utah from Italy. This seemingly innocuous material carried the alfalfa weevil. People noticed weevils in the vicinity of Salt Lake City in 1904; by 1916 the insect had reached southern Idaho. By the 1920s, it had infested fields through much of the state's irrigated landscape, reducing yields as much as 50 percent. It thrived in particular in old alfalfa stands, some of which had been in place for as long as twenty-five years. Ironically, the same clean conditions that temporarily sheltered Idaho's alfalfa fields also favored the alfalfa weevil. Emerging from the furniture packing, it entered a world that lacked the enemies that preyed upon it in lands across the Atlantic.[37]

But not all insects that wreaked havoc in Idaho's irrigated landscape were from Eurasia. There were, for example, native grasshoppers and Mormon crickets. Foothills and rangeland and the plants that grew there constituted these insects' preferred habitat and food. But periodically, the grasshoppers and Mormon crickets moved off the range to ravage irrigated crops. Exactly why they did so is unclear. Farmers and entomologists speculated that overgrazing by livestock deprived the crickets of their normal food and drove them to eat crops, especially grain and alfalfa. Scientists also believed that the insects may have moved into cropland when range vegetation dried up. But whatever the cause of the migrations, the sight of millions of oily black crickets slowly crawling and hopping toward irrigated fields stirred feelings of fear, revulsion, and hatred among farmers.[38]

The danger from grasshoppers and crickets was their voracious appetites; other insects, were dangerous because of the crop-destroying diseases that they carried. A combination of ecological factors, for example, drew the beet leafhopper and the curly top virus to the irrigated landscape's sugar beet fields. During World War I, high wheat prices and ample rainfall induced settlers to attempt dry farming and open new land for irrigation on thousands of acres around Twin Falls and on the Snake River plain west of Rexburg, Blackfoot, and American Falls. When prices fell and precipitation lessened at the close of the war, the settlers abandoned their lands. Tumbling mustard, green tansy mustard, pigweed, red scale, Russian thistle, and other weeds soon covered soil once devoted to crops. Leafhoppers thrived on the fresh vegetation. But each year when summer came and the weeds matured and died, the leafhoppers had to search for a new habitat. They found one in nearby sugar beet fields. As they migrated they spread the curly top virus, which by the 1920s had devastated thousands of acres of sugar beets in Idaho and other parts of the American West. Dying beets encouraged a major restructuring of the sugar industry; Utah-Idaho Sugar and other companies moved factories and closed others, in part because of losses from curly top.[39]

Insects were not the only wild creatures to inhabit the irrigated fields. Various birds and mammals also moved in. Farms ruined the land for sage grouse and other upland game birds but provided excellent habitat for species such as magpie, killdeer, and the English sparrow. Ground squirrels dug burrows within fields, where they gorged on cereals and other crops. Pocket gophers consumed root crops, especially potatoes. In some places, squirrels and gophers literally ate farmers out of house and home and drove them off the land. In 1926, the general manager of the Aberdeen-Springfield Canal Company reported that pocket gophers—apparently their holes—interfered with irrigation. "I have in mind one instance right now," the manager wrote, "in which a good renter is leaving a good farm because the place is so badly infested with pocket gophers that it is impossible to irrigate it." Then there were the jackrabbit populations that periodically irrupted and forced the animals by the thousands from nearby range into irrigated hayfields in search of food. "It wasn't uncommon to see the jackrabbits lined up as thick as they could stand around the haystack," Alice Crandall recalled of a neighbor's farm in the upper Snake River valley. "They grazed in the fields right along with the stock."[40]

Squirrels, gophers, "jacks," and mice in turn attracted predators to the irrigated fields. Charles Clifford Robinson, an upper Snake River valley farmer, recalled that during the early 1900s he had once "climbed atop our haystack and counted 16 coyotes in our fields." Farmers such as Robinson might have been relieved to see coyotes hunting the mammals that threatened their farms. But few felt this way. Robinson's greater concern was the danger that coyotes posed to his family's chickens. More important, coyotes, along with wolves, were the ultimate symbol of the howling wildness that the irrigators were trying to defeat. The good that the animals might have accomplished did not matter. They had no place in a gardenlike landscape.[41]

The presence of so many coyotes and other wild organisms among the crops and livestock suggests a way of categorizing and further analyzing the historical ecology of the irrigated landscape. From their inception, the irrigated fields became the center of a world defined by two interrelated biological regimes, one domesticated and one wild. Under the domesticated regime, the irrigated landscape consisted of private, differentiated spaces—individual farms—where each irrigator carefully tended his or her own crops and livestock. This was the controlled, subdued, static realm of the garden. But in the midst of the garden, indeed inextricably bound up in it, was the regime of wild things, of weeds and insects, gophers and squirrels, jackrabbits and coyotes. The wild regime lived in, fed on, and competed with the garden. And as this competing regime spread, it transcended the abstract boundaries that separated individual farms. Weeds, mammals, birds, and insects moved across property lines and so transformed private spaces into an ecological whole. Much as seeping irrigation water created a hydrological commons, various mobile organisms helped to establish an ecological one.[42]

Usually the two regimes, domesticated and wild, were opposed, but this was not always the case. In some circumstances, clear distinctions between organisms that were wild and those that were domesticated or tame began to dissolve. Crops and other plants that irrigators intentionally introduced—alfalfa, sweet clover, barnyard grass, willows, and other species—could, as Alexander McPherson said, "run wild." Animals, too, moved between the categories. Irrigators loathed coyotes, yet most Snake River valley settlers kept another predator, the common cat. Somewhat feral, cats roamed through farms in search of mice, gophers, squirrels, birds, and other animals. Because geese and other birds ate crops, managers of wildlife refuges sowed grain fields to provide them with food and lure them away from farmers' fields. In effect, the managers created giant bird feeders. Similarly, the state of Idaho bred ring-necked pheasants on game farms. Outside these official facilities, pheasants took cover in actual farms, in fields and along fencerows, canals, and ditches. Some farmers in the hard winter of 1922 provided food and water for the suffering birds. At other times, pheasants ate crops and fought with chickens for the feed that farmers scattered. Although many farmers came to think of pheasants as pests, the presence and behavior of the birds only showed how well they fit into the irrigated landscape. Here, wild and domestic animals sometimes had more similarities than differences.[43]

In most cases the development of the ecological commons and the breakdown of wild and domestic categories posed problems for agriculture. Yet this was not always true. Cats helped farmers. So did bees. These insects ranged across fields irrespective of private property boundaries; along the way, they performed the necessary function of pollinating flowering crops. Some bees were largely independent of human control; farmers in the Sterling vicinity, for example, famous during the 1910s and 1920s for their production of alfalfa seed, may have achieved success in part because of an unusual concentration of leaf-cutting bees. But farmers relied, too, on honeybees, which occupied a more ambiguous position than the leaf-cutters. Like the wilder species, honeybees roamed free in search of nectar; unlike them, honeybees established their colonies in hives that farmers tended. But no matter what kind of insect serviced the crops, pollination took place not so much in private fields as within the broader expanse of the ecological commons.[44]

Farmers had diverted water to fields in an effort to make the desert blossom, yet irrigation had produced more than just crops—it had produced a hybrid biota that countered simplistic images of a subdued, controlled, garden landscape. Beyond the fields was one remaining area where this dynamic mixture of organisms would grow and flourish: the marginal places through which water passed as it moved away from the farms. In seeped areas and drainage ditches, farmers and nature put into place the final pieces of the marvelous, ambiguous, contradictory, and often troublesome new world that was the irrigated landscape.

Seepage first destroyed agricultural habitat, as the history of an area on the south side of the Boise River west of Caldwell illustrates. Originally, this bottomland—the Dixie Valley—was the scene of intense farming. In 1877, journalist W. A. Goulder observed farmers raising hay, grain, and vegetables in the valley's rich alluvial soil. "The settlers here are all delighted with the country," he wrote, "each one thinking that this valley is the best portion of Idaho, and that his own home occupies the best place to be found." Over the coming decades, however, irrigation and seepage from higher ground waterlogged the soil and concentrated salt in it. By the early 1900s, agriculture was in decline in the Dixie Valley. In 1909, Henry A. Wallace, on assignment for *Wallaces' Farmer,* his family's agricultural journal, reported "a good many ragged spots in the grain fields along the river bottom land" caused by patches of salt. By 1940, around 35,000 acres were wet and salty.[45]

But, as it did in other parts of the irrigated landscape, creation accompanied destruction. As crops disappeared from places such as the Dixie Valley, an uncultivated flora advanced. This included plants that tolerated water and salt: greasewood, a spiny desert shrub; rabbitbrush, another arid land bush, distinguished by its pale green foliage and bright yellow flowers; foxtail barley, with its spiky seed head; and saltgrass, a low-growing, lush, dark green grass. With these plants came wild birds, mammals, and insects. But the new fauna was not just wild. It also comprised animals that the farmers themselves introduced: livestock. If the land could no longer sustain vegetables, it could at least support horses, sheep, and cattle. To be sure, seeped lands did not make the best of pastures. Mostly the farmers' animals grazed on the saltgrass, which Henry Wallace called "a short, wiry grass of but little account." Still, it is important to understand that Wallace was making more of an economic judgment than an ecological one. The agricultural poverty that he described did not necessarily indicate an impoverished environment—quite the contrary. Seeped, salty land produced plenty of plants. They just were not the sorts of species from which farmers reaped large surpluses and profits.[46]

While seeping irrigation water was transforming farm fields into rough pastures, it was creating aquatic and riparian habitats nearby. Where the rising water filled depressions and formed ponds, untended plants appeared, as did wild birds, insects, mammals, amphibians, and other kinds of fauna. On the Minidoka project, seeping water filled borrow pits from which the Reclamation Service had extracted material to build elevated canals; eventually the borrow pits had their own biota. Alvin Holmes, who grew up on a Minidoka project farm, recalled the biological transformation that took place:

> In the beginning of the pond era, the little lakes of water were welcome for their recreational value. In the summer the kids went splashing and wading. The water was clear and clean. In due time tadpoles arrived to the amusement of the kids. Then more biological life developed. The water became full of water bugs, snails, algae, moss and minute organisms. The kids' legs got red and inflamed. The ponds were becoming swamps with a disagreeable odor. Cattails and willows began growing everywhere. Muskrats became numerous. Snipes and ducks moved in. Killdeers were running and chirping along the water edge, eating insects. The bordering land was turning white with alkali. Foxtail grass grew everywhere. The land was being ruined for agriculture.

As the ponds filled, some people introduced game fish to them, and they tried to manipulate the habitat to attract game birds. During the early 1900s, for example, a hunting club from the town of Emmett in the Payette Valley sowed rice in ponds to attract ducks.[47]

Seepage led to the creation of another form of aquatic and riparian habitat when irrigators excavated drains to carry the excess water away. Drains made even better habitat than canals because the water flowed more slowly through them and irrigators did not maintain them as well. Aquatic vegetation and terrestrial plants such as weeds, willows, and grass grew lush and thick in the drains. All kinds of wild animals—squirrels and gophers, beavers and muskrats, birds, insects—lived here, too. So did fish, including species that people introduced: during the 1920s, the Ada County Fish and Game League stocked Boise Valley drainage ditches.[48]

Seepage likewise fed ponds and tanks built for the production of commercial trout. Losses from canals and fields substantially augmented groundwater that flowed from the volcanic rock walls of the Snake River canyon. These springs offered fish propagators a source of clean water that remained at a relatively constant temperature, ideal conditions for raising trout. Aquaculturists established a commercial fish farm in the canyon in 1909. From the 1910s through the 1930s, several more fish producers established operations in the canyon and in the vicinity of Twin Falls. Far in the future, during the 1970s, irrigators living west of the town would tap canals and "springs"—places where seepage flowed from the ground—to fill ponds in which they raised trout under contract to food fish companies.[49]

The flow of water through the irrigated landscape thus traced the expansion of a hybrid habitat that was far more complex than the static fields and pastures that were so central to the garden myth. Because this dynamic biological world often challenged the garden myth, because it had the potential to disrupt and even destroy controlled agricultural production, the people of the irrigated landscape had to respond to it. They had to comprehend and explain what was happening to the land, and they had to act accordingly.

A few Idahoans, at least, believed that many of the reviled animals were merely evidence of broader ecological relationships that farmers should actually respect. This was the message of D. W. Fouch in 1907, when he addressed a Farmers' Institute meeting in Parma in the Boise Valley. Fouch decried the "innate instinct in the ordinary mortal" to kill living things indiscriminately without regard for "their usefulness or [the] place they occupy in nature, in sustaining that just equilibrium, or balance of power without which some species become so superabundant as to become a menace to the agriculturist." Fouch defended all kinds of animals that farmers usually considered pests. He contended that coyotes, starlings, crows, and snakes were valuable because they kept various weeds, insects, rodents, and reptiles under control. He expressed especially high regard for hawks: "One of these majestic birds making your farm his habitat is fully as valuable as your best cat—yet his great white breast is too often the target for some unthinking person." Although Fouch's lecture ultimately rested on the dubious assumption that nature exhibits a balance, his ecological awareness did offer a counterpoint to the idea that irrigators could conquer and manipulate the land exactly as they wanted.[50]

Some inhabitants of the irrigated landscape extracted humor from the anguish of pest infestations. In 1921, the *Idaho Farmer* published a poem by Julia Martin, "On Humby's Place," a parody of modern, systematic agriculture. The narrator celebrated Humby's prosperous homestead but

1 Jackson Lake dam, 1911. Engineers often sited irrigation works, including dams, to take advantage of natural topography. The U.S. Reclamation Service built Jackson Lake dam so that it would raise the level of a natural lake, in effect transforming it into a reservoir. Water released from Jackson Lake then flowed through the Snake to downstream irrigators. *(Idaho State Historical Society [ISHS] No. 77-127.5)*

2 *(above)* Twin Falls South Side project headgates, 1905. Water from behind Milner Dam passed through these headgates and into the Main Canal. *(ISHS No. 79-68.1)*

3 *(opposite, top)* Milner Dam, built in 1904, diverted water into canals that carried water to large irrigation projects on the north and south sides of the Snake River. *(E. C. La Rue, USGS No. 919)*

4 *(opposite, bottom)* Brush wing dam on the Snake River, upstream from Idaho Falls, 25 July 1910. Irrigators built dams of brush, rock, and canvas to direct water into their small canals. As summer advanced and the flow diminished, they extended the dams farther into the river. The high water of spring destroyed the structures, and so each year the irrigators rebuilt them. *(E. C. La Rue, USGS No. 900)*

5 Headgates to American Falls Canal, upper Snake River valley, 1905. Workmen turned the wheels to raise or lower the gates, thus controlling the flow of river water into the canal. The Aberdeen-Springfield Canal Company has since replaced this wood structure with a concrete one. *(E. C. La Rue, USGS No. 905)*

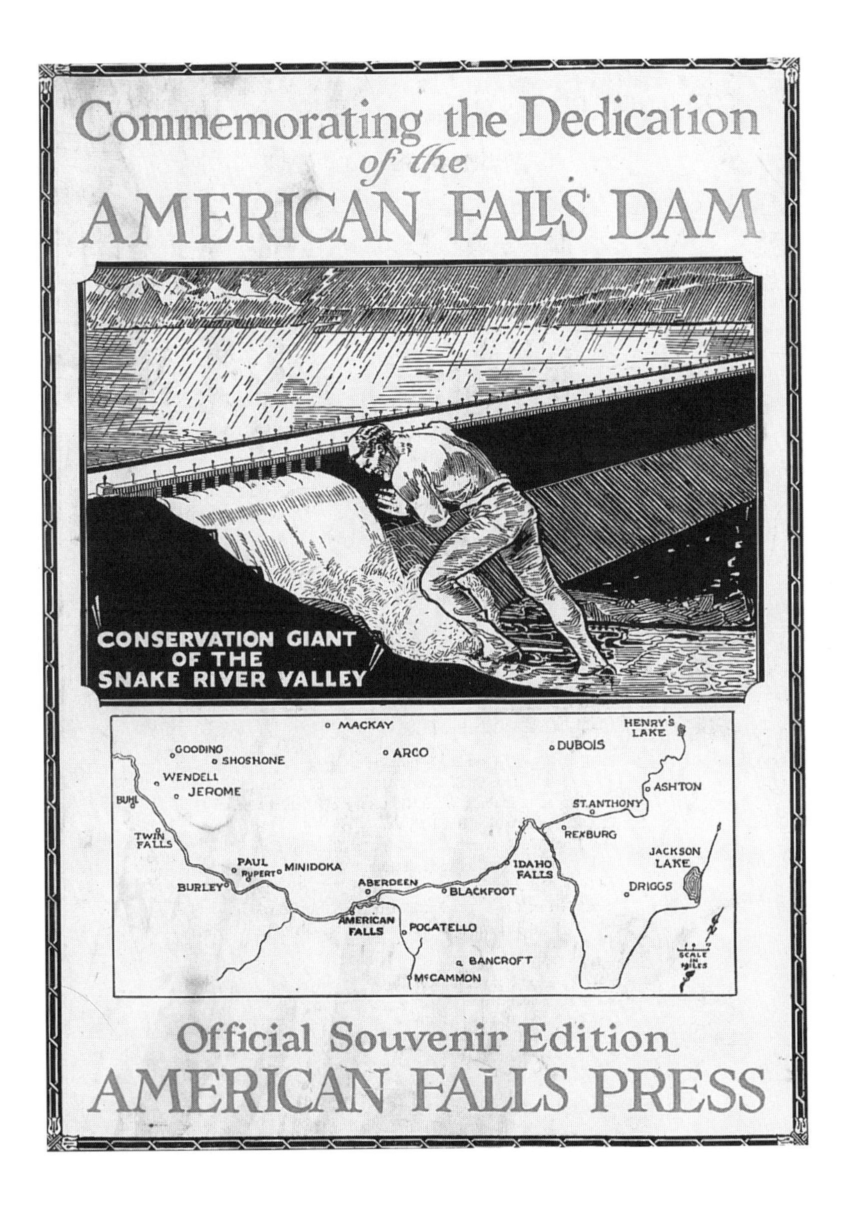

6 "Conservation Giant of the Snake River Valley," about 1927. To some Idahoans, the control of the Snake required masculine, brute force. *(ISHS No. MS 544, Box 5/Pam 18)*

7 Canal headgate, Snake River in Idaho Falls vicinity, 25 July 1910. Many irrigators in the upper valley used wing dams to direct water through crude canal headings. The photograph does not reveal how this particular device actually functioned. One can imagine the problems such loosely built, leaky structures created when officials demanded the precise allocation of water among many canal systems. *(E. C. La Rue, USGS No. 901)*

8 *(opposite, top)* Headgates, Parks and Lewisville canal system, upper Snake River valley, 25 July 1910. Rough but serviceable hydraulic technology. *(E. C. La Rue, USGS No. 897)*

9 *(opposite, bottom)* Dry channel of Snake River, about one mile below Blackfoot, 9 August 1905. Water scarcity exposed problems in allocation systems, provoked conflict, and challenged irrigators to create new distribution methods. *(E. C. La Rue, USGS No. 904)*

10 Irrigation canal near Idaho Falls, 1914. Large canals carried enormous quantities of water. *(R. W. Stone, USGS No. 690)*

11 *(opposite, top)* Control gates, Twin Falls North Side project, about 1912. Irrigation systems consisted of branching canals. Gates such as these controlled the flow of water between conduits. *(ISHS No. 60-176.123)*

12 *(opposite, bottom)* Irrigation ditch, Twin Falls North Side project, early 1900s. Clogged with silt, cattails, willows, and other vegetation, irrigation conduits became unique new riparian habitats. *(ISHS No. 60-176.102)*

13 Construction work, Twin Falls North Side irrigation system, early 1900s. Gangs of men and draft animals carved extensive canal systems out of the sagebrush-covered Snake River plain. Note the horse-drawn slipscrapers, simple but effective earth-moving devices. *(Clarence Bisbee, ISHS No. 73.221.1300)*

14 Perrine Coulee, near Twin Falls, 1912. Irrigators often used natural conduits to deliver water. Engineers incorporated about 400 miles of creeks and draws, which they called coulees, into the Twin Falls Canal Company system. *(Clarence Bisbee, Twin Falls Canal Company collection)*

15 Aberdeen-Springfield Canal Company system, about 1940. Hydraulic technology became a common pasture as irrigators grazed sheep to control the proliferation of vegetation, some of which they had deliberately introduced to stabilize earthen canal banks. *(Aberdeen-Springfield Canal Company, File 4.08)*

16 Vegetation removal, Twin Falls Canal Company system, about 1922. Unwanted flora and fauna thrived in the water as well as on the banks of canals. Here workmen and their teams drag a modified disk harrow through a canal to dislodge the vegetation ("moss"). The men then removed the moss with pitchforks. *(Twin Falls Canal Company collection)*

17 Modified disk harrow for "mossing" of canals. *(Twin Falls Canal Company collection)*

18 Boise irrigation project, probably during the early 1900s. To grow crops, irrigation farmers first had to strip the land of its native sagebrush. Boosters delighted in juxtaposing the sagebrush—the desert wasteland—with the domesticated, agricultural landscape that superseded it. *(ISHS No. 60-77.6)*

19 Mary Hallock Foote, "The Irrigating Ditch," 1889. Inhabitants and observers of the irrigated landscape idealized its tranquil, pastoral qualities.

20 Irrigated field, probably near Twin Falls during the early 1900s. Opening the supply ditch with a shovel, an irrigator allowed water to flow into the furrows. Idahoans admired the skillful use of water and found the clean lines of the furrows aesthetically pleasing. *(Clarence Bisbee, ISHS No. 73.221.344)*

21 Potato harvest near Twin Falls, early 1900s. Idaho's eventual supremacy as a potato region resulted in part from the resistance of a single variety—the Russet Burbank—to disease. *(ISHS No. 73-221.533)*

22 Agricultural fair at Emmett, Idaho, about 1910. When Idahoans referred to the irrigated landscape as a garden, they associated their lives and environment with powerful biblical themes and national myths. At fairs, they often exhibited giant produce (note the cornstalks) that symbolized the garden's incredible fertility. *(ISHS No. 79-95.34)*

23 *(opposite)* A. G. Seiler, "The Story of an Irrigated Farm," 1915. Superficially, the making of the irrigated landscape seemed to be a straightforward process. But the ecological, hydrological, and economic realities of irrigated agriculture belied simple stories of progress.

THE STORY OF
AN IRRIGATED FARM
CONGRESS VOTES THE MONEY
THE PLANS FOR WORK ARE MAPPED OUT
PODUNK
STATION
THE ENGINEERS ARRIVE ON THE JOB
THE LAND IS STAKED OFF
CONSTRUCTION BEGUN
THE DAM COMPLETED
U.S.
LAND OFFICE
APPLYING FOR FARMS
THE SETTLERS ARRIVE
HOMES ARE BUILT
WATER IS TURNED ON
CULTIVATION BEGUN
CROPS FLOURISH
HAULING PRODUCE TO MARKET
CASHIER
THE RETURNS
THE RESULTS OF LABOR

24 Idahoans frequently posed children next to irrigated crops, as in this image by noted Twin Falls photographer Clarence Bisbee. The salubrious environment—the garden—yielded outstanding apples, wheat, corn, alfalfa, peas, sugar beets, potatoes, and livestock; and it yielded vigorous children. *(ISHS No. 73-221.425)*

25 Alfalfa became one of the irrigated landscape's most important crops. Farmers celebrated its versatility and the profits that it brought. But abundant yields, as on this Twin Falls farm, eventually drove prices down. *(Clarence Bisbee, ISHS No. 73-221.257)*

26 Pasture, cow, and seepage below Egin Bench in the upper Snake River valley, 1921. In low areas, irrigation raised groundwater levels, formed pools, concentrated salts in the soil, and created habitat for cattails, saltgrass, and other uncultivated flora and fauna. Unable to use "seeped" land for intensive agriculture, farmers turned it over to pasture. *(H. T. Stearns, USGS No. 241)*

27 Electric Studio portrait of Payette valley orchards, about 1910. The convergence of modern technology and irrigated agriculture—the industrial Eden—fascinated Idahoans. Here, the parallel lines of road, fence, telephone and electrical systems, and fruit trees suggested that nature and culture could be united in efficient industrial production. *(ISHS No. 70-95.12)*

28 *(opposite)* Farmers, especially those who lived in proximity to range land, often organized communal drives to kill the jackrabbits whose populations periodically irrupted. *(ISHS No. 73-221.123c)*

29 Seepage water, probably in the Boise River bottoms in or near Dixie Valley, early 1900s. Note the white, salty patch between the fence and the orchard). *(ISHS No. 63-203.15)*

in the process stood the garden myth on its head. Martin's poem began, "Ever been to Humby's place? / Well you ought to go / Keeps more head of live stock than any man I know. . . ." Of course, readers of the poem were intimately familiar with farms like Humby's, their own included. So they probably smiled ruefully as Martin reminded them of the "livestock" that thrived here—borers and aphids on the fruit trees, midges and rabbits in the clover, beetles in the potatoes, grasshoppers in the garden, and ground squirrels in the wheat: "Humby has the combination—live stock and wheat! / Finds his squirrels the fattest when the straight grain they eat."[51]

But most people did not see much humor in the many unwanted organisms that swept into their fields. Nor, for that matter, were they inclined to view their changing land ecologically. Rather, as crickets, weeds, gophers, squirrels, jackrabbits, and other hated organisms overran the irrigated landscape, Idahoans described what was happening in terms of social conflict. In particular, the metaphor of war—especially during and after World War I—seemed most appropriate to the situation. Hordes of pests were invading their fields; they must combat and destroy the enemy. In myth, irrigators conquered the wilderness so that they could create a new Eden; now, ironically, they had to defend their fields against attack from some of the very organisms that they had been trying to defeat in the first place. Thus the irrigated lands of Idaho's Snake River valley could seem less a garden than a place of trouble and conflict: a battleground.

Weeds were among the most hated opponents in this violent drama. They were "evil," a "menace," a sort of "criminal" that robbed farmers of their livelihood. The plants were an invading army, and Idahoans must wage war against them. In 1918, a resident of Sugar City, Fred Schwendiman, told a typical war-against-weeds story. A farmer hired a group of boys to cut weeds growing in his wheat field. To motivate the boys, the farmer appealed to their collective imagination of the war in Europe. "We are the American army," the farmer told them. "The weeds are the Germans." The boys enthusiastically went after the weeds. After cutting for awhile, one youth suddenly ran ahead of his cohorts and cut down and trampled a "monstrous big weed." Returning to the group, the boy announced that he had "killed the kaiser."[52]

Idahoans similarly interpreted the threat from unruly fauna through the metaphors of crime, invasion, warfare, and extinction. The English sparrow, the *Idaho Farmer* editorialized in 1919, "is one of the outlaws of the feathered tribe, and ought to be put out of business." The bird had so many "bad qualities," the journal elaborated, that "it may be safely classed as 'undesirable,' and should be booked for extinction." That same year, Professor R. H. Smith of the University of Idaho told a meeting of seed farmers at Twin Falls that "nothing short of the most vigorous kind of warfare against insect pests" would prevent the collapse of agriculture in the state. In 1928, managers of the Aberdeen-Springfield Canal Company expressed the feelings of many when they condemned pocket gophers. "There is no one thing that is such a menace and danger to the safety of the canals as this animal," they stated, "and we cannot be safe against costly breaks until the gopher is completly [*sic*] exterminated."[53]

The battle against enemy flora and fauna took many forms. Farmers and their allies plowed, cut, mowed, and burned the weeds. They trapped, shot, and clubbed mammals. They introduced new organisms to prey on the ones that were giving them trouble. They scattered an array of chemical poisons across the land. But no matter how they killed, the farmers usually did it together. Russian thistle, crickets, gophers, jackrabbits—these sorts of flora and fauna occupied the ecological commons, not just private farms. Because these things lived in the commons, farmers most efficiently battled them in common.

Much of the campaign went on in the canals and ditches, the conduits that served as habitat and corridors of travel for weeds and wild animals. Irrigation companies and districts, which the farmers owned and operated together, regularly burned and mowed vegetation that grew on the banks. These organizations also authorized farmers and trappers to run traplines and often paid small bounties for squirrels, muskrats, and other small mammals caught in their systems. They also hired people to keep the canals free of unwanted organisms. E. L. Davis recalled that in the spring of 1911 he and three of his friends took temporary jobs patrolling the Aberdeen-Springfield Canal for muskrats. Ditchriders employed by irrigation companies and districts distributed water to farmers but also looked out for weeds and mammals that could disrupt the flow of water through canals. Some ditchriders received lessons in weed identification from county extension agents, and their tool kits often included a "weed hook" for clearing aquatic vegetation from headgates and weirs.[54]

Through the early 1900s, irrigators suppressed plant life and destroyed mammal habitat by pasturing sheep along canal banks. Irrigation companies and districts erected fences to confine the sheep and invited farmers and stockmen to bring in their animals to graze. The sheep cropped unruly vegetation, thereby depriving rodents and other small mammals of cover and food. Meanwhile, the sheep's hooves trampled and destroyed rodent burrows. Irrigators recognized that sheep were one of their most effective tools in combating the biological menace. In 1922, Twin Falls Canal Company manager J. C. Wheelon stated his belief that, with grazing, the irrigation system "would become a model of neatness and convenience instead of an unsightly jungle of willows and a harboring place for all kinds of rodents and noxious weeds."[55]

While they were battling terrestrial vegetation, irrigators turned on the aquatic plants that often fouled canals and ditches. When the so-called moss proliferated in the summer sun, irrigation system managers drained the canals, killing the plants by depriving them of water. Irrigators also used various mechanical means, such as the "Ziemsen submarine saw," to clean aquatic weeds out of canals and ditches. In 1918, Twin Falls Canal Company workers modified disk harrows so that horses could drag the implements through the canals and uproot the moss. One of the most popular methods of weed removal was "chaining." Horses or tractors on opposite sides of a canal pulled the ends of a heavy chain, dragging it through the water and tearing loose the offending vegetation. After cutting and ripping the moss, workers then scooped it from the water with pitchforks and heaped it in great piles, where it putrefied and reeked in the desert heat.[56]

The same struggle that took place in the canals simultaneously went on in adjacent fields. In a few circumstances farmers and scientists attempted to fight pests through biological means. Through the twentieth century, experiment stations and corporations isolated crop varieties that resisted disease. Occasionally the farmers and scientists used one organism to destroy another. At first the farmers approached this technique with trepidation, however. In 1897, fruit growers debated and then abandoned a plan to enlist a German bird, the Kohlmeise, in their fight against the codling moth; farmers feared that the bird, like many introduced organisms, might itself become a pest. Eventually, though, Idahoans began to release exotic organisms that they believed would help them. Around 1900, Mart Patrie, a resident of Market Lake in the upper Snake River valley, obtained fungus cultures from South Africa. Patrie and a University of Idaho entomologist, J. M. Aldrich, then spread the pathogen among grasshoppers. In 1920, another University of Idaho entomologist, Claude Wakeland, introduced the ichneumon fly to alfalfa fields. The wasplike insect parasitized and destroyed weevil larvae by laying its eggs on them.[57]

More frequently than they employed biological methods, farmers resorted to cultural controls in which they attempted to destroy weeds and insects through techniques such as rotation and tillage. By rotating crops, farmers eliminated the crop habitat of particular insects. The farmers even chose crops because they helped them to destroy weeds. Alfalfa and other legumes, for example, not only restored soil fertility and brought farmers some cash but also smothered certain weeds. Farmers planted crops such as sugar beets or beans mainly for their high market value, but tillage between the rows of these plants also suppressed weed growth. Exactly when insects and weeds forced Idaho irrigators to practice tillage and rotation is difficult to assess, but most farmers probably adopted these methods after World War I. High prices during the war encouraged farmers to prolong the cultivation of lucrative crops. Gradually, however, weeds cut into profits. And when prices collapsed after the war, weed and insect pests finally became intolerable. By the 1920s and 1930s, most farmers in the irrigated landscape had implemented a regimented system of rotation and "clean cultivation."[58]

But the most popular method of battling biological enemies was the use of chemical pesticides, which farmers applied to fields and canal systems. From the nineteenth century through World War II, farmers sprayed fruit trees and crops such as alfalfa with arsenic-based compounds in an effort to rid the plants of insect pests. After World War II, the farmers adopted even more lethal chemicals: chlorinated hydrocarbons like DDT and organic phosphates like parathion. Irrigators and their allies carried out a parallel poisoning campaign against mammals, especially gophers and squirrels. Most often, the exterminators placed poisoned oats and other strychnine-laced bait around burrows. The animals ate the poisoned bait, then retreated into their holes and died. After 1915, various public agencies—the U.S. Department of Agriculture's Bureau of Biological Survey and local chapters of the Farm Bureau, among others—helped farmers purchase and spread poison.[59]

Chemicals provided a powerful weapon with which to fight weeds. Herbicides offered an especially effective method for destroying perennial weeds and for checking weed growth among crops that could not be intertilled. Idaho farmers and agricultural scientists experimented with chemical weed control during the early twentieth century but did not make widespread use of herbicides until the 1920s. This was the era when weed infestations and collapsing prices prompted the growers and their allies to take vigorous action against all floral and faunal pests. The chemical fight against weeds began in 1920, when farmers throughout Idaho applied sodium arsenite to some of the toughest perennial weeds ever to march across their lands. One of these was the dreaded bindweed (also known as morning glory), a vine with funnel-shaped flowers, seeds that remained viable for up to fifty years, and a taproot that went down as far as ten feet.[60]

After this initial battle, the chemical fight intensified. By the end of the decade, farmers had added carbon bisulphide, sodium chlorate, and other herbicides to their weed-killing arsenal. They even used sodium chloride, the same kind of salt that they consumed at the dinner table. In 1926, for example, farmers applied about seventy carloads of salt to weed patches in Twin Falls, Cassia, Jerome, Minidoka, and Bonneville counties. Using salt to kill weeds was ironic, given that soil salinity was a problem for irrigated agriculture. After World War II, farmers, scientists, and government officials assaulted weeds with still more potent herbicides, such as 2,4-D. They also poured compounds with names like xylene, aqualin, and acrolein into canals to eliminate the aquatic plants that grew there.[61]

Whatever method the farmers used to destroy biological enemies, they often carried it out together. This cooperative approach represented a social adjustment to the ecological commons that mobile organisms had created. In this realm, farmers most efficiently fought unwanted plants and animals by pooling resources and coordinating their efforts. Furthermore, it made no sense for one farmer to fight pests if an adjacent landholder did nothing. Indeed, a common complaint in the irrigated landscape was that weeds or gophers multiplied on the land of negligent property owners and then spread to the fields of responsible farmers. This negligence was not acceptable; if irrigators were to overcome unwanted organisms, they must cooperate. As University of Idaho agronomist R. S. Bristol said in 1926, the fight against weeds "must be an organized battle. Growers who try to combat the weed menace on their farms find themselves in an impossible situation when their neighbors, the canals, highways, and other disturbing agencies undo their most persistent efforts."[62]

But the cooperative fight against the pests that invaded the irrigated landscape was not just a response to the exigencies of the ecological commons. To battle unwanted organisms, farmers drew on several familiar forms of cooperative social organization. They based their efforts in part on habits of mutual assistance common to rural Americans. They received financial, organizational, and technical assistance from modern bureaucratic organizations, including the Farm Bureau, the Bureau of Biological Survey, and the University of Idaho experiment stations. And they took inspiration from the cooperative ethos that flourished in the patriotic climate of America's World War I effort. Americans had worked together to defeat their human enemies abroad; now all citizens must cooperate to destroy their biological enemies at home.

The mutual struggle against pests took a variety of forms, from local efforts among a few farmers to sophisticated campaigns led by technical experts and government officials. Fences constituted one cooperative method of countering disruptive, unwanted animals. Normally fences were structures that divided people and distinguished one private parcel of land from another. In the ecological commons, fences also brought people together. During the 1870s, a group of farmers in the Boise Valley attempted to ward off jackrabbits by constructing a "joint enclosure" of woven sagebrush around their farms. In the early 1880s, Mormons in the upper Snake River valley adapted their communitarian socioreligious order to the struggle against ground squirrels—"rats." One Mormon official, for example, instructed "the brethren to fence in fields and farms together, so that they can preserve their crop from drought and rats."[63]

Farmers in the upper Snake River valley also erected fences to fight migrations of Mormon crickets. In 1905 and 1906, crickets migrated off the Snake River plain and headed due east toward irrigated farmland in the vicinity of Egin Bench and the town of St. Anthony. Horrified residents of the area described the crickets as an enemy invasion. "When an army of them is on the move," the *Rigby Star* observed, "the earth they travel over also appears at a distance to . . . be in motion." On Egin Bench, farmers banded together and reportedly erected two low fences of wood and sheet metal, each about ten miles in length, to block the migration. Pits dug at intervals along the fences served to trap the crickets. Fremont county officials called upon all residents to help the embattled fighters bear the cost of the fences. The Egin Bench farmers, they stated, had taken up the fight alone, and now all the people must assume responsibility for defending the county. In each precinct, the officials organized a Local Finance Committee to solicit funds.[64]

Rabbit drives were one of the most popular cooperative responses to an animal threat. Farmers usually organized rabbit drives on the periphery of the irrigated landscape, either on infested farms or in the sagebrush land adjoining them. On foot, horseback, and sometimes with the aid of dogs, farmers banded together and drove the jacks into wire enclosures and then clubbed them to death. On a winter day in 1908 the people of the Market Lake area held a rabbit drive. Five hundred men and boys from around the upper Snake River valley took part. They formed a huge semicircle in the sagebrush and then drove the rabbits toward a V-shaped enclosure. As the semicircle converged, the panicked jacks passed through the V into a slaughter pen, where the bellowing men and boys finished them off. The total estimated kill was between 3,000 and 4,000.[65]

Paradoxically, although farmers viewed jackrabbits as pests, the rabbit drives evolved into important community events. Farmers loved the bloodlust of the hunt; the drives superseded their purely ecological purpose and functioned as ritualized sport. The drives also served as the focus for community gatherings and celebrations. After the 1908 slaughter at Market Lake, the participants headed into the village for sandwiches and coffee or to visit the saloons. That night the town sponsored a ball. After Market Lake abandoned its annual rabbit drive, Louis Shadduck, a local resident, recalled the various purposes of the activity: it killed rabbits, brought people together socially, raised money for worthy causes, and attracted notoriety to the town. At Market Lake and similar locales, rabbit drives at once exterminated jacks and provided people with a forum in which to create and maintain a sense of themselves. Killing reinforced community.[66]

Fences and rabbit drives were simple expressions of rural mutuality. But in some cases, inhabitants of the irrigated landscape applied this principle to modern, sophisticated technology. Irrigators, for example, might handle their chemical sprays together. In many of Idaho's agricultural counties, neighboring farmers formed cooperative spray associations. In 1921, the *Idaho Farmer* reported that "a number of men who are congenial and whose places are located close together" typically formed such a group. They purchased and operated equipment together and elected one of their number as manager of the association.[67]

While small-scale, informal groups and conventional rural mutuality provided the focus for much of the battle against pests, the fight also received the sanction of state law. Through the early 1900s, Idaho instituted a series of statutes that tacitly recognized the ecological commons and provided for the destruction of pests by government officials. In 1907, the state passed a measure that empowered county officials to levy a tax for the purpose of raising funds to pay for the extermination of crickets, grasshoppers, rodents, and rabbits. The 1911 weed control act empowered county commissioners to identify "noxious" weeds and require their destruction within designated "weed extermination areas," sometimes called weed districts. In 1919, the Idaho legislature passed a rodent control measure much like the weed control law.[68]

All of these measures provided an organizational basis for the struggle against pests. Officials set up pest districts (usually along county boundaries), raised money to pay for extermination, and worked in conjunction with an array of groups—farmers, irrigation system managers, the Farm Bureau, county extension agents, the Reclamation Service, and the Bureau of Biological Survey—to organize rabbit drives or purchase and distribute pesticides.[69]

Yet the importance of the pest control statutes was not just the formal structure that they brought to the cooperative war on weeds or rodents. There was an even deeper significance. Both measures gave county officials unique authority: if a landowner failed to take action against pests, the officials could enter that person's land to destroy the offending organisms. The cost of the work would then be levied as a tax on the landowner. By authorizing this public transgression of private space, the pest control statutes essentially acknowledged that property rights in the ecological commons must be qualified. True, the measures sanctioned such action in the interest of protecting the property of neighboring farmers. But by allowing such a transgression, even momentarily, the pest control laws demonstrated the power of the ecological commons and the lengths that people would go to try to combat its worst effects.[70]

How successful, then, was the war against the biota that threatened the dream of the garden? How effective were plowing, tilling, crop rotation, moss removal, trapping, hunting, biological controls, the chemical onslaught, weed and rodent districts, and the cooperative fight? Did the farmers and their allies in science and government triumph? Irrigators were caught up in ecological relationships that they did not fully understand. Although they won many battles and maintained agricultural production, they failed to destroy their biological enemies. These organisms were integral to, and therefore ineradicable from, the irrigated landscape.

Weeds proved amazingly tenacious. Cultural methods had limited effect on the plants. Mowing kept them down but did not destroy them. Burning killed them, yet created open ground into which new weeds moved and thrived. Mechanical removal in some cases only spread weeds further. Tillage chopped them, but even when broken into bits the perennial roots regenerated; and by picking up tiny pieces of root, the farmers' plows, harrows, and other equipment spread weeds between fields. Aquatic plants responded similarly: cutting or chaining shook their seeds loose, and the seeds then germinated; root pieces took hold in the moist soil and sprouted stems and leaves; loose stems grew roots.[71]

Even powerful herbicides failed to eliminate aquatic and terrestrial weeds completely. As the chemicals wore off—as plants absorbed them and water washed them away, or as they percolated deeper into the soil—weeds again moved in. In 1935, the manager of the Twin Falls Canal Company reported that "the progress made in killing noxious weeds on canal banks is not very encouraging." Each year, he said, the company killed one patch of weeds only to find a new patch growing somewhere else. "The total area" infested with weeds, he stated, remained "about the same from one year to another." Irrigators noticed, too, that the plants developed resistance to chemicals and that new, unusual species often took the place of old ones. Sometimes these new weeds resisted herbicides, which suited them to places once occupied by species that chemicals had destroyed. In effect, the killing of one weed opened ground for another. Irrigators thus looked on as a strange and bewildering array of weeds continued to invade their canals, ditches, and fields. In 1954, the Twin Falls Canal Company manager observed the appearance in the system of "new types of moss," including "a very bushy type" and other types that grew "as much as 3 feet long," clung tenaciously to rocks, and required "more frequent mossing."[72]

Not only did the irrigators fail to drive weeds from the land, but their extermination techniques actually created new environmental problems. By continually disturbing the soil, tillage contributed to erosion. Tilled row crops required additional water, as did the more frequent seedings inherent in shorter rotations. Mechanical moss removal wore down canal and ditch banks. Draining canals in midsummer not only killed aquatic vegetation but also deprived crops of water—probably one reason irrigation companies and districts adopted aquatic herbicides. Yet these chemicals created problems, too. Irrigators sometimes experienced difficulty controlling herbicides released into canal water; too great a concentration might enter a farmer's field and poison crops. Herbicides sprayed on terrestrial weeds also had harmful side effects. Carry-over, the persistence of chemicals in the soil, could hinder the growth of the very crops the herbicides were supposed to protect. Chemicals, moreover, infiltrated groundwater and posed potential dangers to people and animals.[73]

Farmers had no better success in wiping out the fauna that threatened agricultural production. Extermination efforts did not rid the irrigated landscape of insect pests—they only changed the conditions of insect life. The beet leafhopper still migrated from weedy lands to sugar beet fields after scientists in the 1930s isolated beet varieties less vulnerable to curly top. Rotation checked insects but did not eliminate them from irrigated fields. Indeed, rotation amounted to a tacit acknowledgement that insects (and weeds) were ineradicable. Nor could pesticides keep insects out of the fields. Insects reproduced rapidly, and their populations eventually developed genetic resistance to the poisons, even DDT, which persisted in the environment for years. There was no solution to the problem; no matter what farmers and scientists did, insects found a place in the irrigated landscape.[74]

Gophers, squirrels, beavers, muskrats, and other small mammals were no more easy to exterminate than insects. But the problem with these animals was not just that they were tenacious; it was that control programs had inherent limits. Irrigators would allocate a certain amount of time, money, equipment, and personnel to killing mammals, but they were unwilling or unable to go any further. To step up mammal extermination—in addition to continuing the other work necessary to keep the canals functioning—probably would have cost more than the irrigators thought it was worth. Thus they took enough action to keep their systems in operation, to keep pocket gophers or muskrats from overwhelming the canals, but that was all. And so mammals survived. Irrigation system maintenance practices similarly gave mammals room to live. When irrigators cleaned the canals and drove out muskrats and beavers, the animals retreated into drainage ditches, which the irrigators did not maintain as well. The drains thus served as refuges for beavers and muskrats—and from there the animals eventually migrated back into the canals.[75]

Economic factors, too, limited mammal control programs in irrigation systems. When fur prices fell, trapping declined. Furthermore, trappers had no incentive to take all the mammals along a particular section of canal or ditch. They took furbearers only so long as animal populations rewarded their efforts. Once they had reduced the populations to a certain level, the trappers moved on to more densely inhabited places. Even broad economic trends influenced trapping and hunting. During the Great Depression of the 1930s, farmers and other local people augmented their incomes by taking animals for their fur and for the bounties that irrigation companies offered for ground squirrels and pocket gophers. During the 1940s, however, gasoline rationing, better-paying jobs, and the movement of young men into the armed forces reduced hunting and trapping in irrigation systems. Finally, the market influenced the sheep flocks that checked vegetation and rodents on canal banks. When crop prices rose during the Second World War, farmers gradually abandoned diversified agriculture, including their herds.[76]

Even legal and bureaucratic measures intended to promote the fight against mammals had weaknesses. Implementation of the rodent control act and the establishment of a rodent control district required petitions from farmers, identification of habitat, coordination between the public and government agencies, and, as always, the allocation of limited resources. While bureaucracy lumbered along, gophers, squirrels, and other troublesome mammals spread across the landscape. Perhaps the greatest indication that laws and government programs had problems was that gophers and squirrels still managed to find refuge on lands of property owners who took no steps to eliminate them. In 1966, decades after the passage of the rodent control act, the manager of the Aberdeen-Springfield Canal Company was still reporting that pocket gophers and muskrats were migrating into the system from adjacent areas. The company had "good luck" in keeping gophers off the canals, he stated, but the animals "are continuously moving in on us from infested areas off of our project and from some of our farms where nothing is being done to control them."[77]

Economics and a cumbersome bureaucracy limited the killing of pest mammals, but perhaps the greatest restriction was imposed by human values. Not all Idahoans, including farmers, viewed muskrats, beavers, and pheasants as problem animals that should be destroyed. Because people valued them for furs, sport, and other reasons, the state legislature passed laws that regulated the trapping and hunting of them. Some farmers sought to amend the wildlife statutes to make it easier to take animals, and in the 1920s and periodically thereafter the legislature relaxed the game laws in response to the farmers' complaints. Ultimately, however, the statutes were not completely removed from the books, and they continued to provide beavers, muskrats, and pheasants a measure of security in and around irrigation systems.[78]

The impossibility of exterminating pests forced irrigators to rethink the notion that they could precisely order the land and retain only desirable organisms. They persisted in advocating eradication, believing that the fight should be carried on to the bitter end; but despite their exterminationist rhetoric they began to admit frankly that they could never rid themselves of the weeds, insects, and mammals that they detested. If this was a war, it was a permanent one, a struggle that they could never ultimately win. In 1921, Claude Wakeland, the University of Idaho entomologist, told farmers that the alfalfa weevil "has come to stay." He warned them that only a diversified agriculture would allow them to absorb losses from the weevil and avoid bankruptcy. Two years later, University of Idaho horticulturist C. C. Vincent made a similar acknowledgment. "We can never hope to stop fighting" orchard pests, he said, "for they are always with us."[79]

Such statements echoed through the coming years. "The problem of handling 'moss' is one we always have with us," wrote the manager of the Aberdeen-Springfield Canal Company in 1936. Another irrigation system manager reported in 1938 that despite ten years of ground squirrel "extermination work" the animals "seem to be as numerous as ever." He believed that without such a program the "rodents would soon destroy the canal system." In 1945, Ival Gooding, manager of the Twin Falls Canal Company, concluded that "noxious weeds, muskrats, pocket gophers and ground squirrels" would "continue to be a problem in canal maintenance for years to come," so the company was developing a "systematic plan to combat these menaces." Twenty years hence, in 1965, Ada County farmer Carl Huntsinger lamented his weeds. "It's getting worse—not better," he said. "We've pretty well controlled the white top and morning glory but not Canadian thistle. I spend more on treating weeds than I do on taxes." It was a hard lesson, but an important one. Unwanted living things, farmers such as Huntsinger learned again and again, could be counted among life's few certainties.[80]

Farmers, engineers, scientists, government officials, and journalists had once anticipated that the irrigated landscape would be different. At times they had idealized the Snake River valley environment as a clean, empty place, a construction site of sorts, filled with malleable resources or materials awaiting manipulation by human hands. They had hoped to create the world that D. W. Ross, W. F. G. Thacher, and no doubt R. H. Loomis imagined, a world that functioned with an almost mechanistic precision. And in their most optimistic moments they imagined that they had conquered the desert and turned it into a garden. As the poet Irene Welch Grissom said, they had vanquished the chaos of the wilderness and created an ordered and fruitful landscape, a new world that redeemed nature and humanity alike.

In practice, nature and the irrigated landscape belied their dreams. A new world replaced an old, but it was not the one they had imagined and desired. Rather than simply work their mastery over the environment and create a static, subdued landscape, irrigators engaged nature in a kind of dialogue in which they continually attempted to reconcile agricultural practices with hydrological and biological processes. Ironically, their own actions produced new environmental conditions that in turn weakened their efforts to achieve control and stasis. Farmers had initiated irrigated agriculture, but they could not regulate it the way they did water in a plumbing system or potted plants in a greenhouse. A dynamic nature influenced the functioning of canals and the growing of crops, even to the extent that the boundary between what the irrigators considered artificial and natural, domesticated and wild, grew hazy, indistinct, and sometimes disappeared altogether.

When that boundary blurred, a new landscape, a new world, both human and natural, was born. Water seeped from canals and fields into the ground and finally emerged in springs that delighted tourists, provided water for fish farms, and replenished the Snake and other streams. Seepage water pooled in low places and ditches and provided habitat for cattails, muskrats, beavers, and many other living things. In some places water rose to the surface and formed small lakes that attracted fish, fowl, and irrigators alike. Birds also congregated at reservoirs, leading humans into the supreme irony of establishing wildlife refuges on bodies of water that were man-made. Carp—aquatic swine—escaped from ponds and, running wild, remade the environment of Idaho streams. Species such as pheasants thrived wild in farmlands yet like domesticated chickens could not live without farms. Bees, both wild and domestic, pollinated alfalfa and made honey. Hydraulic technology—ditches and canals—formed unique habitats for aquatic and terrestrial plants and all manner of fauna, including sheep. Farm cats, nominally domestic but partly feral, roamed through the fields in search of mice and birds. Insects such as the beet leafhopper flourished on acres of weeds growing rampant on abandoned farms and then spread into cultivated lands where they gorged on sugar beets. Weeds grew all over the irrigated landscape but thrived when closest to human activity, in fields and along irrigation ditches. And at times, irrigators themselves seemed unable to agree on what constituted a weed, simultaneously cultivating and trying to eradicate the same type of plant.

This new, sometimes chaotic landscape compelled irrigators to act cooperatively. In one way, the irrigated landscape reflected the conventions of market agriculture: it was a landscape of individual farms and private property. But to uphold the integrity of this arrangement, irrigators joined forces and worked together. Though they might differentiate their individual farms with fences, with the cricket or rabbit fence they made common cause to protect fields from inscrutable and sometimes frightening natural forces. Another sort of fence, the wire enclosure of the rabbit drive, also brought farmers together; as bellowing men and boys converged on the slaughter pen, the bonds of community grew tighter. In some instances, however, the ecological commons rendered fences irrelevant. Seeping water, downy weed seeds that floated on the breeze, little gophers scurrying across the ground—these forces of nature transcended the fences that objectified abstract survey lines. In a sense, groundwater and organisms transformed private spaces into a public arena where a type of community prevailed over individualism. Normally irrigators defended private property rights, but in the ecological commons they exercised a sort of collective capitalist mandate to force individuals to pay for drainage or kill weeds.

Much as the irrigated landscape challenged institutions and compelled social changes, it countered the irrigators' belief that they could in fact control nature precisely. On one level, they continued to trumpet their mastery over the environment. Such boasts, however, would not hold up in daily life, in the grubby landscape where farmers, scientists, and engineers fought weeds, battled muskrats, and struggled to regulate water. When they reflected on the new world they had helped to create, Idahoans acknowledged that they could not completely banish unwanted flora and fauna, that they could not always prevent seepage, and that streams were not like pipelines. At such moments they sometimes groped for more appropriate ways to define their relationship to the land.

In trying to understand the irrigators and their beliefs, we should not analyze only their dominant or controlling metaphors and myths; rather, we should examine the differences among the metaphors and myths that they used. For these multiple and often incompatible images—laboratory science, housekeeping, garden idyll, battlefield, crime, taxation—illuminated a complex world in which the irrigators always had to test their ideals against nature. When their interaction with the land produced environmental consequences that denied their mastery, they searched for new constructs that more effectively explained their predicament. The irrigators' many visions thus revealed their reciprocal relationship with nature, the creation and re-creation of their landscape, the unfolding of their history.

3/ **Dividing Water**

Conflict, Cooperation, and Allocation on the Upper Snake River

Fundamental to the irrigated landscape's unfolding history were the laws and conventions that irrigators adopted to govern the allocation of water. Irrigators sought to divide the flow of rivers and canals among themselves; accomplishing this division in a just, cooperative, and, within limits, equitable fashion required rules and regulations. But in a landscape in which human systems had become inextricably entwined with complex and disruptive hydrological processes, the orderly distribution of water could easily break down. Such a breakdown took place on a July morning in 1919.

Early that morning, Joe Koury rose from his bed and went outside to tend his irrigation water. Farmers such as Koury were used to this task. They were accustomed to slipping on rubber boots, grabbing shovels, and tromping across the landscape to perform one of the most routine but important chores on the farm. If necessary, the farmer shifted the irrigating stream from one field to another. And he looked for malfunctions in the system. Debris might have plugged a ditch. Perhaps a gopher had been busy. Maybe the current had begun to erode a bank or furrow. Somewhere, no doubt, the crumbling earth would need a little work, some patching and filling, to keep the water flowing on its way.

Yet Koury also scrutinized the landscape for signs of a more serious problem. Koury went out not just to change the stream or check on the ditches but to make sure that none of his neighbors had taken water that he believed rightfully belonged to him. In Idaho and the American West during the late nineteenth and early twentieth centuries, such a violation of an irrigator's rights almost always led to conflict, even violence. Drought usually precipitated these altercations. In the driest years, canals simply did not carry enough water to supply each farmer with a continuous stream, as they usually did. To compensate for the scarcity, water users took turns and irrigated their farms in rotating shifts; for a limited time, each irrigator on a canal or lateral would use the entire supply. But while drought led water users to cooperate, it also increased the potential for conflict. Misunderstandings and disagreements over water allocation easily developed. Whose turn was it? How long was the turn? As crops withered, tempers flared; an impatient, angry, or aggrieved farmer was likely to stomp out to a headgate and simply commandeer the stream. Desperate or dishonest irrigators even resorted to theft; under cover of darkness they opened their headgates and took water at someone else's expense. Thus when Joe Koury left his house, shovel in hand, he probably knew that his work might amount to more than just routine maintenance. The summer of 1919 was one of the driest on record in the upper Snake River valley, and farmers everywhere—Koury included—patrolled their ditches and monitored their water with extreme vigilance. Indeed, so anxious was Koury that he went out that morning before breakfast. It was his turn, and nobody was going to cut into it.

He realized his worst fears when he reached the ditches that crossed his farm. They were running dry. Furious, Koury went to the headgate that shunted water from the community lateral into his ditches. There stood William Grover, Jr., his neighbor. To Koury's dismay, Grover had closed Koury's headgate and opened his own. Koury accused Grover of taking the water; Grover countered that it was now his turn to have the stream. Ironically, local irrigators had recently given Grover responsibility for preventing just the sort of argument in which he had now become embroiled. They had chosen Grover to serve as lateral watermaster, the man responsible for overseeing peaceable water rotation during this droughty season. But this morning at the headgate, cooperation and amity broke down into verbal conflict, then violence.

The shovels that Koury and Grover carried, tools for directing life-giving water to crops and livestock, now became weapons, instruments of injury and death. Both men struck blows. Grover's blade hit Koury's temple; Koury collapsed into the water. Grover immediately pulled him out, but Koury was dead. Later, prosecutors would argue that Koury, a battle-hardened Great War veteran trained in bayonet drill, could not have lost the fight had he actually brought his shovel into play. But no matter; Koury had survived the Meuse-Argonne and the horrors of trench warfare only to meet his fate in an Idaho irrigation canal.[1]

Joe Koury's death starkly illustrated the strange mixture of conflict and cooperation that surrounded—and still surrounds—water management in Idaho and other arid parts of the American West. In the arid West, conflict and cooperation existed together like two parts of a schizophrenic personality. In one way, irrigation was largely a cooperative enterprise. Some irrigators worked alone, but most did not; building diversion dams, excavating lengthy canals, and allocating water were tasks most efficiently accomplished collectively. Cooperative social behavior, motivated by religious ideals or the habits of mutual assistance common among rural Americans, also compelled irrigators to join forces. And to further their shared hydraulic objectives, irrigators often formed cooperative institutions: the irrigation district and the aptly named mutual company, for example, gave farmers the organizational means to finance and maintain their canal systems. In certain respects the Corbett Slough, the conduit that carried water to the farms of Koury and Grover, exemplified the cooperative ideal. To meet the technical challenges of moving water out of the Snake and onto land, farmers in the 1880s had come together to build, operate, and maintain an irrigation system. The drought-inspired rotation program that Grover administered, moreover, represented an extension of that neighborly cooperation.[2]

Yet the killing of Joe Koury demonstrated that conflict could destroy the harmony that the irrigators prized. Usually this conflict stemmed from efforts to divide a stream of water and distribute it among many irrigators. For most irrigators in the American West, dividing water ultimately meant privatizing it, turning the common property of a stream into private property on a farm. Although irrigators, those on Corbett Slough included, might share the burdens of procuring water, they still lived in a society and nation that idealized private gain and the economically independent man. Koury and Grover, after all, each wanted water for himself, for his own farm, for his own profit. And as their fight showed, reducing a stream to private property could easily lead to violent conflict. Disagreement, misunderstanding, or thievery could upset the process. Such problems often simply derived from the nature of water itself. Dividing this fluid, mobile, scarce substance into privately owned quantities—measuring and physically separating it—was difficult and created ample opportunities for error, misunderstanding, and disagreement. Dividing water, which almost always depended on the ability of irrigators to cooperate with one another, thus frequently spawned conflict. And it was no coincidence that, as in the Koury-Grover fight, the struggles frequently broke out at headgates, the devices with which irrigators physically apportioned water. Not merely an instrument of hydraulic technology, the headgate also stood as the locus and symbol of conflict.[3]

The ditch bank duel between Joe Koury and William Grover on that warm July morning in 1919 was not an isolated incident; rather, it represented but one small fight in a broader, ongoing struggle over Idaho's water. This struggle encompassed not just single canal systems but large portions of rivers and watersheds; it spanned not just lone years such as 1919 but entire decades. Beginning in the late nineteenth century, farmers, canal companies, irrigation districts, and various government entities across Idaho's irrigated landscape battled over the allocation of scarce water. The conflict usually played itself out on two interconnected levels. Within canal systems, farmers disputed water allocation, took water from each other, and, like Koury and Grover, fought with shovels. Concurrently, irrigation companies and districts fought over the division of the rivers themselves. On the rivers the struggle usually entailed lawsuits, but irrigators found intimidation and threats of violence useful tools as well. But conflict was not an end in itself for irrigators; throughout this period, water users on both levels sought to establish—or reestablish—the cooperation that made for stable irrigated agriculture.

Conditions on the upper Snake River between the 1880s and 1920s provide a case study of the interplay of cooperation and conflict that typified water allocation in Idaho and much of the American West. Water allocation on the Snake evolved in two periods, roughly before and after the completion of Jackson Lake dam and reservoir. In the first period, from the 1880s to the early 1910s, irrigators from the Blackfoot area to the vicinity of Rigby struggled to divide the scarce waters of this hydrologically complicated river. The earliest irrigators simply diverted water into their canals with little regard for rules or laws. But drought and increased irrigation eventually so depleted the Snake that farmers resorted to the legal doctrine of prior appropriation to divide the shrinking river. Theoretically pragmatic and functional, and suited to the ideal of individual self-interest, the law in practice failed to provide for smooth distribution of water. To overcome the flaws of prior appropriation, ensure that needy irrigators received water during extreme drought, and forestall conflict, irrigators created extralegal, cooperative allocation methods.

In the second period, from about 1910 to the mid-1920s, irrigators around Minidoka and Twin Falls vied with upstream irrigators—primarily those between Blackfoot and Rigby—for control of the Snake. The downriver farmers drew much of their water supply from Jackson Lake dam and reservoir, completed in 1907 and expanded in 1910 and 1916. This stored water had to flow several hundred miles through the Snake's channel, past the people in the Rigby-Blackfoot vicinity, to get to the Minidoka–Twin Falls irrigators. But public officials responsible for allocating the Snake had trouble differentiating the stored water from the regular, or natural, flow of the hydrologically complicated river. Conflict broke out between the upstream irrigators and the Minidoka–Twin Falls water users. Once again, a social solution based on informal, cooperative allocation methods seemed the answer. This cooperation, which culminated in the creation of the Snake River Committee of Nine in 1923, combined all elements of water management—prior appropriation, dams, reservoirs, irrigation companies and districts, state and federal agencies, the river itself—into one loose, composite system. Snake River irrigators hoped that this ad hoc cooperative organization would prevent the sort of conflict that led to farmers bludgeoning each other with shovels.

The common thread in the history of water allocation on the upper Snake is the river itself. The story begins with the river and its tributaries in the late nineteenth century, when farmers first settled along the stream in substantial numbers. In Wyoming, the Snake trickled off the Continental Divide, ran west and south through mountainous terrain, and drained into Jackson Lake on the east side of the Teton mountains. From this alpine lake the Snake plunged south through rocky, pine-covered gorges. Then the river turned west and emerged from the mountains onto the plain to which it lent its name. At the canyon mouth the Snake slowed somewhat and flowed over a huge alluvial fan that it had built up over millennia. After taking in Henrys Fork and other streams, the Snake curved south and west and passed over a series of falls: American, Cauldron Lyn, Twin, Shoshone.

Along its twisting course, the upper Snake fluctuated in volume. Each spring, snow in the headwaters melted and the river rose in its bed, spilled over its banks, and scoured new courses for itself in the underlying alluvium. As summer approached, the snow disappeared and the Snake's volume fell, until by early autumn the flow was but a fraction of what it had been during the late spring flood. Within these broad seasonal cycles the Snake's flow varied according to hydrological conditions. The amount of winter snowfall determined spring and summer flows; light snowfall meant less water in the spring. Groundwater tables along the Snake influenced its volume, too. Where the table was low, river water seeped into the gravel. Where the table was high, groundwater percolated through the gravel into the stream. Disturbance and change characterized the Snake; from this writhing, pulsing, complex river the first irrigators would divert water to their fields.[4]

Those early settlers began arriving in the upper valley in the 1880s and 1890s. During the 1880s, Mormon farmers left Utah's crowded farming areas and relocated in the Snake River forks country, around where several rivers (Falls, Teton, Henrys Fork) and creeks (Willow, Moody) flowed into the South Fork of the Snake. They arrived in small groups of friends or kin, carrying with them the sanction but not the organizational and financial support of their church. By 1890, the Mormons had covered the forks country with a complicated network of small canals, each drawing water from the Snake River or one of its tributaries. Through the 1890s and into the early 1900s, still more settlers, many of them non-Mormons, came to the upper valley. Between 1896 and 1900, the area of irrigated land in Fremont and Bingham counties rose 225 percent, from 65,000 to 211,111 acres. Large projects, such as the 16,000-acre New Sweden Irrigation District, accounted for much of the added farm acreage.[5]

Through the late nineteenth century, irrigators in the upper Snake River valley adjusted their agriculture to seasonal hydrological variations. The farmers often extended temporary rock, brush, and canvas dams into the Snake and its tributaries, capturing the flow and drawing it into their canals; as summer approached and the streams fell, they extended the dams farther, until at last the water ran below the level of the canals, leaving them dry. The farmers also grew crops according to the availability of water. In 1890, engineer Frederick Haynes Newell observed that Cassia County irrigators cultivated greater or lesser amounts of land each year "depending somewhat upon the appearance of snow among the high peaks." Hay and grains, the predominant crops in Idaho before the early twentieth century, were suited to the fluctuating water regimen. Hay and grain required relatively little water compared to other plants, and then only in the spring and early summer when melting snow swelled the Snake and other streams. In 1900 state engineer W. D. Ross reported that because the upper Snake River valley "is very largely a grain country, the maximum use of water" took place "during the period of maximum supply," the months of May, June, and July. By August and September, when the river fell in volume, the harvest of hay and grain was already under way.[6]

The legal doctrine of prior appropriation ostensibly provided irrigators with a system for dividing the Snake and other streams. Prior appropriation was the basis for most western American water law by the late nineteenth century. Californians first implemented it in the gold rush camps. Gradually, water users in other territories and states adopted it. By 1890, it was firmly established in Idaho statutes, supreme court decisions, and the state constitution. Two essential features characterized the doctrine. First, irrigators claimed rights to distinct quantities of a stream's flow, the earliest, or "prior" rights taking precedence over later ones (thus the time-honored western phrase, "First in time is first in right"). Second, to retain water rights, an appropriator had to exercise them by actually diverting water from a stream and continually putting it to "beneficial use," most importantly, domestic consumption and agriculture (thus, "Use it or lose it").[7]

Deceptively simple in theory, prior appropriation embodied the individualist and capitalist values of a society intent on exploiting and developing the western public domain. Under the law, irrigators attempted to treat streams like "natural resources," raw materials that people could manipulate, possess, and transform into wealth. Prior appropriation and water rights ostensibly divided a stream into units that belonged to farmers, who then turned the water into food and dollars. The doctrine allowed irrigators to negotiate between nature and the market, to translate a wild substance into private property. In a sense, the doctrine attempted to reconcile two entities—the real stream that actually flowed through the landscape and the abstract, legal river belonging to many individual irrigators. Water rights allowed irrigators to imagine a stream as the sum of many individual quantities, each soon to be privatized and consumed. In addition, the "first in time is first in right" maxim endeavored to make water ownership conform to the hydrological reality of creeks and rivers that fluctuated in volume. In theory, when a stream ran low, owners of prior rights received water before irrigators owning junior rights. A legal river, in short, had to swell or diminish in accord with the real one.[8]

Prior appropriation supposedly provided irrigators in the upper valley with an uncomplicated, utilitarian system for dividing the Snake River and other streams among many parties. Yet for all its theoretical virtues, the law in actuality did not work smoothly. In the beginning, prior appropriation and water rights had little practical meaning for upper valley irrigators: the Snake and its tributaries supplied all the water they needed, without regard to priority. Furthermore, this apparent abundance provided little incentive for farmers to register their rights accurately, if at all. "Many of the early settlers" in the upper Snake River valley "did not record their claims in priority," Byrd Trego later recalled, "because they did not believe that there would ever be a water shortage on [the] Snake River due to its great volume." Conversely, still other farmers—ignorant of how much water the rivers actually carried and the techniques of measuring it—claimed enormous amounts, far more than ever flowed there. In 1900, Ross, the state engineer, reported that farmers on Henrys (North) Fork of the Snake had appropriated 74,460 cubic feet per second (cfs) of water, thirty-five times the minimum flow of the river at St. Anthony.[9]

Nineteenth-century Idaho law also failed to provide irrigators with clear, compelling rules for precisely registering their claims. Not until 1881 did they have an official method for recording water rights with county officials. Even then, the instructions for noting the date, location, and amount of the claim were vague and imprecise. Because of these legal ambiguities, prior appropriation in Idaho first functioned, according to state engineer James Stephenson, as "an extension of the right of 'squatter sovereignty.'" Irrigators simply took water as they needed it, seldom measuring it accurately and almost never keeping a clear, up-to-date account of their claims.[10]

Irrigators in the upper Snake River valley persisted in their "squatter sovereignty" as long as creeks and rivers, agricultural practices, and their own numbers accommodated them. In the absence of drought, with continued hay and grain cultivation and with few farms and ranches, irrigators blithely took all the water they wanted. So long as water remained relatively plentiful, they had little practical use for the law of prior appropriation. The amount of water available to upper Snake River valley farmers, however, eventually began to diminish. In August of 1901 and 1902, a reach of the Snake in the vicinity of Blackfoot ran dry for the first time in local memory. In part, decreased precipitation—drought—at the Snake's mountain headwaters caused the shortages. Additionally, irrigators in the upper valley diverted increasing amounts of water as the population grew and as they brought more land into production.[11]

Changes in agricultural practice in the upper valley also helped to shrink the Snake and other local streams. By the early 1900s, farmers in Fremont and Bingham counties were switching from hay and grain to sugar beets, which required water not just in spring and early summer but throughout the summer and early autumn. In 1905, one perceptive resident of the upper valley identified beet culture as the essential reason for the decreasing flow of the Snake each summer around the town of Blackfoot. Farmers, he wrote, irrigated "the cereals and hay" early, when the river carried enough water for all. By the time "extreme low water" conditions set in, they had finished most of their irrigating. Sugar beet cultivation altered this basic pattern of water use. To irrigate "the vast acreage of beets now planted," farmers needed water much longer, even after the river flow had fallen. Consequently, "the greater the amount of beets raised" upstream, "the shorter the water will be at the Blackfoot end of the valley."[12]

For farmers in the Blackfoot area, hydrological conditions in the Snake River exacerbated water scarcity. After emerging from the mountains on the eastern edge of Idaho, the Snake alternately gained and lost water in relation to surrounding water tables. To the dismay of irrigators, the Snake lost water about eight miles upstream from Blackfoot, before it reached their canals. The water then seeped through the river's alluvial bed and finally rose about ten miles downstream from town, where it collided with an underground formation of volcanic rock. A local newspaper reported that "when Snake river gets at a low stage nearly all the water sinks [upstream from Blackfoot] and comes out again a short distance above American Falls."[13]

As drought, intensification of irrigation, and natural hydrological conditions made water increasingly scarce, many irrigators were no longer willing to tolerate unregulated use. When the Snake ran dry in the early 1900s, previously complacent irrigators suddenly became competitors for water. The Blackfoot irrigators felt the pinch of scarcity more than others, and they in particular desired order in water distribution. The prior appropriation doctrine should have provided that order; it should have stood ready to regulate all the claims, to fill rights in chronological sequence until no more water remained. The legal river suddenly had importance, and it had to conform to the fact of a real stream severely diminished in volume.

But in practice the system did not work; as the Snake flowed to a trickle, prior appropriation failed. In the absence of a clear, readily available record of water rights, nothing dissuaded farmers from believing that they, not their upstream or downstream neighbors, owned the prior rights. In 1906, a Blackfoot resident observed that all along the upper Snake, groups of farmers believed that "they had the oldest rights on the river and that it was the other localities that were late rights and should go short of water." He noted that many people had settled the valley almost at the same time, making "the rights of the various localities . . . practically identical from one end of the river to another." In practice, the irrigators had no system for allocating water.[14]

The failure of prior appropriation compelled irrigators in the upper Snake River valley to try to clarify the water rights record. A lawsuit involving large numbers of water rights seemed to offer them the best method of resolving the problem. In the summer of 1901, the Rexburg Irrigation Company filed suit against the Teton Irrigation Canal Company for the purpose of clarifying title to water rights on the Teton River, a tributary of Henrys (North) Fork of the Snake. As news of the lawsuit spread, dozens of irrigation companies and hundreds of individuals throughout the entire upper valley joined as plaintiffs and defendants.[15]

Like many water lawsuits throughout the American West, *Rexburg Irri-*

gation Company et al. v. Teton Irrigation Canal Company et al. served as a formal adjudication process to determine the status of all the water rights on a stream and its tributaries. The judicial decree that decided the lawsuit attempted to rectify the deficiencies in the record of claims and to provide for the enforcement of priority. A judge collected testimony and other evidence necessary to establish an accurate chronological list of appropriations. With the information at hand, he determined as precisely as possible the date, amount, and quantity of each appropriation. He then issued the decree, and a watermaster physically distributed the stream among the claimants.[16]

But *Rexburg v. Teton,* despite its promise and to the dismay of irrigators, provided no immediate relief for water conflicts in the upper Snake River valley. Adjudication was a complicated, laborious, time-consuming process; indeed, as one observer noted, the case had "as many twists and curves as the Snake river itself." Judge James M. Stevens, who oversaw the case, did not deliver his decision until December 1910; and not until August 1911, ten years after the case began, was the decree, with its prioritized list of water rights, published and made available to irrigators. During the interim, Judge Stevens and the referee he appointed, F. J. Cowen, heard testimony, the state engineer surveyed and mapped canals and farms, and Stevens and Cowen studied the evidence and prepared the final decree. Meanwhile, irrigators chafed at the slow progress.[17]

As the legal apparatus lumbered along, water distribution on the upper Snake continued in a haphazard fashion. Some irrigation companies, fearing the legal loss of water they did not divert, "would take out all [the water] their canals would carry and in many cases, turn it out upon arid plains to run to waste while crops worth millions of dollars farther down the valley were suffering for water to mature them." Small dams, built in response to water shortage, complicated water distribution. When the Snake fell in volume during the driest part of the summer, it often left canal headgates high and dry; to get water back into their canals, irrigators extended temporary dams of rock, wood, and canvas into the river's channel. With each dam, the Snake fell still farther; as it decreased, irrigators built more and tighter dams. "Dams and counter dams" eventually left parts of the Snake around Blackfoot totally dry and farmers in that area without enough water.[18]

One can imagine the scene around Blackfoot during the summer of

1905 as farmers in the area ran out of the water necessary to mature their precious sugar beets. They gazed at the dry riverbed and contemplated the exposed cobbles baking in the relentless sunlight. Anguished, they walked through their beet fields and inspected the limp, withered leaves. Their labor, the money spent on seed and hired hands, all of the things they had invested in this year's crop, they now stood to lose. Hope evaporated like leftover puddles in the Snake's dry bed. And to think that the farmers but a few miles upstream around Idaho Falls had full canals.

What, the Blackfoot farmers wondered, could they do to give their crops a reprieve? What could they do, in the absence of a court decree, that might ensure them at least a minimal amount of water? In this hot dry summer, they would try what farmers all over the American West would try when conventional, individualistic ways of securing water failed: cooperation. The Blackfoot irrigators came together as a group, as a community united by deprivation, to devise a collective solution to their plight. They gathered at a mass meeting in the town and there contrived a plan of accord: they would reach out to the upstream irrigators and, in a spirit of civic-mindedness and fair play, ask for "some kind of division of the waters by mutual consent." This would be the first attempt to divide the upper Snake cooperatively, outside the morass of prior appropriation. The Blackfoot group appointed a committee of three to travel upriver to make the appeal.

It almost succeeded. At Idaho Falls the committee met with canal managers who expressed sympathy for the Blackfoot farmers' situation. Two operations, the Idaho Canal and the Burgess Canal companies, agreed to close their gates and "turn out" some water so that it would flow downstream to the deprived beet growers. But at an Idaho Falls meeting, other local irrigators began to balk. Some canal managers said they would first have to secure permission from all of their farmers. Other water users fell back on self-interest and prior appropriation. They had the oldest water rights, they claimed; they had no obligation to forgo their water to help someone else. Quite probably these intractable irrigators feared that giving water to the Blackfoot farmers might jeopardize their own rights to it. Under the beneficial use mandate of prior appropriation, irrigators had to use it or lose it. With the decree still pending, stubborn Idaho Falls irrigators believed that they could not risk temporarily giving up some water. Thus failed the Blackfoot irrigators' cooperative mission.

With time running out on their shriveling crops, the Blackfoot people now saw no other recourse but to consider the upriver irrigators adversaries. Accordingly, they asked a court to issue a legal injunction ordering the upstream irrigators to leave water in the Snake for the downstream canals. They got what they wanted; Judge Alfred Budge (filling in for an absent Judge Stevens) appointed D. P. Pratt, a Blackfoot man, to distribute Snake River water and regulate canal headgates in the Rigby–Idaho Falls vicinity. Pratt in turn selected two other Blackfoot irrigators to help him with this task. Pratt was supposed to distribute the water as best he could with regard to priority while making sure that the Blackfoot irrigators got enough water to save their crops. In practice, Pratt and his two assistants rotated water between farmers in the Idaho Falls–Rigby vicinity and those around Blackfoot on a weekly basis. But Pratt and his allocation methods did not sit well with upriver irrigators. They looked upon Pratt and his deputies as interloping agents of the Blackfoot farmers.

As usual, conflict flared at the headgates, at the places where Snake River water became the property of irrigation districts and canal companies. Apparently, no violence occurred, but the upstream farmers made it difficult for Pratt to perform his duties. One canal watermaster opened a headgate, locked it in place, then stood by and defied Pratt and his men to open it. At the Farmers' Progress Company headgates, a duel ensued. Pratt and his assistants lowered the gates; a Farmers' Progress crew promptly raised them. Back and forth they went, lowering and raising the apparatus. Finally Pratt ordered a retreat. He then had the three Farmers' Progress men charged with contempt of the court order.

In subsequent years, irrigators in the Idaho Falls–Rigby area continued to rail against what they saw as the illegal expropriation of water by the Blackfoot irrigators. After 1905, Judge Stevens tried to defuse the volatile situation by avoiding the mistake of appointing local men to the position of watermaster. Conflict between upstream and downstream users, however, continued, and Idaho Falls–Rigby irrigators kept up their complaints. "Of course it is understood," sneered an Idaho Falls newspaper, "that when the canal owners of the southern portion of the county want water they ask the court to give them an order to take it without any reference to priority of rights, or notice to the canals in the north end of the county." The upstream irrigators, many of them original settlers in the upper valley, did have some reason to protest, because in general their rights were indeed older than those of the Blackfoot irrigators. In the absence of a decree or a firm record of priority, however, Judge Stevens awarded the Blackfoot farmers water because they needed it, not necessarily because they had any clear right to it. Only in relatively brief periods of ample precipitation, like those that occurred in late summer 1906, did tensions ease.[19]

Persistent conflict over water prompted upper Snake River valley residents to look to hydraulic technology to solve the problems of shortage and distribution. If the doctrine of prior appropriation and the legal system that supported it could not solve water problems, perhaps reservoirs could. With storage dams, some irrigators believed, they could garner additional water. Each spring, they watched the Snake flush massive volumes of water through the valley and on toward the sea. Large dams could interrupt the course of nature and hold this flood for later use. By storing water in reservoirs, irrigators could overcome the tricky hydrology of the Snake, surmount the problems of prior appropriation, forestall litigation, and ensure an abundant water supply for all.[20]

One Blackfoot resident, W. D. Trego, portrayed cooperative dams as a panacea for water conflicts. Even with the decree, he said, irrigators, encouraged by greedy lawyers, would continue legal action against one another, "but will that make any water for us to grow beets, potatoes, wheat, oats and hay, or any other crop?" Furthermore, said Trego, "the Snake never can be regulated so as to give every canal just the water that [it is] entitled to, on account of the supply continually changing." No matter how precisely officials distributed the water, even owners of "the oldest rights on the River" would not always receive their full complement. To Trego, the solution to water scarcity and conflict lay beyond the law, in jointly operated dams and reservoirs. "Let us all look at the matter in a business-like manner," he urged, "as well as in the light of what is right between man and man, and go together and get water and plenty of it, and not spend our wealth fighting over priority of rights." In Trego's view, cooperation should transcend individualism and the conflict that accompanied it.[21]

There was, however, the problem of marshaling the organizational and economic resources necessary to construct a large dam. Upstream residents, confident in their water right priorities and in their advantageous geographical position, did not necessarily agree that they should all join in common cause to build a dam. In addition, irrigators who wanted a dam lacked the expertise and financing necessary for such a work.[22]

Trego and like-minded people of the upper Snake River valley thus looked to the federal government for assistance. In particular, they desired to secure a water supply far upstream at Jackson Lake reservoir in Wyoming, in the Teton Range. Arthur Foote, an engineer with the United States Geological Survey, had identified Jackson Lake as an ideal storage site during the late 1880s. Under the 1902 National Reclamation Act, federal engineers initiated construction of a dam at the lake's outlet. Jackson Dam, completed in 1907 and rebuilt in 1910 and 1916, raised the lake's surface elevation, creating a store of water for large new irrigation projects at Twin Falls and Minidoka, which lay downstream from the farmers around Idaho Falls and Blackfoot. The Twin Falls developments included two projects established under the Carey Act, in which federal and state governments cooperated with private corporations to build irrigation systems and induce settlement. Opened in 1905 and 1908, the Twin Falls projects eventually encompassed about 330,000 acres. The Minidoka project, established under the National Reclamation Act and opened in 1907, totaled approximately 120,000 acres.[23]

In 1906, W. D. Trego arranged for engineer D. W. Ross, now of the U.S. Reclamation Service, to come to Idaho Falls and talk to irrigators who had "asked for the privilege of getting an interest in the [Jackson Lake] reservoir." Ross addressed a gathering, explaining the details of the new dam and reservoir and "waxing enthusiastic" about upper valley agriculture and its potential. He made clear, however, that Reclamation Service assistance for upper Snake River valley irrigators, in the form of reservoir water, lay in the future. He told the assembled farmers that "the government had no propositions to make" and that, for now, the best way for them to get more water was to reduce their wasteful use of it. At least in the near future, only the farmers at Minidoka and Twin Falls would have access to the water kept in Jackson Lake.[24]

Ad hoc allocation of water on the upper Snake River thus continued, although not without challenge from aggrieved upstream irrigators. These people still believed that the court-appointed watermasters, even if from outside the valley and therefore ostensibly neutral, overlooked their water rights when allocating the Snake. To strengthen their case, farmers around Rigby and Idaho Falls formally organized. In August 1910, representatives from canal companies on the South Fork of the Snake met in Rigby to form the Farmers Protective Irrigation Association (FPIA). Like the less formal Blackfoot group of five years before, the FPIA was a cooperative organization. Initially, however, it had no intention of working with the Blackfoot irrigators and dividing the waters of the Snake "by mutual consent." The FPIA was strictly defensive. The court-appointed watermasters were taking the upstream farmers' water; through their new organization the farmers aimed to get it back.[25]

But the upstream farmers' defensiveness lasted only a short time. Events soon made them more willing to cooperate and share water with the downstream users. In December 1910, a few months after the FPIA's birth, Judge Stevens finally issued the decree that settled *Rexburg v. Teton*. By August 1911 the decree appeared in print; now all the irrigators along the upper Snake could see just where their rights stood in the list of priority. No doubt to their delight, the farmers of Rigby and Idaho Falls saw that their rights indeed were the oldest in the upper valley. No one could take their water from them now; they could be as selfish with it as they wanted. Yet selfishness was not what the Rexburg decree induced in the upstream irrigators; on the contrary, it actually made them more generous. The decree apparently convinced them that they could release some of their water to Blackfoot without jeopardizing their rights to it. If their rightful control over the donated water were challenged because they had failed to use it themselves (use it or lose it), they could invoke the unassailable decree. Thus by confirming prior appropriation and its individualistic basis, by figuratively chiseling the system in stone, the Rexburg decree actually made upstream farmers feel secure enough in their rights to move beyond self-interest and send water to needy farmers with junior rights. As early as 1913, upper valley irrigators began to make this extralegal practice—which they called "temporary transfers"—a regular part of upper Snake River allocation. And the organization that coordinated the temporary transfers was none other than the FPIA. Born out of antagonism, the FPIA matured into an agency of community accord.[26]

Cooperative organization thus became essential to stable water allocation on the upper Snake. The Rexburg decree and prioritization of water rights alone would not have established peaceable or equitable water distribution. A watermaster still had to reconcile an idealized system with an erratic, complicated river. Prior appropriation recognized that a stream's total available water supply fluctuated, but beyond that, the list of prioritized rights represented water abstracted from nature, from its spatial context and from its hydrology. The Snake as a whole did not simply rise and fall in volume; because of groundwater levels, rainfall on various parts of the watershed, and "return flow" (surface or underground water that flowed into the river from canals and fields), the river fluctuated in particular reaches of its channel. Thus a watermaster still had to match the priority list with this dynamic stream, something that he could never accomplish perfectly. To the extent that real water did not correlate with water rights, there would be cause for quarrels. In particular, the most vulnerable irrigators—those with the latest rights, and those in the Blackfoot vicinity especially—would have reason to challenge the allocations. But on the upper Snake River, the FPIA and the extralegal system of temporary transfers forestalled such conflict. During crisis, irrigators did not necessarily have to resort to judges or lawsuits to get water.

Upper valley irrigators did not find themselves living completely without conflict, however. Even as they worked out differences among themselves they had to confront a more serious threat to amicable water distribution. This threat came from Jackson Lake dam, the waters of Jackson Lake reservoir, and the recipients of that stored water: the farmers of the Minidoka and Twin Falls projects.

Water managers intended the dam to regulate a dynamic river: the structure captured the Snake's spring flood and created a tremendous water reserve that, when finally released, counteracted the July or August drop in the river's volume. But *regulate* is not the best word to describe what Jackson Lake dam and reservoir did to the Snake, because the stored water actually complicated the river's hydrology and thus the irrigators' water allocation systems. In the Snake's channel, reservoir water intended for Minidoka and Twin Falls mixed with water that normally flowed there. Water distribution now became even more difficult than before, because upper valley irrigators—those around Rigby, Idaho Falls, and Blackfoot—had to distinguish not only between chronological rights but between two kinds of water: stored water and natural flow. The problem of physically allocating the two intermixed categories of water led to conflict between irrigators around Rigby, Idaho Falls, and Blackfoot (natural flow users) and those in the vicinity of Minidoka and Twin Falls (stored water users).

Abstractly, the differences between stored water and natural flow were quite stark. Stored water—the reservoir—resembled a resource that people stockpiled; relatively motionless, it sat within a sort of container. Groups of downstream irrigators held prioritized rights to the stored water, but these rights did not function the same way that natural flow rights did. Irrigators exercised their natural flow rights at the specific places where they actually diverted water from a stream into ditches or canals. In contrast, users of stored water legally acquired that water at the reservoir, not at the headgates of their canals. But the strongest distinction between the two categories appeared in the different ways that irrigators measured them. Irrigators expressed their stored water rights in terms of a static measure, acre-feet; they represented natural flow rights as cubic feet per second (cfs) a figure that accounted for motion as well as volume.[27]

The problem for water managers—state, local, and federal officials—was keeping track of the two types of water as they flowed together in the Snake. The officials had to distinguish stored water from natural flow, both in the reservoir itself and in the river's channel; they had to distribute water to upper valley irrigators with natural flow rights and at the same time shunt water through the river for more than 300 miles to stored water users at Minidoka and Twin Falls. The engineer A. E. Robinson summed up the problem in a 1912 report. "On the face of it," he wrote,

it would seem to be a simple problem to divert a certain amount from a reservoir into the head of a stream, and take it out down below undiminished except for losses due to natural causes, and at the same time allow persons holding rights to the natural flow to draw the water to which they were entitled. But actual solution of the problem depends upon so many unknown quantities that the equitable distribution of water becomes one demanding a great amount of hard work and clear thinking.

The challenge that Robinson and other water managers faced was nothing less than a variation on the basic problem of reconciling the ideal, schematic, legal river and the real, complicated stream.[28]

The difficulties began at Jackson Lake. Here reservoir water sat within a container, but the container was not exactly like a bottle or tank—it was not closed and impervious. Rather, the reservoir water was subject to uncontrollable, and to a certain extent unmeasurable, hydrological processes. Pooled in Jackson Lake, water both evaporated and seeped into the ground. When the Reclamation Service drew the reservoir down, the "bank storage" seeped back in. The precise quantities of evaporation and bank storage, moreover, varied according to hydrological conditions. Cloudy weather inhibited evaporation; clear skies boosted it. Drought and low water lessened bank storage; abundant precipitation and a full reservoir increased it. These conditions presented water managers with nearly irresolvable problems. Did evaporation and bank storage come from stored water or natural flow? If both, what percentage did each type of water contribute? What was the evaporation and bank storage in the natural Jackson Lake, the lake prior to the dam? How did the reservoir affect this natural evaporation and bank storage? Such questions were virtually impossible to answer because both stored water and natural flow mixed together in the reservoir.[29]

Transporting stored water through the Snake's channel magnified the problem of water allocation. One difficulty typically developed around July. At this time, the natural flow of the river began to decrease. Irrigators in the vicinity of Rigby, Idaho Falls, and Blackfoot now had to confront the fact of scarcity. Their Snake—the natural flow—was receding in its bed. Yet this diminution was not obvious. Hot July weather usually compelled the downstream Minidoka–Twin Falls irrigators to request formally their reservoir supply; just when the natural flow was decreasing, the Reclamation Service opened the gates at Jackson Lake dam, and stored water surged down the river. Consequently, around Rigby, Idaho Falls, or Blackfoot, the Snake's total volume might not be falling at all. Indeed, the stored water might actually cause the river to rise. But the irrigators of Rigby, Idaho Falls, and Blackfoot did not understand the complex relationship between stored water and natural flow. They looked at the full Snake and refused to believe the water managers who told them that their portion of the river—the natural flow—was decreasing, that priority would have to be implemented and headgates closed. It simply did not make sense to these natural flow irrigators that they should be deprived of water when the river was so full.[30]

And the problems did not end here. Water managers had to account for fluctuations in stored water and natural flow as the two kinds of water ran through the Snake. Along its course, the river lost water to evaporation and seepage. The amount of loss depended on an array of factors, including quantity and speed. The greater the volume of water, for example, the greater the loss. These fluctuations baffled water managers and made allocation difficult. To which category—stored water or natural flow—did the loss apply? What percentage of loss could be attributed to each?[31]

Water managers had an equally difficult time assessing gains in the Snake's volume. Gains came not simply from springs or a high water table but from return flow, the source of which was unclear. Return flow consisted largely of irrigation water (originally natural flow) that seeped back into the Snake from canals, ditches, and fields in the area surrounding Rigby, Idaho Falls, and Blackfoot. But return flow also consisted of stored water that, after its release, was lost in transit and that eventually trickled back into the river's channel. Water managers faced enormous problems tracing return flow to its source and then crediting it to stored water or natural flow. But to make matters worse, geography and time further confused the task. For example, did return flow from stored water loss seep back into the river in time for the water managers to credit it to the stored water users, or did it finally make its way back into the river after the irrigation season had ended? Or did return flow from upstream canals enter the river too far downstream for the natural flow irrigators to make use of it? Water managers could not readily reconcile return flow in two columns on a ledger sheet. They could not easily match the legal river's abstract quantities with the mysteries of the real Snake.

Nevertheless, they had to make the attempt. Each year water managers compiled measurements from the limited number of gauging stations on the Snake River and then made informed approximations of the losses and gains to stored water and natural flow. Their calculations generally favored the natural flow users. During the 1910s, water managers typically estimated an overall 5 to 10 percent loss of stored water between Moran gauging station, just below Jackson Lake dam, and Blackfoot station, downstream from the headgates of the last upper valley canals.[32]

Natural flow irrigators in the upper valley had no objections to the water managers' methods—as long as the river carried ample water. But with drought and water scarcity, they changed their minds and began to find fault with the system. In particular, natural flow irrigators with the latest rights, especially those in the Blackfoot area, offered the greatest objection to the official accounting of the Snake. When water grew scarce and the natural flow of the river fell, water managers began to close headgates of canals with late rights. Starting with the last water right, they cut off each successive older right, closing headgates as necessary to supply the canals with the earliest rights. Irrigators with late rights now had reason to question the calculations, and they argued that the Snake actually carried more natural flow than the managers had figured. Anger increased in proportion to the diminishing water, and once again headgates became points of contention.[33]

Controversy over stored water and natural flow involved numerous groups, organizations, and overlapping government agencies. The Reclamation Service (renamed the U.S. Bureau of Reclamation in 1923) operated Jackson Lake dam and reservoir for the Minidoka–Twin Falls irrigators. The U.S. Geological Survey maintained stream gauges along the Snake, which it used to collect data on the river's flow. Beginning in 1909, the Idaho state engineer appointed a special deputy to control and allocate stored water that flowed from Jackson Lake. Until 1910, a court-appointed watermaster regulated the natural flow of the Snake in the area covered by the Rexburg decree; starting in 1911, water users in this area (designated Water District 36 in 1915) elected their watermaster. Meanwhile, irrigators around Minidoka and Twin Falls had their own watermaster. In addition to the official water agencies and administrators, grass-roots organizations formed. The FPIA was one such group; there would be still others. These entities gradually worked toward the creation of an extralegal, cooperative management system—eventually embodied in the Snake River Committee of Nine—that attempted to resolve the technical problems of dividing and distributing stored water and natural flow.[34]

The conflict that led to the formation of the Committee of Nine developed as early as 1911. In late July of that year, the Reclamation Service adjusted the gates at Jackson Lake dam to flush 10,000 cfs of water from the reservoir. As the reservoir level fell, the discharge gradually decreased to 1,800 cfs, until at last the delivery to the Twin Falls and Minidoka projects was complete. The dam tenders then partially closed the gates, allowing only the calculated natural flow—1,000 cfs—to pass over the dam. As the Snake drastically fell in volume from the peak stored water discharge, "reports . . . circulated the upper valley," said D. G. Martin, "to the effect that the Government had closed its gates at Jackson Lake entirely and was not allowing any water to escape into the stream." The upper valley farmers' perception, however mistaken, revealed their fear that Jackson Lake dam and its operators deprived them of their natural flow rights.[35]

A similar scenario unfolded in the summer of 1914. In July the river fell as it usually did, but this time it dropped lower and more suddenly than in other years. On July 19, the Reclamation Service opened the gates at Jackson Lake dam, allowing stored water to pass into the Snake. Shortly thereafter, the state engineer's special deputy and his assistants began to close upper valley canals holding late rights. Natural flow irrigators now observed, and rebelled against, the apparent paradox of headgate closures while the river still carried plenty of water overall. Special deputy W. O. Cotton reported that these irrigators "could not understand why they were cut so much more this year than ever before, and so sure were they that there was some error somewhere that they were protecting their rights with military force." Cotton did not explain what this military force entailed, but his statement probably referred to angry farmers guarding headgates along the Snake.[36]

As the crisis deepened, water managers appealed to the FPIA for help. Cotton reported that he arranged with the FPIA officers to "lay this matter before them and see if we could not secure their cooperation in holding some of the canals to the amount to which they were entitled." At the meeting, representatives of canals holding late water rights—the Farmers' Friend, Rudy, Snake River Valley, and others—voiced their belief that the natural flow was not falling as rapidly as Cotton had calculated.

The FPIA backed up their claim, protesting the Reclamation Service's "interference" with the decreed rights of upper valley irrigators. The FPIA asked the Reclamation Service to determine accurately the return flow and credit it to the users of natural flow.

Determining the return flow was virtually impossible, so the FPIA and the Reclamation Service worked out a temporary compromise. For ten days, canals in the upper valley received a greater amount of return flow than the Reclamation Service had credited them, enough to quiet the concerns of irrigators with late rights. Through cooperative means, the irrigators and the Reclamation Service had averted disastrous conflict, perhaps even violence.

And they had set a precedent for additional cooperation in Snake River allocation. Beginning in 1915, the Reclamation Service, working with the FPIA, made limited amounts of reservoir water available to upper valley irrigators. In August 1915, water shortage again threatened the natural flow users who owned late rights. John Empey, the FPIA president, in conjunction with the Reclamation Service, other government agencies, and prominent figures such as Senator Fred Dubois, arranged for upper valley irrigators to receive water from Jackson Lake reservoir. The FPIA acquired 1,000 cfs of stored water for ten days (a figure mathematically converted from acre-feet) and agreed to have upper valley canals with the oldest water rights transfer some of their natural flow to canals with junior rights. After ten days, the irrigators with the late rights returned the borrowed water (presumably by allowing natural flow either to collect in Jackson Lake reservoir or to pass on to Minidoka Dam and its reservoir, Lake Walcott). The arrangement thus allowed upper valley irrigators with late rights to avert disaster and mature their sugar beets. Upper valley irrigators again secured water from Jackson Lake reservoir several years later. In 1916, the Reclamation Service rebuilt Jackson Dam and expanded the reservoir by 400,000 acre-feet. The following year, thirteen upper valley irrigation companies and districts contracted with the Reclamation Service to receive 100,000 acre-feet of water from the expanded Jackson Lake. These upper valley irrigators still relied primarily on their natural flow water rights, but the dam now gave them a small reserve.[37]

New organizational innovations and planning processes contributed to cooperative water management on the upper Snake. In August 1914, following the July water allocation crisis, all natural flow irrigators in the upper valley—including Blackfoot canal companies holding late rights—joined together under the leadership of the FPIA. The FPIA further consolidated its position in 1916, when irrigators elected its president, John Empey, to the position of watermaster for Water District 36. As watermaster, Empey had responsibility for allocating the Snake's natural flow between Milner Dam near Twin Falls and the Wyoming border. In his position, he strove to consolidate and streamline management of natural flow and stored water. He coordinated his duties with those of D. G. Martin, the state engineer's special deputy in charge of stored water, even working out of Martin's headquarters in Idaho Falls. Empey in 1917 further proposed that all irrigation organizations and government agencies hold a meeting each spring "for the purpose of having a clearer understanding of general matters" related to the Snake River and for "promoting friendly relations" among stored water and natural flow irrigators. In April 1918, following his suggestion, representatives of the state engineer, the Reclamation Service, the Twin Falls projects, and the FPIA met at Idaho Falls to lay out "tentative plans" for water management in the upcoming months.[38]

Water users and agencies soon took steps to institutionalize this cooperative management of the Snake. In April 1919, representatives of irrigators met with state engineer Warren Swendsen to prepare "a plan which would provide a more stable organization and one better able to cope with the difficulties [of water management] as presented from year to year." A committee of nine men, three each from Henrys Fork, the South (main) Fork, and the Minidoka–Twin Falls projects, worked out "the details of this plan." This was the genesis of what would become the Committee of Nine. Acting on the committee's recommendation, Swendsen appointed engineer G. Clyde Baldwin, a USGS official responsible for stream measurement in Idaho, to be special deputy in charge of stored water. Swendsen then selected three men, one from each of the three areas, to serve as an advisory committee to assist Baldwin and make recommendations for Snake River management.[39]

Thus by spring 1919, irrigators and government officials had erected a loose network of organizations, agencies, committees, and allocation practices intended to distribute stored water and natural flow smoothly and to contain conflict. As a whole, however, these arrangements had yet to be tested by severe drought. Indeed, irrigators and government officials had worked out many of their cooperative programs in a wet period. The years 1916–18 were the time of Idaho's great dry farming boom, a brief moist era when the skies showered water on the usually arid Snake Plain and the river's spring flood rose to its highest level since 1894. With water so abundant, perhaps irrigators could afford to be more affable and cooperative. Yet this surfeit of water soon came to an end. When the irrigators met with Swendsen in April 1919, Jackson Lake reservoir had accumulated only about 70 percent of its capacity. Snake River valley irrigators did not know it yet, but the coming summer would be one of the driest on record.[40]

Farmers in the upper valley began to experience the effects of the drought by early June, the time in wetter years when the river attained its peak flow. On 11 June, working down the list of priority, the watermaster cut all water rights to 1903; by 1 July, he had shut off all post-1890 water rights; by 1 August he had worked his way down to the 1885 rights. As the headgate closures continued and crop failures appeared imminent, irrigators with the latest natural flow rights grew restive, and conflict developed. Most desperate irrigators refrained from fighting with one another or obstructing watermasters and government officials, but a few did not. In late June, the District 36 watermaster cut off the Corbett Slough's 1892 water right of 130 cfs, leaving irrigators with only an 1889 right of about 109 cfs; days later, on 7 July, Koury and Grover met in their fateful headgate battle. Some irrigators raised canal headgates on the Snake River after water managers had lowered them. Even chains and padlocks on the headgates failed to stop these water-hungry farmers. The USGS's Baldwin, who was special deputy for stored water, reported that he and Empey had posted guards at the headgates of three canals to prevent farmers from tampering with them. Throughout the upper valley, a few irrigators found themselves arrested and facing criminal charges for interfering with water distribution.[41]

As the situation worsened, upper valley irrigators together took steps to allocate water efficiently, provide water to those with late rights whose crops were in danger of failure, and prevent further conflict. They gathered in Blackfoot and Idaho Falls and devised plans for securing stored water from Jackson Lake. Upper valley canals with rights to the reservoir, however, had already exhausted their small supply. And the drought had drastically reduced the stored water of the Minidoka–Twin Falls farmers, who felt unable to lend any of their reserve as they had done in 1915. Thus the only relief measure left for upper valley irrigators was the system of temporary transfers, first implemented in the early 1910s, in which farmers with senior rights and more ample water transferred their natural flow to canals with junior rights cut short by drought. According to Baldwin, "More than one hundred of these temporary transfers were made. So far as known the water delivered . . . was always donated." Irrigators in the upper valley could not avoid some crop losses, but the temporary transfers allowed these people to minimize the worst effects of the drought. Farmers at a disadvantage, such as those in the Blackfoot vicinity, received at least a minimal amount of relief during the crisis.[42]

On the whole, the cooperative water management arrangements had achieved the objective of efficient, amicable water allocation during the 1919 drought. Yet signs of tension and conflict had been evident. Some violence did occur. And clearly, incidents of headgate tampering on the Snake involved farmers who were not just desperate for water but who refused to believe that what they saw flowing in the river was mostly stored water, not natural flow. On 9 July the natural flow of the river had reached its lowest point a month earlier than usual, "but owing to the stored water passing down the channel it did not present the same appearance." To the fearful and angry irrigators, the Snake did not seem drought starved; their eyes told them otherwise. Other than the word of the river managers, these farmers had no proof that the Snake was predominantly stored water. They could not see, and therefore could not accept, the abstract divisions of water that made up the legal river. They suspected that the stored water users at Minidoka and Twin Falls were benefiting at their expense.[43]

Some of the farmers believed that the inequities originated in the measurement and control of water at Jackson Lake reservoir. Consequently, John Empey and a committee of irrigators from the FPIA traveled to Jackson Lake in early August 1919 to investigate. The committee found no evidence of mistakes or wrongdoing. "After inquiring into and examining the methods now being used to determine the natural flow and stored water," the group concluded, "we are satisfied that the water is being handled with justice to all as nearly as it is humanly possible."[44]

The FPIA's findings failed to satisfy the aggrieved upper valley irrigators, who in coming years continued to believe that they were not receiving all of the natural flow to which they were entitled. Water shortage, which continued through the 1920s, exacerbated their discontent. In August 1920, George Hill of Rigby perhaps summed up the feelings of many of his fellow irrigators in a letter to the *Idaho Falls Times*. Hill argued that the Snake carried plenty of natural flow even though the special deputy and the watermaster had begun to close upper valley headgates. His proof was that Minidoka–Twin Falls farmers had more than enough water, so much that it was flowing over Minidoka Dam and Shoshone Falls, downstream from their main headgates. In fact, the special deputy and watermaster occasionally did miscalculate losses and gains to the extent that the river at Minidoka or Twin Falls might have more water than farmers there could use. Hill, however, believed that "the people of the lower country" had "a well-laid plan" to "undermine our water rights." Eventually angry irrigators like Hill—especially ones with late water rights—reopened the issue of stored water and natural flow and challenged the FPIA, the Reclamation Service, and the Minidoka–Twin Falls farmers. Snake River irrigators would have a final conflict to overcome in their efforts to achieve stable, cooperative water allocation.[45]

In 1921, discontent among Hill and other upper valley irrigators coalesced into a political challenge to Empey, the FPIA, and Baldwin. In November of that year, representatives of twenty irrigation companies and districts met in Rigby to form the Upper Snake River Union of Water Users. Although open to all upper valley irrigators, the Union of Water Users clearly reflected the interests of those with late rights or a geographical or hydrological disadvantage. Blackfoot irrigators, including H. N. Smith of the Peoples' Canal Company, led the move to form the organization, which elected him its president. Other members of the Union of Water Users included the Idaho Irrigation District, the Enterprise Canal Company, the Farmers' Friend Irrigation Company, and others, all of which depended, at least in part, on vulnerable late water rights.[46]

The Union of Water Users signaled the fragmentation of the coalition of irrigators that had formed the FPIA. Leaders of the new group "stated that the . . . association is in no way antagonistic to the Farmers Protective Irrigation [Association], but rather an additional assistance to the canals of the valley in securing a square deal on water." Although the FPIA had done much to advance the interests of upper valley irrigators as a whole, the Union of Water Users clearly believed that it had not fully lived up to its purpose, especially its role of defending the interests of irrigators with late natural flow water rights.[47]

The Union of Water Users thus attempted to change water allocation practices on the Snake. First it tried to unseat Empey in the election for District 36 watermaster. Failing at that, the group then decided that it must hire an engineer "to check up on the water ratings of the various canals, independently of the reclamation service or the Farmers Protective Irrigation association." The Union of Water Users hired William O. Cotton to perform the work and to represent the organization in its dealings with the state engineer, his special deputy, and the Reclamation Service. Cotton, a principal in the Idaho Falls engineering firm of Cotton and Wilson, had extensive experience in hydrology and irrigation. For years he had managed the Idaho Irrigation District, and at one time he had served as special deputy in charge of stored water.[48]

During 1922 and 1923, Cotton presented the Union of Water Users' case to government officials and the special deputy's advisory committee. Using a detailed analysis of the Snake River, he attempted to show the inadequacy and unfairness of water allocation practices, which he said slighted natural flow irrigators in the upper valley in favor of the stored water users downstream. First, Cotton addressed water management on the Snake below Jackson Lake dam. The special deputy should charge a greater loss to stored water in transit between the dam and Blackfoot gauging station; that official, moreover, should credit natural flow users with more return flow. Cotton then turned to accounting procedures at Jackson Lake reservoir. The Reclamation Service and the special deputy failed to credit upper valley irrigators with a natural lake storage that had accrued to them prior to the dam; furthermore, the natural flow users should receive an additional amount of water in exchange for increased evaporation on Jackson Lake caused by the reservoir's greater surface area. Thus, considering all factors (transmission loss on stored water, return flow, natural lake storage, and evaporation), natural flow irrigators should receive about 400,000 acre-feet more water each year than the special deputy had been allocating them. Cotton's claim was substantial, given that the storage capacity of Jackson Lake itself amounted to about 850,000 acre-feet. A technically proficient, respected engineer, he had mounted a serious challenge to official water distribution practices on the Snake River.[49]

Yet the conflict threatened to become more intense when government officials—primarily state engineer Swendsen, special deputy Baldwin, and the Reclamation Service's Frank Banks (chief engineer in the design and construction of Jackson Lake dam)—denied the validity of virtually each claim that Cotton made and defended existing water distribution practices. In the matter of natural storage in Jackson Lake, for example, Baldwin asserted that he and the watermaster delayed cutting water to upper valley canals on the assumption that the benefit to natural flow users from natural storage usually came in the early part of the irrigating season. Furthermore, he said, the existing procedure assumed that additional bank storage caused by the reservoir offset natural lake storage and excess evaporation. Baldwin and the other government engineers also contended that Cotton based his calculations on insufficient data.[50]

Finally, Baldwin pointed out that distribution of the Snake's water would always be potentially flawed. Although more information on the river's hydrology could be gathered, the problem of determining "proper storage transmission losses is a very complicated one . . . which . . . can never be solved exactly." Too many factors—variations in river level, groundwater, climate, return flow—affected the exact determination of stored water and natural flow. This complexity had important consequences for water management: "While the present system is a workable one and is believed to be reasonably fair," he stated, "the fact that it is not final in any sense leaves opportunity for continual complaint and dissatisfaction." Baldwin approved of any additional investigation of the Snake that would lead to a more acceptable management program, but he stated that "any such schedule [of water allocation] will be more or less arbitrary in character." Baldwin's comments clearly addressed the virtual impossibility of making the Snake conform to an abstract system that legally divided the river into stored water and natural flow.[51]

The implications of Baldwin's observations were clear: an arbitrary answer to a technical problem must be decided through negotiation, compromise, and cooperation. Only when irrigators suspended questions about exactly how the Snake functioned would they arrive at an allocation policy acceptable to all. Consequently, Swendsen set up a meeting in January 1923 at which irrigators would once again attempt to work out their differences. It was important, he told them, that they come up with a solution to the problem "without recourse to expensive and drawn out litigation." At the meeting, representatives of more than sixty canal companies appointed a special committee to prepare a cooperative plan for distributing water in 1923. R. E. Shepherd of the Twin Falls North Side Canal Company proposed that the committee consist of two men from the North Fork, four from the South Fork and the main river, and three from the Minidoka–Twin Falls area. Upper valley farmers accepted the plan, probably because it gave the South Fork irrigators, many of whom owned late rights, greater representation. In a subsequent meeting, irrigators voted to replace the old advisory committee with the new group, now formally called the Committee of Nine.[52]

For help resolving the dispute over stored water and natural flow, the Committee of Nine looked to an outside engineer. Following a suggestion from Shepherd, the committee members decided to hire an engineer of "disinterested expert judgment" to study the problems of water distribution on the Snake. As an Idaho Falls newspaper explained, the committee sought a technician with "broad experience in the control of 'intermittent' rivers—rivers that disappear and rise again; rivers that belong to different states or counties or different valleys [through] which they flow—through which they flow and sink and rise and sink again, serving perhaps in different settlements over and over as they travel on and rise as return flow into the parent stream." After perusing a directory titled *Who's Who in Engineering,* the group chose Ralph Meeker of Denver, an engineer who appeared able to comprehend the Snake.[53]

But Meeker's report did not solve the problem of water allocation as irrigators, particularly the Union of Water Users, had hoped. Delivered to the Committee of Nine in February 1924, his study basically approved prevailing practices. In Meeker's judgment, the controversy stemmed less from unfair allocation than from water shortage caused by drought and further expansion of farming in the Snake River valley. Echoing Baldwin's earlier comments, Meeker took pains to remind irrigators that they could never exactly reconcile water rights and water, the legal and real rivers. Flowing water, he stated, "cannot be measured with the same precision" as "an acre of land . . . or a ton of wheat"; it could not be apportioned "with the same accuracy as a quart of milk." Meeker was pointing to the critical gap, the gray area, between an idealized system and water, between precise quantities and unbounded nature. Knowing that irrigators usually filled that hazy space with conflict and litigation, he called upon them instead to bridge it with toleration. "At times a 'give and take' must prevail among water users," he averred. "Above all, a spirit of tolerance will do much to arrive at a place of mutual understanding." Meeker reinforced what some irrigators and water technicians already knew: ad hoc water allocation, acceptable to all and reached through compromise and cooperation, must replace conflict.[54]

Cotton and the Union of Water Users had no choice but to negotiate. They might have continued the fight, perhaps through the litigation that Baldwin said they should avoid. But circumstances, especially nature, were against them. The Snake, as the watermasters and engineers had learned, was a capricious entity, mutable and irregular within the confines of its broad seasonal cycles, a living river that humans could neither measure exactly nor regulate fully. Because of its complex hydrology, the Union of Water Users would be unable to demonstrate precisely and unassailably the amount of natural flow of which they had been deprived. More important, the upcoming season promised a water dearth as great as in 1905, 1915, or 1919, and upper valley irrigators needed at least a minimal amount of water security. In the face of the drought, the cooperative side of western irrigated agriculture triumphed.[55]

The compromise agreement of 1924, and agreements worked out in subsequent years, represented an extralegal, somewhat arbitrary method of distribution that was technically imprecise but generally acceptable to all. The irrigators did not actually solve the technical problem of distributing water: they did not discover a means to divide the Snake into quantities that exactly corresponded to water rights. Instead, the arrangements they established reflected political compromises as much as expertise in hydrology. Each year, the Committee of Nine attempted to shape the "compromise schedule" to meet irrigators' shifting perceptions of what constituted a fair distribution of water. In a sense, quantities assigned for variables such as stored water loss and return flow became bargaining chips in each year's negotiations. In 1924, for example, the natural flow users accepted a minimal percentage loss to stored flow and agreed that bank storage at Jackson Lake reservoir equaled the natural bank storage prior to the dam's construction. In return for these concessions, the Bureau of Reclamation (successor to the old Reclamation Service) allowed upper valley farmers to keep their natural flow in the reservoir for use later in the summer, when their most valuable crops—sugar beets and, increasingly, potatoes—needed it. It was a complicated, abstract arrangement: natural flow left in the Snake would be credited to stored water users; an equivalent amount of stored water would be counted as natural flow.[56]

Despite its complexity, the 1924 pact was a momentous event: it marked the institutionalization of the Committee of Nine and its yearly agreements, and it brought to a close a tumultuous phase in the development of irrigation in the upper valley. Snake River irrigators had at last devised a system of pragmatic, cooperative allocation practices. Now they had a method of curbing the tendency to become embroiled in irresolvable technical and legal arguments. The Committee of Nine's ability to resolve differences also encouraged upper valley irrigators to mend the FPIA–Union of Water Users split; in 1925, the two groups merged to form the Upper Snake River Valley Water Users Protective Union.[57]

In coming decades, these cooperative arrangements would give farmers the means to deal with still more changes and fluctuations in the river. There would be more dams (American Falls in 1927, Palisades in 1958, and others) and stored water to allocate. And as always there would be the droughts that mocked the reservoirs and threatened to disrupt the orderly distribution of water. In society as in nature, stability would be fleeting. But institutions like the Committee of Nine would allow the irrigators to respond flexibly to a world in motion.[58]

Cooperative allocation practices on the Snake were the outgrowth of the irrigators' long interaction with the river and the environment that surrounded it. The earliest irrigators adjusted their agricultural activities to meet the demands of making a living from this complex stream. They banded together to channel water to their farms. And the river's pulsations—spring rise, then a fall in volume as summer advanced—shaped their crop choices and planting habits. These farmers, in sum, addressed their activities to the real river, much as they oriented their lives to diurnal rhythms and the changing seasons.

At the same time, irrigators were capitalists, members of a society that sought to divide nature—even rivers—into individual, privately owned parts. Prior appropriation provided them with a system for translating moving water into discrete, measured quantities that had economic value. The law functioned as an intellectual instrument for transforming the real river into its legal, quantified counterpart.

But prior appropriation was imprecise. The legal river, after all, was only a representation of a stream that was much more complicated and unpredictable than the mere cubic feet per second that it carried. The real river remained part of the earth, flowing over and through it, often irregularly and erratically. Engineers knew that the Snake held water, but they did not know *exactly* how much, when, and where the river carried it. Privatizing the stream, turning its flow into exact quantities at the headgates, was not an easy task. In practice, prior appropriation proved to be a poor method of transferring water across the shifting, ambiguous zone that separated nature and culture, environment and economic system.

In the upper Snake River valley, the problems of prior appropriation appeared most evident after the construction of Jackson Lake dam and reservoir. Here, engineers sought to wrest water from nature, from the hydrologic cycle. They captured the flood caused by solar heat melting the snow and ice of winter, and then they stilled it, dissipating the kinetic energy that it had gathered as it tumbled down the peaks and crashed through the rocky bed of the Snake. In the reservoir, engineers refined water, stored it, made a surplus of it. Thus contained, it was like capital, a necessary component in agricultural production.

But the engineers' attempt to separate water from the hydrologic cycle was both incomplete and temporary, and it complicated water distribution under the system of prior appropriation. In the reservoir, water was still subject to evaporation and seepage. Even greater instability arose in the delivery of the stored water. To send it to irrigators at Minidoka and Twin Falls, engineers had to reintroduce it to wildness, move it from its refined condition back to a state of nature. Before the farmers at Minidoka and Twin Falls could use it, the stored water had to travel 300 miles through the Snake River's channel. In the reservoir and then in the Snake itself, the two abstract categories of water, stored and natural flow, mingled. Engineers could generally distinguish between them, but not perfectly. The disjuncture between actual water, the real river, and categories of water rights invited conflict among irrigators.

The introduction of stored water to the Snake had still more consequences for water management: it conflicted with the upper valley farmers' long-standing experience of the river. In dry years, as the Snake fell in late June or early July, they watched the seeming paradox of watermasters closing their headgates while the stream still flowed at a relatively high stage. They could not accept, or did not want to accept, that the rushing water actually represented a decrease of natural flow and an increase of stored water being released from Jackson Lake reservoir. Indeed, irrigators probably did not think of the Snake in terms of discrete categories; a farmer who dipped a bucket in the river could not identify which portion of water was stored and which was natural flow. The failure of water managers to distinguish precisely between the two only intensified the upper valley irrigators' fear that someone was cheating them. These farmers were capitalists, but their faith in that ultimate of capitalist abstractions—the legal river—proved limited.

To overcome the conflicts, especially legal conflicts, that often ensued from water distribution problems, irrigators turned to extralegal forms of cooperation. A basic instrument of cooperation was the temporary transfer, in which irrigators with early water rights lent water to farmers whose late rights left them vulnerable to scarcity. In a sense, a moral economy of water underlay the transfers: the practice rested on the basic belief that it was wrong to let a neighbor or nearby farmer fail for lack of water. Temporary transfers showed that irrigators could both uphold the individualist, competitive system of prior appropriation and periodically agree to set that system aside in the interest of helping each other.[59]

In the Farmers Protective Irrigation Association, the special deputy's advisory committee, and the Committee of Nine, extralegal cooperation took institutional form. When these organizations functioned most effectively, they bridged tremendous differences among irrigators and between irrigators and government officials. Like the temporary transfers, these organizations reflected not the abandonment of prior appropriation but rather the effort to work out compromises when the law failed to address the complexity of the Snake and its tributaries.

When assessing the Committee of Nine, it is perhaps tempting to look beyond its cooperative basis and see the outlines of an authoritarian agency, a kind of Hobbesian Leviathan that ended anarchy and established order in the allocation of a scarce natural resource. In certain respects, the gradual formation of the committee was part of a trend that indeed concentrated power in the hands of a few technical experts. G. Clyde Baldwin, the USGS engineer and special deputy in charge of stored water, once likened his position to that of the "general manager of a corporation" and the advisory committee—precursor to the Committee of Nine—to the "board of directors." But to conclude that the Committee of Nine and the lawyers and engineers associated with it constituted an authoritarian structure of water management is to overlook its fragility. Baldwin and the committee were only as strong as the support they received from Snake River irrigators. When some of them, like the Union of Water Users, threatened to withdraw their support, the entire system stood on the verge of collapse. The Committee of Nine thus is best understood not as a Hobbesian Leviathan but as an ad hoc, tenuous organization built upon compromise, consensus, and finally cooperation.[60]

That cooperation had deep roots in American civilization. Mormon farmers in the upper valley, with their communitarian ethos and cooperative irrigation methods, no doubt encouraged and reinforced the tendency of all farmers along the Snake to work together to allocate water. But non-Mormon irrigators—in the Blackfoot vicinity, for example—also implemented cooperative practices, by themselves and in conjunction with Mormon farmers. The FPIA and other organizations resembled claims associations and clubs common throughout the West and Midwest. Furthermore, the FPIA, temporary transfers, and the Committee of Nine developed during a time when farmers across the nation seized on the idea of cooperative organization as a means to survive in a complex industrial economy. Finally, the patriotism and government economic programs associated with World War I created a climate that encouraged citizens to work together for the common good.[61]

Most important to the formation of cooperative practices was the encounter between the irrigators and the river itself. The irrigators simply had no easy means to transform the Snake into precise quantities flowing onto the fields of individual farms. Hydrological conditions were too complex and erratic, and the opportunities for disagreement, conflict, and violence too great. The irrigators could achieve their objective only by fashioning methods and organizations that allowed them collectively to negotiate the movement of water from its uncontrolled state to its final destination in their fields. In their efforts to use and alter the river, the irrigators had to confront conditions that in turn forced them to change. They had to build new social institutions through which they adapted to the complicated river that they, in part, had helped to create.

Thus the development of water management on the upper Snake River was part of the larger process of landscape formation, of human systems simultaneously transforming and adjusting to the earth. And this pattern manifested itself not just in water management but in an even more basic human activity: work. To work the irrigated landscape, to use mind and muscle to manipulate and respond to nature, required fundamental modifications in the irrigators' society.

4/ **Labor and Landscape**

Irrigated Agriculture and Work

In May 1877, journalist W. A. Goulder traveled through the Boise Valley to survey its agricultural development. His dispatches mostly depicted small family farms in a landscape of irrigation ditches, hay and wheat fields, verdant pastures, and groves of willows and Lombardy poplars. According to Goulder, cultivated acreages ran from 100 to 300 acres; individuals or families generally worked the small spreads alone. "Like most of the farmers in the valley," Goulder wrote of David Heron, "he does the greater portion of the work with his own hands." But these people were not just social isolates, for they also shared certain tasks: "Passing along down the road I saw Mr. J. B. Pierce, who, with several of his neighbors, was hard at work cleaning out and enlarging a water ditch." Indeed, Goulder observed that kinship, familiarity, and mutual agricultural burdens united the valley's settlers. He noted that many of the inhabitants were related and had lived among one another for years, and that a shared pioneer experience—including "common labors"—had "caused a fraternal feeling which makes them seem as members of one family."[1]

Yet, as Goulder's dispatches revealed, communities of small farmers were not the only people who tilled the valley's rich soil. About six miles from Boise, the journalist came to the W. L. Thurman farm—also called Thurman's Mill—and here he found a contrasting landscape and a different pattern of work. Thurman farmed on a large scale: 1,280 acres, 600 of them in wheat. The farmyard, actually the core of a mill village, included a wagon shop and a store filled with groceries and other merchandise. Nearby stood the Thurman family residence, "separately inclosed and surrounded by shade trees." A flour mill, powered by water from an irrigation canal and capable of producing fifty barrels of flour per day, formed the centerpiece of the village. Thurman, of course, did not perform most of the work with his own hands, as did farmers like David Heron. Rather, Thurman's "agent and man of business, Mr. Harrison," oversaw the entire operation, including many laborers. "This large estate and business forms a little empire of itself," Goulder reported, "and requires quite an army of employees to keep the work going on."[2]

Family farmers and an army of employees: Goulder had described the two predominant types of labor that shaped Idaho's irrigated landscape between the 1860s and the 1920s. Each reflected a particular pattern of human interaction with the biophysical world; each produced distinctive landscape forms. One type, the sort that Goulder observed among the common farmers of the Boise Valley, can be called the *family mode.* Under this arrangement, family, friends, and neighbors used simple technology to build and maintain irrigation systems collectively. Although these people generally worked their crops by themselves, they helped each other at harvest and other times. Their world, their land, bore the imprint of their small-scale labor: rock, brush, and canvas wing dams that extended into rivers; small, crude canals that wound through river bottoms; little farms and patchwork fields.[3]

A second, contrasting type of labor was the *industrial mode:* the use of managers, engineers, and large numbers of wageworkers for agricultural production. Businessmen such as Thurman mobilized this type of labor on an elementary level; during the late nineteenth and early twentieth centuries, modern, bureaucratic corporations—with the assistance of government agencies—expanded and perfected it. Like the family mode, the industrial mode produced corresponding landscape elements. Technological expertise and massed labor generated large dams, vast networks of canals, sizable farms, mills, and eventually huge processing factories that towered over the surrounding fields.

To travel through the irrigated landscape, as Goulder did, was to encounter these two forms of labor and the structures and fields that each produced. Yet each form was, of course, not completely separate. In everyday life, the two patterns often overlapped or merged, as the Boise Valley in Goulder's time demonstrated. Thurman, after all, did not operate his mill in isolation from the surrounding small farmers; Thurman's workers milled the farmers' wheat. Indeed, the small farmers' prosperity, even existence, in part depended on dealings with operations such as Thurman's Mill. By the same token, Thurman profited from his association with the nearby small producers. Thus family and industrial labor can be described in isolation, as distinct categories, but at certain places and moments they exhibited important—and sometimes surprising—symbiotic connections.

Each mode, family and industrial, at its most basic level manipulated water with dams and canals. Between the 1860s and early 1900s, family farmers labored on dozens of small irrigation systems along the Snake River and its tributaries. Relatives and neighbors who lived near each other often came together to build these diversion structures and conduits. In 1864 and 1865, Seth Bixby, John Carpenter, John Patterson, and other nearby farmers excavated the New Union Ditch Company Canal in the Boise Valley. R. H. and Smith Stockton built the Parma Canal in 1877 to carry water from the Boise River a short distance to their farms. William Bubb, M. E. Payne, M. E. Pratt, and Charles and Jacob Rein began work in late winter 1889 on a canal (called the Bubb or Payne Canal) that also channeled water from the Boise.[4]

Religion sometimes reinforced the family and neighborhood ties that formed the basis of small-scale canal labor. Mormon communities in the upper Snake River valley, for example, blended kinship and a patriarchal religious order in the establishment of irrigation systems. The Salem Union Canal, built by the farmers of Salem around 1890, typified the Mormon system. Augustus Belnap, a church leader later recognized as the "father" of the canal, used his religious authority to exhort and organize local farmers to construct the system. The first group of six men to work on the project included Belnap and his brother Amasy, and the brothers Pete and Hans Hansen. Spiritual ideals, a symbolic father, and actual kinship connections blended in the toil and sweat that produced the Salem Union Canal.[5]

The actual process of canal construction reflected the grass-roots labor methods of the pioneer irrigators. Often these early farmers formed legal corporations to excavate and maintain irrigation systems. In Idaho and other parts of the American West, these entities were known as mutual or cooperative companies; Idaho farmers also called them home companies. Such terms indicated the local, cooperative basis of the operations. Each irrigator purchased shares in a company, and each year the company assessed a fee on the shareholders to pay for maintenance. In theory, the farmers would buy shares and clear their assessments in cash, and the company would use the money to employ workers to dig and repair the canal. But in practice this is not what usually happened. Early Idaho farmers had little cash—often they had yet to harvest a crop—so they purchased their shares and paid their assessments in labor. As W. A. Goulder remarked, canal companies that operated in this fashion represented "more muscle than money."[6]

Because they had more muscle than money, and because that muscle was limited, the early irrigators tended to build canals where it was physically easiest: in and around bottomlands in the Snake River forks country or along tributary rivers such as the Boise or Payette. Here the small groups of kin, friends, and neighbors took advantage of natural conditions to maximize their limited skills and labor power. In the relatively flat, gentle topography of the bottoms, the farmers diverted and routed water with relative ease. In such places they did not usually have to overcome steep elevations or other obstacles. Some early irrigators linked their systems with natural sloughs. In 1885, the farmers of the Burgess Canal and Irrigation Company began to construct a conduit to their farms. "Very little work was needed," wrote M. J. Hammond, "as they chose one of the old channels left years before by the flood waters of the Snake river."[7]

But building a canal was not always so easy—the process often involved far more than just routing water through an old river channel. Unlike the Burgess Canal and Irrigation Company conduit, most canals required at least a degree of planning and intensive physical work. To put in an irrigation system, farmers used their knowledge of simple construction techniques, their own collective effort, and the power of their horse and mule teams. Sometimes they hired an engineer or surveyor to locate a canal route, but on many conduits they performed the task themselves. To determine where the canal would go, farmers might use a simple survey device made from a level and a board. In some instances, though, they just situated the canal in the general direction they wanted it to follow. "It is purported that the Harrison Canal," wrote S. Reed Andrus of an upper Snake River valley conduit, "was routed by the natural eye without instrumental survey."[8]

After determining the canal route, irrigators excavated it. Together the men tore at the earth with picks and shovels. They hitched plows to horse and mule teams and made furrows to loosen the ground. They scooped soil and rock with small horse-drawn scrapers. One kind of scraper was a crude device made of boards fitted with handles and metal edges. Another type was a bucket made of riveted iron, called a slip scraper or Mormon scraper, that held about one-fifth of a cubic yard. When scraping the earth, the farmers worked in pairs; one man drove the horses, the other emptied the scraper. Sometimes the men diverted a small quantity of water into a half-completed ditch to soften hard ground; perhaps this is why they also called the slip scraper the slush scraper. Women also participated in the work, further making canal construction a family endeavor; wives and daughters cooked meals for the men, sometimes at the construction site, and tended crops and livestock.[9]

With limited resources and few members, the home companies usually built small, simple irrigation systems. Canals were narrow, shallow, seldom more than ten or fifteen miles long at most, and they delivered water to farms that together encompassed only several hundred or a few thousand acres. By 1905, the New Union Ditch Company Canal irrigated 689 acres through a ditch that ran about two miles. The Parma Canal carried water roughly one mile to just 440 acres. The Bubb Canal ran for two miles and by 1903 served 770 acres. Often the physical structure of the early ditches was quite crude. The Bubb Canal measured six feet across on the bottom and two feet in depth. When completed in 1865, the Siebenberg Co-Operative Ditch Company Canal was but four feet wide at the bottom and only a foot and a half deep. Joseph Jensen recalled that in the late 1890s, the Riverside Canal, a conduit in the upper Snake River valley, was "nothing but a 'pig trough.'"[10]

Despite such rough beginnings, the home company irrigators worked each year to improve and maintain their systems, and so the need for cooperative labor continued long after the initial completion of a canal. As the demand for water increased—as the population grew, as irrigated acreages increased, and as farmers adopted water-intensive crops—they made their canals wider, deeper, and sometimes longer. And they performed annual maintenance on their systems. They removed debris, shored up banks, and dug out silt that had settled in the canal bottoms. They restored wood, stone, and concrete headgates. They rebuilt the flimsy rock, brush, and canvas diversion dams that the spring floods destroyed. At times the work was extremely taxing and dangerous. Floodwaters surged through the Hillside Canal in the upper Snake River valley one spring, washing out a portion of the canal bank and threatening to inundate fields. The farmers scrambled to repair the breach, but the water carried away each sandbag that they dropped into the opening. One man came up with a solution to the problem: he lowered himself into the cold water and stood on the sandbags, holding them in place with his weight. He labored in this fashion for an hour and a half, as the bags piled up. Then he mounted a horse and rode three miles for a change of clothes. For the early home company irrigators, the canals were not just structures that conveyed water; they were monuments of pioneer toil.[11]

The names that farmers attached to canals expressed their pride in their labor and the sense of ownership they derived from the work. Irrigation systems sometimes featured titles with words such as *cooperative, mutual, peoples', farmers',* and *union.* The Siebenberg Co-Operative Ditch Company, for example, which originated in 1865, delivered water to a small group of farmers in the Boise Valley. In 1894, a group of irrigators in the upper Snake River valley began work on the Peoples' Canal near Blackfoot. A group of farmers along Falls River, a tributary of the Snake, established the Farmers' Own Ditch Company in 1896 to convey water to their fields. Many early irrigation canals in Idaho carried the names of an individual farmer or a pair of farmers, another indication of the small-scale labor that built such systems. Besides the Bubb or Payne Canal and the Andrews Canal, the Boise Valley had the Mace-Catlin, Davis and Hart, Conway-Hamming, and Graham-Gilbert ditches, and others. The upper Snake River valley featured the Nielsen-Hansen and Woodmansee-Johnson canals, the Lowder and Jennings Canal Company, and the Wolfe, Hill-Pettinger, and Nelson-Corey ditches, and more.[12]

The hard physical work that initially built and maintained the canals, however, gradually became a thing of the past for many farmers. By the late 1890s, the basis of canal labor began to change. Money, not muscle, increasingly became the foundation of irrigation companies. More profitable crops, including sugar beets and potatoes, brought the farmers handsome returns. As the money rolled in, farmers paid their annual canal assessments in cash instead of labor. Some also adopted a new organizational form, the irrigation district, that allowed them to finance canal improvements. Home companies now had the means to hire others—engineers, contractors, and wageworkers—for construction and maintenance. Moreover, the farmers could hire these backs, shoulders, arms, hands, legs, and brains in relatively large quantities. Canal construction and maintenance, once undertaken on a family and neighborly basis, now proceeded in the industrial mode. With this new arrangement, the farmers generated important environmental modifications: some home companies used their expanded labor power to fashion larger, more technologically sophisticated canals.[13]

The farmers of Teton, a Mormon village in the upper Snake River valley, constructed such a conduit. When they founded the settlement in the early 1880s, the farmers first established a small irrigation system. They organized the Teton Irrigation Company and built a canal to carry water a short distance from the Teton River to their farms. Like their counterparts in other home companies, the farmers excavated the canal themselves. "Much of the labor," recalled Goldie C. Thomson, "was donated because money was scarce." With water the Teton farms began to flourish, but the supply eventually proved inadequate. By the late 1890s and early 1900s, drought and the expansion of irrigation encouraged the villagers to look for an additional water source. They found a new supply to the northeast, in Falls River. Yet the project posed enormous obstacles; the canal had to traverse more than twenty miles of rugged terrain and would require a pipeline across the Teton River. The Teton farmers simply lacked the skills and numbers of people necessary for timely completion of the work; this was not the sort of project that home companies normally built.[14]

Only by raising funds to hire industrial expertise and labor could they construct the new conduit. In 1903 they organized the Enterprise Canal Company, through which they hoped to arrange a bank loan to pay for the canal. The corporation failed to come up with the necessary funds, so the farmers turned to another form of organization and in 1904–1905 established the Enterprise Irrigation District. Idaho's irrigation district law, first enacted in 1895, allowed farmers to issue bonds for canal construction and improvement, the security provided by a lien against land within the district. After selling $44,000 in bonds to a Denver investor, the Enterprise Irrigation District in 1908 hired the necessary labor and moved forward with canal construction. A bookkeeper kept a financial record of the project; a surveyor mapped out a route; a construction superintendent oversaw the work of 242 men, who drove teams, operated scrapers, and blasted rock with dynamite. The district contracted with the Pacific Coast Pipe Company to build a 2,076-foot wood-stave pipeline across the Teton River. Even the name of the new irrigation system suggested how far the Teton farmers had departed from the family mode. "Enterprise Canal" made no reference to agrarian cooperation or family farmers; indeed, it rendered anomymous the many wageworkers whose aching limbs had actually moved soil and rock.[15]

But it is important to keep in perspective the Enterprise Canal and the labor that produced it. Despite the Teton farmers' adoption of the industrial mode, their operation still was much smaller and simpler than the largest canal projects in the Snake River valley. In many ways their new irrigation system remained rooted in the world of the Mormon village. Apparently they hired no professional engineer to design the canal and direct the work; a local man, J. R. Thomson, served as construction superintendent. They indeed employed a sizable number of men, but fewer than many Idaho irrigation projects and for less time. And although larger than most home company systems, their twenty-two-mile canal served only about 7,000 acres of irrigated land.[16]

Modern corporations and government agencies were responsible for irrigation systems that fully exemplified the industrial mode. These projects dwarfed the home companies' hydraulic works. A few were speculative ventures in which investors planned to deliver water to farmers for a fee. Most, however, involved state and federal governments. The 1894 Carey Act authorized corporations and public agencies to build irrigation works for farmers. The federal government allotted land to Idaho and other western states for settlement by irrigators; Idaho then contracted with developers to excavate the canals. Under the National Reclamation Act of 1902, the U.S. Reclamation Service built dams and canals to supply water to settlers. The purpose of all these projects was to serve as many farmers as possible, so bureaucrats and engineers designed canal systems and storage structures of massive, breathtaking size. Arrowrock Dam on the Boise River, built in 1916 by the Reclamation Service, was at 349 feet the tallest in the world. The Minidoka project on the south central Snake, another federal endeavor, moved water from

Minidoka Dam through large canals to 130,000 acres on both sides of the river. The Carey Act, too, produced vast projects. The Twin Falls Land and Water Company, backed by the Pennsylvania steel magnates Frank Buhl and Peter Kimberly, in 1905–1907 completed an immense irrigation project on the south side of the Snake River canyon. The system featured a massive network of primary canals that extended for more than eighty miles and irrigated some 200,000 acres.[17]

To build such great hydraulic works, businessmen and government officials had to mobilize labor on an industrial basis: tasks were diverse, were highly specialized, and required hundreds of people. Engineers designed dams and conduits and oversaw their construction. Surveyors mapped the topography and staked canal routes, while draftsmen rendered construction drawings. Hand drillers bored holes in bedrock and packed them with dynamite. Muckers wielded picks and shovels; skinners drove teams and operated slip scrapers. Equipment men ran steam shovels, pile drivers, and concrete mixers, and carpenters built forms. Bosses supervised each task. Nurses and doctors tended the injured. Cooks prepared meals. Clerks, timekeepers, and accountants kept records. Elliot Paul, a Reclamation Service employee, described the commotion of construction work at Jackson Lake dam on a typical day during the winter of 1910–1911:

> At eight o'clock the whistle of the steam shovel blew, the muckers began shoveling, the skinners started up their horses to drag the slips around, and the hard-rock men, those who held drills sitting on gunnysacks to keep from freezing to the bed of the Snake, began their poetry of motion. Machines chugged and spluttered, pumps sloshed and gurgled, and the members of the surveying party set up the level or the transit—or at least, I did, since that was my job—and we drove stakes with tacks in them to indicate where the dirt and rock should be excavated, where and how high it should be piled on the dykes and waste banks, where the wooden forms for the concrete should be built, and Scotty made notes from which Torrance and his crew in the drafting room could compute the yardage and estimate progress.[18]

Many of the employees on such industrial projects came from "the army of manual laborers," as Carlos Schwantes has called them, that roamed the West in search of jobs during the late nineteenth and early twentieth centuries. A number were Americans, but a construction site might also include workers of various other nationalities: Greeks, Austrians, Basques, Danes, Bohemians, Italians, Canadians, Irish, Japanese, Bulgarians, French, and many more. Supervisors at dam and canal sites typically grouped these men in gangs, according to their ethnicity. At the Jackson Lake dam construction site, engineer Frank Crowe believed that certain nationalities were specially suited to particular tasks—for example, in his estimation Bulgarians made superior muckers. Crowe even quartered common laborers according to ethnicity: Americans and other English-speaking workers, Scandinavians, and Frenchmen bunked in one building; "Bohunks," which included east and central Europeans along with Spaniards, occupied another.[19]

Although these men moved on after they completed the great irrigation systems, the industrial mode of labor continued during maintenance and expansion. According to the law, settlers eventually assumed control of federal and the public-private Carey Act projects, but the farmers alone simply could not mobilize the labor necessary to maintain and operate the canals that businessmen, government officials, and engineers had devised. These huge hydraulic systems were simply too big and complex. So the farmers hired staffs of laborers to perform the work for them. The Twin Falls Canal Company (TFCC), the farmers' organization that superseded the Twin Falls Land and Water Company, took over an irrigation system that included Milner Dam and about 1,000 miles of canals, laterals, and coulees, all of which covered some 400 square miles. To operate and maintain the system, the TFCC board of directors employed approximately 100 people, including a project manager (often a professional engineer), heavy equipment operators and other workers, and ditch riders. Much like other industrial occupations, canal work was often dangerous. Alfred Peters, a longtime TFCC employee, recalled that seventeen men died working on the company's system, in tunneling accidents, ditch cave-ins, rockfalls, and mishaps involving dynamite and machinery.[20]

The profoundly industrial character of such labor, however, did not preclude the family mode from playing at least a minor role in Idaho's large irrigation projects. Common laborers, after all, could be family farmers, too. Mormon farmers from the home companies, in need of cash or merely wanting to augment their incomes, found employment as wageworkers and contractors on big canal construction projects in the upper Snake River valley. Often they mobilized their labor according to kinship and community connections. George Wintle and his neighbors, who dug the small Parsons Ditch in the late 1880s, took their horse teams and scrapers and worked on various canals, including the Aberdeen-Springfield system, a large Carey Act project. The Wadsworth family excavated small canals to their farms in the upper valley during the same period and later rounded up their teams and worked on the Great Western Canal, a speculative development. John Watson recalled his work on the Peoples' Canal and noted that he and his friends—Peter Fjellstad and the twins Herbert and Robert Adams—"took contracts to build other ditches."[21]

In some cases family farmers worked as wage and contract laborers and used their earnings to purchase a farm or augment a homestead on a federal or Carey Act project. During the early 1900s, the Mormons John and Johanna Josephson homesteaded eighty acres of Carey Act land on the Aberdeen-Springfield project and also worked on the construction of the canal system. John "obtained a job tending the big steam boilers that operated the jack hammers used to blast the lava rock in the canal bed"; Johanna, with the help of her two young daughters, cooked for canal workmen for eighty dollars per month. Henry A. Wallace, while visiting the Minidoka project in 1909, inspected the construction of the Minidoka Dam powerhouse and learned that "the big crew" working at the site consisted of farmers. John McQuiston, Jr., a local settler and Wallace's host, informed him that the men "haven't been able to get enough to live off of their places yet, so they come and work for the government on the ditches."[22]

The Reclamation Service even attempted to organize farmers to build parts of the Minidoka system on roughly the same basis as the home company canals. Beginning in 1906, the agency required Minidoka project settlers to excavate and maintain lateral ditches (also called sublaterals or community laterals) that carried the water from the primary canals directly to the farms. The handful of farmers on each lateral were supposed to band together to dig the conduit. In a sense, the industrial mode of labor would end and the family mode begin at the places where the primary conduits and laterals intersected.

But much to the dismay of the Reclamation Service, the farmers did not always cooperate with one another. They often lacked the community spirit that typified the work of the home company systems. A farmer at the head of a lateral might get all the water he needed and feel no obligation to ensure that the settler farther down the ditch got any. Some settlers took outside jobs to pay for their farms, leaving others to build the ditch on their own. Some "settlers" were merely speculators waiting for property values to rise so that they could sell; these people were content for someone else to do the work of building the lateral. Eventually the Reclamation Service took over the task of excavating the laterals, and industrial work superseded the faltering efforts of the neighboring farmers. Thus it was appropriate that these laterals almost never carried the names of any people but were designated by a numbering system (e.g., 56-3 or 128) reminiscent of that used for tracks in a railroad yard or assembly lines in a factory.[23]

Family and industrial modes thus generated the many dams and canals that covered the Snake River valley and fundamentally defined the irrigated landscape. Yet the labor that produced these hydraulic structures was ancillary to the work that shaped an adjacent and related realm, that of the farm. Irrigation water made its way through ditches and onto fields, and here human effort joined it to soil, seed, and sunlight to produce abundant crops. In late-nineteenth- and early-twentieth-century Idaho, this labor usually (but not exclusively) followed the family mode; farm families "worked the land."

Working the land began with the establishment of a farm. From the 1860s to the 1920s, thousands of people migrated to Idaho and purchased tracts that ranged from 40 to 640 acres. Making these plots productive took an enormous amount of effort. First, farmers cleared the land, which usually involved tearing up sagebrush with horse-drawn rakes. In conjunction with the construction of supply canals, settlers then excavated their own farm ditches and prepared their fields for irrigation. Then they plowed the soil, planted crops, and seeded pastures. And in the meantime, they cut logs, sawed lumber, and put up fences, houses, barns, and other structures. By the early twentieth century, some settlers were able to purchase farms already built, but even these people frequently had to make improvements to their new homes.

Occasionally farmers did not have enough food and money to support themselves while they completed their farms, so they sought outside work. In such circumstances, the two primary modes of labor in the irrigated landscape, family and industrial, again overlapped. Some settlers, as we have seen, found employment on irrigation projects. Others took jobs on the railroad, perhaps the greatest agent of industrialization in the American West. Mormon folk, among the most family-oriented and agriculturally inclined people in the region, worked on railroads and used their wages to support their irrigated Idaho farms. Richard Dye and his wife started a farm near Firth in the upper Snake River valley in 1885, but not until five years later did they actually begin irrigating. During the intervening years Dye worked as a section hand for the Union Pacific in Ogden in the winter and cleared sagebrush and dug ditches in the summer. After laying claim to a homestead in the Iona vicinity in 1883, the brothers Rufus and Leander Norton took jobs as tracklayers. In 1884, Cyrus Rostan and a group of five other young men claimed homesteads around Lorenzo. "They then went to Missoula, Montana to work on the railroad for awhile" before returning to their farms.[24]

When a family was finally able to devote all of its energy to farming rather than outside industrial employment, it faced a crucial challenge: to coordinate its capacity for work with the labor requirements of its various agricultural activities. The goal was to perform each task as efficiently as possible, thereby sparing time and effort for the other farm chores. Allocating labor to field irrigation was central to this overall process. Ideally, each farm had to develop a system of fields and ditches that most efficiently used the labor power of family members, especially its male head. E. B. Darlington said that farmers had to practice an "economy of labor" in irrigation. According to the engineer, "land should be so prepared and such methods of application adopted as will accomplish [irrigation] with the least expenditure of labor and water." If they did this work efficiently, farmers could devote an adequate level of labor to other farm chores.[25]

To maximize their labor in irrigation, family farmers could choose from several types of ditch and field systems. In border flooding, the farmer laid out fields in narrow, graded strips, 30 to 100 feet wide, that followed the slope of the ground. Dikes, roughly one foot high, separated each strip. The farmer then fed water into each strip, either through a small headgate or by digging an opening in the supply ditch. The dikes kept the water at a uniform level as it flowed downward through each strip.

In furrow irrigation, also called the corrugation system, the farmer sent shallow streams of water from the supply or feed ditch into the furrows or corrugations that ran between crop rows. Basin flooding involved the construction of square basins around orchard trees, into which the farmer channeled water. Subirrigation required the farmer to keep the underground water table up to the level of crop roots, usually by allowing water to seep from ditches. With simple flood irrigation, the farmer dug openings in the supply ditch and directed a sheet of water across a graded, sloped field.[26]

Each of these landscape arrangements had advantages in terms of labor. Complex systems such as the furrow method or border flooding needed a relatively large amount of labor to lay out, but once built the structures made irrigation easier. The simplest practice, flood irrigation, involved little initial work to shape the land, but directing the sheet of water evenly across the ground was difficult and required a great deal of skill and effort. None of the systems, however, functioned well if the farmer did not build properly or apply the water carefully. Darlington complained that some irrigators were sloppy in their workmanship and so tended to waste water. To compensate for poorly made irrigation structures or their own lack of care, they spread excessive amounts of water over the ground. "In such cases," he stated, "the farmer is trying to make the water do the work that he himself has neglected."[27]

Yet field irrigation involved more than just choosing a particular landscape system and operating it properly. In building their ditches and spreading water, beginning irrigators also had to learn to take into account topography and soil characteristics. Failure to consider these environmental conditions fully could result in lost labor, as the Swedish immigrant Carl Sealander discovered. In 1902, Sealander acquired a small tract in the New Sweden Irrigation District and dug a ditch from the main canal system to his new farm. In one place, he had to erect a six-foot-high embankment to convey the water across a depression. Construction required special care—the soil that made up the embankment fill had to be wetted and allowed to settle before it could hold a full stream of water. Sealander thought that he had done the job correctly, but to his dismay water turned into the ditch washed out the fill. He then rebuilt the embankment but later discovered that once more the water had broken through, carrying away even more earth than before. "I felt so bad

I lay down on the ground and wept," Sealander said. "All that work had to be done over again."[28]

Some irrigators, including the boys Floyd and Jake Eilers, developed special work routines to manipulate water on difficult terrain. According to Floyd, the sandy soil of his family's Minidoka project farm made irrigating an especially challenging task. When the brothers dug an opening in a supply ditch, they could not precisely control the stream because it tended to erode the ditch bank and make the hole larger than necessary. Filling the breach, moreover, was a problem. To close the opening Floyd and Jake often used chunks of sod along with loose soil. But the two youthful irrigators, as Floyd recalled, "worked out a system for the really tough ones. I would tell [Jake] to take his clothes off and have him sit in the ditch to stop most of the flow while I pitched dirt in to close the gap."[29]

Family farmers such as the Eilers had to adjust their work routines not only to soil conditions but to water scarcity as well. Many irrigation systems supplied water to farmers on a continuous flow basis—every farmer received a full irrigating stream as long as the source would allow. Under such conditions, all the ditches on a farm would be carrying water, and farmers could irrigate whenever they desired. But when summer or drought reduced a river's volume, each farm received less, thereby encouraging farmers to adjust their irrigation techniques and thus the labor that they devoted to sending water through their fields. Rather than have small amounts trickle through each farm ditch and onto the fields, an inefficient and time-consuming use of water and labor, the irrigator kept the water in one stream, or "head," and moved it from field to field as needed. Some irrigation systems had their members adopt rotation, in which neighboring farmers took turns using a single large head of water. Alexander McPherson, superintendent of the Twin Falls Land and Water Company, explained that rotation had the advantage of "relieving one farmer of the labor of looking after water and enabling him to attend to other work while his neighbor, with a large supply of water, is able to go over his field more rapidly, thus economizing time and labor."[30]

Of course, although irrigating with one stream made the farmer's work more efficient and in some ways easier than directing the flows in many ditches, the method required greater discipline from those who adopted it. During rotation, for example, farmers might have to take a turn of water at night. Typically, a farmer opened a supply ditch and started the water flowing onto a field. Then he lay down at the end of the field and took his rest. Eventually the water reached him and wet his feet or clothing. Roused by this hydraulic alarm clock, he would move the stream to a new location. Delbert Fowler remembered using this practice during the 1920s on his family's farm near Shelley, in the upper Snake River valley. "When the water got down to where I was," he stated, "it woke me up and then I changed this stream to a different place."[31]

On certain irrigation systems, rotation might even begin to make the small-scale labor of family farmers resemble industrial labor. Although rotation partly evolved from the work habits of preindustrial irrigators such as Mormons, the method greatly appealed to the modern industrial sensibilities of engineers, government officials, and canal managers—people who occasionally referred to irrigation systems as machines. In such industrial systems, engineers wanted to put water to its most efficient use. The most efficient use was achieved by carefully measuring and allocating the rotating head of water. Much as factory managers regulated machines and workers through the use of the clock and the precise measurement of time, certain canal managers attempted to regiment irrigation and irrigators through the precise measurement of water, using meters, weirs, and flumes. Thus when efficiency-minded experts like Alexander McPherson or E. B. Darlington drew up irrigation schedules and administered rotation, farmers might at least vaguely resemble factory employees.[32]

But no matter how implemented, the disciplined work of irrigation was crucial, because farm families had to spare labor for the tending of crops. Farmers chose crops for various reasons, including the profit they would bring and their suitability to the soil and climate of a given area. But farmers also had to consider what tasks the cultivation of those plants would require. In the irrigated landscape, crops such as wheat and hay grew in dense stands that needed no weeding or tilling and relatively little water: here the labor requirement was comparatively low, except at harvest. In contrast, row crops such as sugar beets or potatoes had a much higher labor demand. These needed careful tilling and irrigation throughout the season and intense labor at harvest. Thus most farm families in the irrigated landscape often grew combinations of crops, such as wheat and beets, that reflected their attempt to juggle relative profitability, environmental constraints, and labor requirements.

Wheat was a popular choice for family farmers in Idaho's irrigated landscape. Small-scale wheat production and the labor of family and neighbors were ideally suited to each other. The overall ease and simplicity of raising the grain made it attractive to a farm family with limited hands. Wheat's minimal labor requirement during the growing season, furthermore, allowed a family to devote some of its effort to the cultivation of more demanding and lucrative row crops. Only at harvest did wheat require greater labor than a family could provide.[33]

At this time, neighboring farm families usually shared or exchanged the labor of cutting and threshing. If adjacent farmers could not individually afford the binders or threshers needed to cut, gather, and thresh their wheat, they sometimes collectively purchased and operated the machines. They took the binder from farm to farm, cutting the wheat and then placing the bundled, sheaved grain in piles, called shocks. After allowing the crop to dry, they threshed it. Often neighboring farmers hired a custom thresherman, in some instances another farmer, and he in turn hired a few wageworkers—at least an engineer and a machine tender (a separator man)—to help him run his equipment. Then the farm families supplied the rest of the labor. Virtually everyone, including women and older children, took part. Men and boys collected the shocks and transported them in wagons to the threshing unit. There the hardiest among them wielded pitchforks and steadily fed the sheaves into the machine, which spewed kernels back into the wagons for transport to a nearby grain elevator.[34]

Meanwhile, women and girls performed what some people believed was the most strenuous task of all. At each farm, the wife and daughters, often with the help of neighboring women, took on the job of cooking meals for the entire threshing crew. "The women would cook up huge amounts of food to satisfy the hungry hands, noon and sometimes at night," recalled Floyd Eilers. Often the task could be onerous. With humor and a trace of bitterness, Annie Pike Greenwood recalled the intense effort that females expended on meals:

At times I had as many as thirty-two men to feed, some of these being agents, ditch-riders, and so forth, for wherever there was a threshing, all the men in the district gathered like flies around a screen door when a rain-storm is threatening. And it was a grand visiting-time for the men, too. They were in a genial mood, and the agents could talk up business to half a dozen at once during the noon-hour or a breakdown of machinery. For there always was a breakdown, and we sagebrush women must expect to feed the threshing crew several more meals than we had been warned about, while someone dashed to town for broken parts, and the men enjoyed a rest.[35]

Geography encouraged the sharing of labor that typified the family farmers' wheat harvest. Given the relatively small irrigated patches of wheat and other grains, and the generally low prices that these crops brought, it was often uneconomical for farm families to purchase costly machinery individually. But jointly purchasing and operating binders or threshers allowed them to overcome the problem. A dense settlement pattern typical of early irrigation also promoted cooperative harvest labor: farmers could easily travel from one farm to the next to help each other. Furthermore, the small wheat acreages made for a brief harvest season that did not demand too much of each farm family's time and effort.[36]

Wheat production on the largest irrigated farms, in contrast to the family operations, functioned in the industrial mode. Here spatial and economic conditions were different from those on the small farms, and so field work became much like factory work. Big farmers were actually less farmers than businessmen, and they cultivated hundreds, even thousands, of acres of wheat and other crops. Large-scale wheat production was an effort to achieve an economy of scale: to maximize profit from relatively low-priced wheat, wealthy farmers grew lots of it. With farms this large, the big producers could not expect small farmers to volunteer to help with threshing. Nor would they have wanted such aid. The big farmers sought intense, heavy production, and the only way for them to achieve this was to mobilize labor on a wage basis. On the large farms, harvesting was not a cooperative exchange between neighboring family farmers. If small farmers worked for a big farmer, they did so as his hired laborers, not as his neighbors.

Such industrial wheat farms appeared throughout the early history of Idaho's irrigated landscape. W. A. Goulder described Thurman's Mill in the Boise Valley, a nineteenth-century version of such an operation. The farm of "Commodore" Tom Lee, 2,200 acres of wheat and other crops near Rigby in the upper Snake River valley, represented a twentieth-century incarnation. A former railroad official, Lee knew how to coordinate people and machinery on an industrial basis, and he applied his skills to his farm, "the Bungalow," which he established in 1906. To help him oversee the farm and its workers, he hired a manager. And he operated "such machinery as seeders, mowers and binders, eighteen hours per day, changing shifts of men and horses every four hours." In 1908, Lee overcame the inefficiencies of steam power and further systematized his operations when he began using a mobile electric motor to run threshing and other equipment. Thus to work on Lee's farm at harvest was to enter a world very different from the world of the family farm: this was a factory-like place of clanking machinery, humming motors, and regimented labor.[37]

It would be a mistake, though, to conclude from Lee's operation that factory farming necessarily mobilized labor on a strictly industrial basis. Certain forms of crop production were just as industrialized as wheat, yet these relied heavily on the family mode. Consider the example of sugar beets. During the early 1900s, the Utah Sugar, Utah-Idaho Sugar (U&I), and Amalgamated Sugar companies introduced beets to the family farmers of Idaho's Snake River valley. Small producers, particularly the Mormons, seemed to offer the corporations ideal conditions for making sugar. They had plenty of irrigated acreage on which to raise beets, and more important, they were an excellent source of the labor necessary to grow and process the crop.[38]

Sugar beets required intensive cultivation—mostly arduous hand work—and frequent irrigation. The beet produced multiple seedballs or clusters; when planted, each cluster in turn sprouted numerous shoots. To mature the beet properly, the sprouts had to be "blocked and thinned," the strongest plant selected from the cluster and given ample growing room. Careful irrigation through the summer made for maximum growth, and two or three hoeings ensured that the beets would have the highest sugar content possible. At harvest, someone had to pull the beets from the ground, cut off or top the foliage with a special knife, and then transport the crop to a nearby company factory for processing. Thinning, hoeing, weeding, irrigating, pulling, topping: these were grueling tasks best carried out, the sugar companies believed, by a stable, experienced, labor source—the family farmers of the Snake River valley.[39]

Thus the operations of industrial corporations merged with the labor of small farmers. In fact, so close was the connection between beet factory and beet field that a remarkable situation sometimes developed: distinctions between family and industrial modes began to disappear. As Charles F. Saylor of the U.S. Department of Agriculture wrote in 1904, "There is such a community of interests between factory and farm in the production of sugar that it is hard to draw a definite line of separation" between the two. The contract between the farmers and the sugar companies initiated this convergence. The farmers could negotiate the acreage they would raise and the price they would receive for the beets—here they preserved a degree of labor autonomy. But the contract also imposed a form of industrial discipline on the farmers: it outlined the methods they must use in raising the crop and stipulated that they must accept the direction of a factory agronomist or agricultural expert. Small farmers labored in the family mode, but they had to follow the dictates of an industrial production routine.[40]

During the winter, sugar production fully transformed some farm family members—usually adult males—into industrial workers. At this season the men took processing jobs in which they refined raw beets into sugar. As the days grew short and cold winds blew flecks of powdery snow across the fields, they made their way from their farms to a nearby refinery, a large brick or concrete complex surmounted by tall smokestacks, flanked by railroad tracks, and governed by the time clock and the shriek of a steam whistle. How many farming men participated in this kind of wagework is unclear, but in the early days of Idaho sugar production there must have been a fair number. In 1903, the Utah Sugar Company opened the first factory in the upper valley; the Idaho Falls facility "employed approximately 200 persons, primarily local farmers who had grown beets." In 1911, a federal government study reported that Idaho sugar factory laborers "are drawn almost exclusively from the farms of the community and from neighboring towns."[41]

Although the sugar corporations at first anticipated that the family farmers would furnish virtually all their labor needs, they soon realized that these people would not be enough. The industry was still expanding, and the family farmers alone could not furnish the labor necessary to raise larger quantities of beets. Thus, almost from the advent of beet cultivation in the Snake River valley, farmers, with corporate assistance, increasingly came to rely on contract field labor. Farmers themselves had signed contracts with sugar companies and worked as wage laborers in factories; now they had to mobilize their own workers according to the industrial mode.[42]

Children were the first laborers specially recruited for sugar beet work. In June 1903, for example, groups of Utah farm boys, already experienced at beet cultivation, traveled to the Idaho Falls vicinity and there found employment thinning beets. At the same time, boys and girls from Idaho Falls went out from the town to labor in the fields. Using children to thin and hoe beets allowed farmers and sugar corporations to avoid hiring "undesirable" transient laborers, a migratory, unstable, alien class of people. Children perhaps seemed familiar and reliable, more like neighbors, extended families, or the farmers' own offspring.[43]

But this arrangement did not last for long; the children were not enough. As early as 1904, farmers in the upper Snake River valley hired Japanese to thin, hoe, and harvest beets. By 1911, between 600 and 700 Japanese were in the fields. Because the Japanese had already found employment in Idaho on railroads and canal construction projects, they offered farmers a convenient labor pool. At first, farmers contracted with labor bosses who controlled gangs of Japanese workers. By 1911, U&I had instituted a formal program of hiring. The company contracted with a Japanese corporation for a supply of workers; U&I paid the firm a per-acre fee that also covered the cost of transporting the laborers to Idaho and the salary of a resident Japanese agent who supervised them. Individual farmers then contracted with the workers directly, on a piece rate basis.[44]

The 1907 "Gentlemen's Agreement" between the United States and Japan limited Japanese immigration, so farmers looked for an additional source of beet labor. They found a good supply among "German-Russian" immigrants, ethnic Germans who had once settled along Russia's Volga River. In 1908, U&I brought in an initial group of these people. They seemed a logical choice; raising beets was already part of their agricultural heritage, so they did not have to learn how to tend the crop. German Russians offered one other advantage: they provided competition for the Japanese already in Idaho. According to a federal government report, sugar companies first imported German Russians "in order to check the demands of the Japanese, who, it is asserted, were becoming 'hard to handle.'"[45]

World War I and the 1924 National Origins Act altered sugar beet labor conditions again and encouraged the importation of yet another ethnic group, Spanish-speaking people from Mexico and the American Southwest. The war fostered agricultural expansion at the same time that military service took young men from the farms and blocked immigration from Europe. Then the 1924 law further closed off the flow of foreign workers, except from Mexico. Because of these developments, sugar companies in Idaho and elsewhere in the American West looked to Mexicans and Mexican Americans to fill the need for sugar beet labor. Although some Mexicans had come to Idaho earlier, most arrived during and after the war. By 1918, U&I and Amalgamated Sugar—which opened its first factory in Burley in 1912—had recruited hundreds of Hispanic people from El Paso, Texas, to work in Idaho beet fields. In 1919, both corporations established labor agents in the city.[46]

In a matter of years, sugar beet cultivation had virtually institutionalized a seasonal movement of migrant workers through the irrigated landscape. Yet these people did not necessarily remain bound to the piece rate and the industrial mode. Some, especially the Japanese and German Russians, sought the opportunity to work the land on their own, in the family mode, as family farmers. Many Japanese had been farmers in their native country, and they came to America with agricultural knowledge and a desire to raise their own crops and livestock. In Idaho, they saved their money, learned irrigated agriculture, and eventually leased land. Some, like the families of Henry Fujii and Henry Hashitani, formed partnerships, pooled their capital, and operated a farm collectively. In other cases single families acquired farms. In 1916, K. Inouye, his wife, and their children cultivated about ninety acres, including thirty-four acres of sugar beets, in the upper Snake River valley near Shelley. And some Japanese eventually hired their own field laborers. Sakichi Ogawa, for example, a farmer near Idaho Falls, employed Filipino workers to tend his crop. Skilled and enterprising, the Japanese by 1909 farmed a surprisingly high proportion—perhaps 25 or 30 percent—of Idaho's total beet acreage. It was not easy for them, for they faced obstacles that other farmers did not. Idaho law prevented them from owning land,

and a 1923 statute denied them the right to lease. But at least some of them found ways to get around these restrictions.[47]

The German Russians, too, had once been farmers elsewhere, so for them the acquisition of a farm also meant the fulfillment of a cultural ideal. Indeed, the German-Russian field laborers usually worked in family units that were already oriented to farming. According to a government report on seasonal agricultural work, "Almost all of them had been farmers in Russia, and they were accustomed to hard work in which all the family members joined." In Idaho, the same labor pattern prevailed; a German-Russian family as a unit took contracts to work the sugar beet fields. German Russians also tended to have large families and so "could contract for large acreages and have a large total family income." Some of them did quite well, even better than the farmers who hired them. Annie Pike Greenwood recalled "a whole family" of German Russians who once thinned beets on the Greenwood farm. Although they wore rough clothing and walked barefoot among the beet plants, Greenwood noted with dismay the contrast between the laborers' "fine car" and the Greenwoods' dilapidated "old secondhand Ford." Like other German-Russian agricultural workers, the Greenwoods' employees were probably saving their money so that they could purchase or lease land and make the transition from family field laborers to family farmers.[48]

Eventually, U&I and Amalgamated Sugar even encouraged and helped the Japanese and German Russians to acquire farms. The sugar companies believed that they could expand production by stabilizing these migrants and turning them into farmers. So they offered land to the laborers through lease (by cash or on shares) and sale (except for the Japanese, who by law could not own land). U&I and Amalgamated Sugar also provided loans to the new immigrant farmers.[49]

U&I continued this trend of creating sugar beet farms and farmers when, in the early 1920s, it established the Hays and Osgood projects on 8,200 acres in the upper Snake River valley. The Osgood project was founded on land that a large agricultural operation, the Idaho Falls Dry Farm Association, had previously planted in grain. Yet neither the Osgood nor the Hays project functioned like industrial agribusiness, as a single, giant farming unit. Instead, U&I leased small, forty-acre tracts to individual families who then grew beets and other crops under company supervision.[50]

Beet labor had now come full circle. U&I and Amalgamated Sugar first introduced beets to family farmers in the Snake River valley. Next the corporations imported field laborers to work the crop. Soon they sought to transform these workers into farmers. And at last U&I established modern beet plantations that it divided into small family farms. But whether U&I attached itself to small Mormon farmers or attracted family tenants to the Hays and Osgood projects, the process of combining family and industrial labor remained essentially the same. As an analyst of beet labor stated, "Sugar beet agriculture and the factory production of beet sugar are inextricably linked in . . . a curious union of family farms and million dollar corporations."[51]

And so it often was with work in the irrigated landscape. Family and industrial labor did function separately, but they could also become quite tangled and even merge. This was a place in which family farmers might tend their own crops and ditches, work cooperatively on canals or the wheat harvest, find temporary employment on a railroad or large construction project, raise sugar beets under the direction of a company agronomist, contract with migrants to tend those beets, and take a job at a sugar factory processing the very crop that they, nominal family farmers, had produced. This, too, was a landscape in which common laborers could turn themselves into farmers. Here, people like Annie Pike Greenwood employed German-Russian field laborers who were really no less a farm family than the Greenwoods themselves.

These interconnections revealed one of the greatest ironies of irrigation in the American West. Irrigated agriculture, as many of its proponents envisioned it, would provide Americans an escape from the worst effects of industrialization and the urbanization that accompanied it. By opening the arid West to farming, the republic could renew its agricultural roots. Yet by the late nineteenth and early twentieth centuries, irrigated farming almost invariably involved railroads, factories, and heavy construction. Thus the work that people performed in the irrigated landscape reflected a convergence of the family farm ideal and industrialization. In Idaho's Snake River valley, labor linked America's agrarian past and its onrushing industrial future.[52]

But no matter what form this labor took, changing conditions gradually reduced the need for it. Irrigation companies and districts eventually adopted bulldozers and other mechanized earth-moving equipment; now one man, a cat skinner (a Caterpillar operator), could do more work than several of the best mule skinners combined. On the irrigated farms, harvester-thresher combines, introduced in the 1930s and adopted incrementally thereafter, lessened the need for community labor in wheat production.[53]

Changing technology similarly shrank the labor needs of sugar beet production. Larger, more efficient factories allowed sugar corporations to close down smaller, dispersed plants and consolidate their operations. Automobiles hastened centralization, because farmers could truck their beets to the new, now distant factories. In turn, centralized factories required fewer workers. Changes in farm technology paralleled the new factory systems and reoriented farm labor. Farmers eventually adopted machines that made the cultivation and harvesting of beets much less dependent on hand labor. The world war labor shortages that compelled corporations to import Mexicans also stimulated the invention and purchase of specialized "labor-saving" field machinery. From the 1910s through the 1960s, farmers gradually came to use mechanical planters, cultivators, harvesters, and loaders.[54]

But as much as technology, nature influenced changes in the labor of sugar beet production. The curly top virus, spread by the beet leafhopper, contributed to factory closures and an overall restructuring of the sugar beet industry in the Rocky Mountain West. Nature, too, affected field mechanization; the multigerm seed obstructed the adoption of machines and so actually delayed the abandonment of field labor. Wayne Rasmussen has written that the "spring and summer operations—planting, blocking, thinning, and weeding—were not to yield so easily to mechanization. This was due largely to the nature of the beet seed." In the late 1940s, sugar companies and the U.S. Department of Agriculture sponsored a search for a monogerm beet seed that would not require intensive hand labor. In 1948, two Russian émigré scientists, Viacheslav and Helen Savitsky, discovered five plants in Oregon that contained monogerm seed. Two of the plants were interbred with other beets to produce a monogerm hybrid suitable for crop production. By 1966, virtually the entire sugar beet crop in Idaho's irrigated landscape and the rest of the

American West came from monogerm seed. When the "nature" of the beet seed had finally changed, the need for intensive labor suddenly vanished. Migrant workers were, at last, completely redundant.[55]

The disappearance of migrants from the sugar beet fields highlighted a central characteristic of the irrigated landscape: the ongoing interplay between social institutions, labor included, and the biophysical environment. Organized on the level of small farms, industrial corporations, or a mixture of the two, labor altered the earth's surface on a massive scale. We can best understand this point if we imagine gangs of sweaty men and draft animals carving vast irrigation systems out of bottomlands and the dusty Snake River plain. Here a great body of workers moved landforms, shifted the course of rivers, introduced new plants and animals, and so remade the world.

Yet this was not a one-way process in which humans simply shaped the land. For the land in turn affected the people who worked it. Irrigated crops best illustrated this. Each crop had biological characteristics and agronomic requirements that demanded varying levels of human effort, and this demand had important consequences for the labor systems of irrigated agriculture. Seed, whether wheat, beet, or other kinds, did not simply contain the potential for plant life; locked in each seed, as well, was a social potential revealed in labor.[56] Of course, exactly what form that labor took—family farmers, industrial employees, or a combination thereof—depended on choices that people made. But those choices were, to a certain extent, always influenced by the needs of the seed, the plant, the crop. Thus the work that people performed and the land in which they worked were extensions of one another. In Idaho's Snake River valley, land and work evolved together; the irrigated landscape was a landscape of labor.

5/ **From Field to Market**

Agricultural Production in the Irrigated Landscape

When inhabitants of Idaho's Snake River valley worked, they both defined themselves socially and shaped the irrigated landscape. But their labor resulted in something more: it involved these people in a productive process that merged nature with yet another human system, the market. Farmers and agricultural workers manipulated soil nutrients, sunshine, water, and seed to produce crops and livestock. Then the farmers sold the fruits of this labor, sometimes to local people but more often to food processors or wholesale merchants in regional trade centers. These middlemen in turn marketed the agricultural produce to consumers. Plant and animal growth and the labor that nurtured it eventually culminated in food on someone's table and cash in a farmer's pocket.

Abstractly, this movement of agricultural produce from field to market could seem highly ordered and profitable. In 1915, A. G. Seiler of the U.S. Reclamation Service published a scenario of irrigated agriculture, "The Story of An Irrigated Farm" (fig. 23), that depicted a smooth, systematic transformation of nature into money. Seiler's story, rendered in a series of cartoonlike sketches, began with a congressional appropriation and the construction of a federal irrigation project. When the infrastructure, including a large dam, was completed, the settlers acquired farms, erected homes, and worked the land. Headgates were opened, the people cultivated their fields, and crops flourished. Then, beneath a harvest moon, a weary farmer hauled his wagon load of produce "to market." The monetary reward was ample, as stacks of money at the cashier's window demonstrated. And at last there appeared the final "results of labor," what the settler had striven for all along: a prosperous farm replete with tidy fields, a substantial house, and a new automobile. With planning and hard work, went Seiler's tale, the land could be ordered and made to produce wealth.[1]

"The Story of An Irrigated Farm," and representations like it, had great appeal for the creators and inhabitants of the irrigated landscape. In moments of optimism, when the future beckoned or when their hard work indeed brought them handsome profits, the story did seem to be one in which they had rearranged nature into a linear, precise (even mechanistic) system that began with water and ended with a pile of cash. Seiler in fact had reason to compose just such a version of events, for 1915 was a relatively prosperous year for farmers across the American West. Irrigators were doing well, the region was filling up with settlers, and the Reclamation Service seemed to be realizing its promise as a modern federal government bureaucracy.[2]

Yet Seiler's story may have served another purpose as well: it may have been an attempt to put a hopeful face on some troubling conditions. Irrigated agriculture, in fact, was seldom as simple, straightforward, and profitable as he suggested, even in the best of times. Much was missing from his narrative. Drought, water rights disputes, weeds, insects, diseases, crumbling canals—these made no appearance in "The Story of An Irrigated Farm." Nor, for that matter, did the perturbations of that abstract thing called "the market": there was no hint of the falling prices or greedy middlemen that might turn a farmer's dream of prosperity into despair. Seiler, in sum, had ignored the complexity, contingency, and hazards that agricultural production actually entailed. Contrary to his portrait, this was not an enterprise in which people just grew things, transported them to market, and then headed to the bank with a pile of cash.

Inhabitants of the irrigated landscape thus at times turned to images and metaphors that more accurately conveyed a sense of the economic and environmental challenges confronting them. Such was the case with E. F. Rinehart of the U.S. Department of Agriculture. In 1915, Rinehart reported with dismay that some settlers on the Minidoka irrigation project tried one commodity after another in a relentless, self-defeating effort to take advantage of each price trend. As Rinehart put it, the farmers were "continually chasing a market rainbow, going into a crop or livestock industry when the prices are high and going out when they are low" and never concentrating on plants or animals that would bring steady income over the long term. Unlike the farmers in Seiler's story, who were creating stacks of money, the Minidoka farmers, Rinehart suggested, were pursuing an elusive pot of gold. Ed Vance, an extension agent, observed a similar frenzy in Twin Falls County in 1917, and he described it in equally vivid terms. A succession of crops and stock—potatoes, apples, clover seed, alfalfa, wheat, dairy cows, sugar beets, and others—made the "whole county" seem like "one big experiment farm."[3]

But the people of the irrigated landscape resorted to an even more troubling metaphor to described their work: farming, they said, was gambling. In 1917, potato expert Eugene Grubb stated that "the average farmer is a guesser and a gambler," impulsive in buying and selling and inadequately informed of proper agricultural techniques. The following year, Earl Irwin of Rupert, Idaho, likened farming to "a gambling game. The farmer bets his time, seed, money and labor against the weather, failure, rodents, and the profiteers." Annie Pike Greenwood in 1934 looked back on her years spent on an irrigated tract in Idaho and mused that "all farmers are speculators—the most reckless gamblers in the world, with all the cards stacked against them."[4]

These contrasting images—order and disorder, systematized farming versus gambling—summarized one of the basic dilemmas of agricultural production in the irrigated areas of Idaho's Snake River valley. Farmers and their associates, primarily engineers and scientists, sought to build and operate a precise and lucrative landscape, a virtual money-making machine. Indeed, this quest expressed some of their deepest aspirations for their land and lives. Yet despite their best efforts, environmental and economic factors worked against them. They always had to contend with a natural world that resisted their attempts to regiment it, and they constantly struggled to sell commodities on a capricious, unpredictable market. Agricultural production in the irrigated landscape thus involved a perpetual drive to impose order on an inherently chaotic, hazardous situation, to smooth the transition of crops from field to market.

How, then, did farmers go about raising and selling crops and livestock? What sorts of production and marketing routines did they devise? How did their systems develop and change? How did their decisions, and the consequences of those decisions, help to generate the irrigated landscape? A close examination of any of the plants or animals that Idaho's irrigation farmers produced could provide answers to these questions. But two crops, alfalfa and potatoes, will suffice. Alfalfa, widely grown in the Snake River valley since the early twentieth century, was a multipurpose plant, both livestock feed and soil conditioner. The potato, which many farmers adopted during the same time, was predominantly a specialized crop sold largely to urban consumers. Together, these plants tell the story of the farmers' efforts to garner profit from an uncertain world.

A central feature of that world was the market in which Idaho farmers operated during the late nineteenth and early twentieth centuries. Early irrigated agriculture addressed a demand for food and forage that came mostly from nearby mining districts and their boomtowns, places such as the Boise River basin or Silver City in the Owyhee Mountains. From the 1860s to the 1880s, irrigation farmers mostly grew hay for draft animals and wheat for miners. But beginning in 1884 with the completion of the Oregon Short Line through the Snake River valley, the situation changed. The railroad now turned the farmers' attention from local to regional and national markets. As this happened, the irrigated landscape became, in effect, an extension of the American dinner table. Connected by rail to economic centers such as Los Angeles, Denver, and Chicago, farmers quickly oriented their production to fill the conventional American demand for beef, potatoes, bread, milk, and sweets. But even with this predictable "beef and potatoes syndrome," as the historian Harvey Levenstein has called it, the market for such standard dietary fare was never steady. Most important, the heavy demand and high prices of the First World War collapsed in the 1920s.[5]

Idaho farmers raised alfalfa and potatoes in relation to this evolving local, regional, and finally national economy. They introduced alfalfa to the Snake River valley during the nineteenth century; by 1910, they devoted 36 percent of the irrigated landscape to the crop. They had good reason to give it such emphasis, for it was a versatile plant that performed several agricultural functions: it added nitrogen to the soil, served as animal fodder, and produced seed for yet more alfalfa. And when the market was right, it brought substantial cash returns.[6]

In Idaho and around the American West, irrigation farmers, agricultural scientists, and government officials praised alfalfa's usefulness and profitability. During the early 1900s, for example, agronomist and alfalfa booster George L. Clothier lauded the multiple purposes that the plant served. Alfalfa, he said, miraculously converted nature's raw materials into valuable things—milk, wool, fat lambs and hogs, fertile soil, and, even more important, money. In his words, alfalfa "brings gold into the farmer's purse by processes more mysterious than the alchemy of old."[7]

There were problems, however, with such pronouncements, for Clothier's alchemical metaphor had much in common with Seiler's "Story of An Irrigated Farm." Both emphasized the smooth transformation of nature into monetary wealth and, in doing so, obscured and even mystified complex biological, agronomic, and economic processes. Neither conveyed a clear sense of the factors that shaped agriculture in places such as Idaho's irrigated landscape. That a farmer could simply work the land and end up with a pile of cash was a fiction, and so was the idea that alfalfa was simply a Merlin that magically turned crude substances into gold coin.

Then how did this plant function? What, precisely, did those "processes more mysterious than the alchemy of old" entail? Alfalfa was, in part, a biological tool with which farmers made the soil fertile for other marketable crops, such as wheat, sugar beets, and potatoes. When plowed, the stems, leaves, and roots decomposed, adding organic matter to the spare desert soil. As a legume, alfalfa also infused the soil with nitrogen. *Rhizobium* bacteria lived in nodules on alfalfa roots and collected atmospheric nitrogen during the respiratory process of the host plant. When alfalfa decomposed, the roots and nodules, along with stems and leaves, passed the nitrogen into the soil.[8]

Idahoans noted increased yields in crops planted after alfalfa. In 1909, a farmer near Twin Falls informed a young Henry A. Wallace that wheat sown in "old alfalfa ground . . . will run forty bushels" per acre, in contrast to the twenty-five bushels on untreated soil. "Were it not for the fact that the cultivation of alfalfa and other legumes adds nitrogen to the soil," Elias Nelson of the Idaho Agricultural Experiment Station reported that same year,

> we should be obliged to purchase expensive commercial nitrogen such as nitrate of soda and guano and apply them to the land. . . . To supply the deficiency of nitrogen is imperative on our new land as there is no cheaper or more practicable way of making our soil fertile. . . . The various field crops

yield nearly if not twice as much after alfalfa has occupied the land as before. A crop of alfalfa is worth ten tons of sugar beets per acre.

Although alfalfa plowed into the soil never sold on the market, its value finally appeared in the beets or other crops that it helped to produce.[9]

Alfalfa's use as a soil conditioner had a parallel in its role as livestock feed. The same nitrogen-fixing ability that made the plant good for the soil and thus for other crops also made it a highly nutritious animal feed, because nitrogen was a crucial component of the amino acids necessary for manufacturing proteins. Here again, alfalfa was a biological tool used for producing something else. Idaho irrigation farmers fed alfalfa to dairy cows that provided milk both for the farm family and for sale on the market. They also fed it to sheep that in turn provided marketable wool and mutton. During the early 1900s, the irrigators' flocks grazed in the mountains in the summer and then moved to the farms in winter, where they consumed alfalfa hay. Alfalfa, too, served as fodder for draft animals. Not unlike gasoline in a tractor or coal in a steam engine, alfalfa fueled the horses and mules that provided the power for plowing, tilling, harvesting, and hauling.[10]

Alfalfa was a soil fertilizer and animal feed for the farm, one among many tools that a farmer used—and it was something else besides, for alfalfa hay also had great value when sold to other farmers or livestock growers. Less miraculous than Clothier implied, how alfalfa became a cash crop in Idaho's irrigated landscape is a lesson in environmental and economic contingency. As irrigation expanded during the 1880s and 1890s, a ranching market developed for alfalfa hay. Although an indirect extension of the larger national demand for beef and wool, this ranching market resulted from ecological conditions in and around the Snake River basin. By the 1880s, overstocking of cattle on the open range was depleting native grasses. As forage decreased, animals struggled to survive the winter. During the cold months of 1889–1890, thousands of cattle perished. Disaster on the open range in turn yielded a boom on the irrigated lands: cattle ranchers and then sheep raisers purchased alfalfa as winter feed to compensate for the deteriorated range. After 1890, farmers grew most of their hay for sale to ranchers. This relationship even adapted itself to the established practice of transhumance in Idaho. Ranchers wintered livestock (especially sheep) on the irrigated farms in the valleys; here the animals browsed on field stubble and were fed alfalfa. In spring the ranchers moved the herds to nearby mountains to graze. The market for alfalfa further expanded once farmers could ship the crop by railroad to livestock growers in nearby states.[11]

But this was not a story of unqualified success, for the magnificent alfalfa boom gradually turned into a resounding bust. Part of the problem was simply economic: overproduction. During the 1910s, farmers more than doubled the state's alfalfa acreage; in Twin Falls County alone, the amount of alfalfa land grew from about 18,000 to 54,000 acres. Increased alfalfa production followed by declining demand at the close of the First World War drove prices down. The market simply could not absorb any more hay.[12]

An abundance of alfalfa with no market was in itself a serious problem, but ecological problems compounded the economic: overproduction had unforeseen environmental consequences that further weakened the ability of farmers to sell their crop. Through the early twentieth century, the demand for hay encouraged irrigation farmers not only to increase alfalfa acreages but to keep their fields in the same stand for long periods—as long as twenty-five years in extreme cases. Old alfalfa fields, however, were vulnerable to the alfalfa weevil that appeared in southern Idaho by 1916. As the weevil infestation worsened, surrounding states imposed quarantines on Idaho alfalfa in an effort to stop the spread of the insect. With this outside market closed off, the value of Idaho alfalfa declined still more.[13]

As the glut worsened, farmers had greater and greater difficulty converting hay into dollars. The problems appeared as early as 1917, when farmers and livestock growers bickered over the price of hay. The *Idaho Republican* labeled the conflict a "rate war," an assessment that conflicted with the notion of farmers easily transforming nature's bounty into cash. The farmers demanded top dollar for the crop, at least fifteen dollars per ton. The stockmen wanted to pay no more than ten dollars, and they threatened to ship their animals outside the region, to the Great Basin or the Midwest, to feed them. When the Idaho Bureau of Farm Markets mediated the dispute, the agency's director, Harvey Allred, supported the higher prices.[14]

The farmers, however, lost their advantage after World War I, and in the 1920s prices hit bottom. On the Minidoka project, alfalfa hay that brought as much as $35.00 per ton during the war sold for as little as $2.50 during the following decade. Annie Pike Greenwood reported that her husband had spent $8.50 per ton to raise alfalfa, but the "sheepmen" would pay only $6.00 for it. The loss was devastating to the Greenwoods and other farmers who had counted on the profits that the hay would bring. The plant that George Clothier said transmuted itself by alchemy into gold now seemed to have turned into dross.[15]

Yet the farmers who survived the agricultural depression of the 1920s did not abandon the crop. It was too useful on the farm, and more important, they found another money-making purpose for the plant: fed to dairy herds, alfalfa boosted production of milk. Fortunately for the farmers, the milk market was expanding just at the time the alfalfa market was collapsing. Increased demand came not only from a growing national population but also from higher per capita consumption. Within this nationwide market, the West Coast, especially California and the burgeoning metropolis of Los Angeles, cried out for more cheese and butter. In response to the trend, especially the West Coast clamor, Idaho irrigation farmers fed hay to more and more milk cows. In 1924, R. E. Shepherd of the Minidoka North Side Irrigation Project reported that "we have got ten thousand cows and are now milking our way out" of the excess alfalfa problem. Between 1919 and 1926, Idaho farmers as a whole increased the number of dairy cows by 38 percent, from 118,000 to 163,000 head. From 1919 to 1924, milk production expanded by 50 percent. Growing numbers of factories, including a large plant operated by Kraft, processed the milk into butter and cheese. In turn, the Oregon Short Line carried greater quantities of dairy products, mostly to Los Angeles: the railroad shipped 177 cars of cheese in 1923 and 471 in 1926; butter shipments in the same years went from 188 to 469 cars. Once again alfalfa brought money into farmers' pockets, although in part because of good fortune and unpredictable economic circumstances.[16]

And the money continued to flow when groups of farmers oriented their alfalfa culture to a growing market for seed. Through the early twentieth century, farmers in the American West and around the nation—particularly the Great Plains and upper Midwest states such as Wisconsin—grew more alfalfa in the interest of improving soil and producing animal feed. Between 1899 and 1909, alfalfa acreage in the United States approximately doubled; from 1909 to 1919, it almost doubled again. This expansion in turn made alfalfa seed an important commodity. The demand grew especially intense when World War I reduced the influx of foreign seed, most of which came from Turkestan by way of Hamburg, Darmstadt, or Vladivostok. Before the war, annual imports averaged over 7 million pounds, roughly one-quarter of the total domestic demand. But by 1918, imports had fallen to 87,000 pounds. With foreign sources closed, domestic production boomed. In 1919, Idaho farmers devoted 5,500 acres to alfalfa for seed; by 1926 the total had risen to 20,000.[17]

Yet just as hay production encountered economic, biological, and ecological problems, so did the growing and selling of seed. Because germination and growth were the ultimate purpose of seed, the market placed the highest value on seed with physical qualities that appeared to indicate its potential vigor: relative absence of weed seeds; smooth, plump appearance and shape; a bright yellow color. Such qualities, however, were often difficult to achieve. Various environmental factors—climate, weather, weeds, irrigation, harvesting, even features of the plant itself—could impart characteristics to the seed that lessened its viability. Seed that was cracked, wrinkled, dirty, polluted with weed seeds, immature, or of mottled color usually germinated less well and so tended to have a lower market value.[18]

Even before World War I, many Idahoans—farmers, government officials, boosters—worried that production of low-quality seed might destroy their efforts to establish Idaho's reputation as one of America's prime seed regions. In addition, they were concerned that poor seed might promote the spread of weeds throughout the state's agricultural lands. Consequently, in 1911 Idaho passed its first Pure Seed Act, a measure designed to control and stabilize the production of alfalfa and other seed.[19]

The Pure Seed Act was an attempt to regulate the two things on which farmers depended but which could undermine their efforts to gain wealth: nature and the market. In part, the law sought to impose a new order on seed production—on nature. It defined weed seeds, established aesthetic and purity standards, and eventually created five grades of alfalfa and other seeds. The grades, somewhat arbitrary categories, provided farmers with agricultural and environmental standards they had to meet in order to sell their product. To meet the grades, the farmers had to keep their fields clean of weeds. At the same time, the Pure Seed Act regulated the market and brought it into conformity with seed production. Grading indicated to farmers and other buyers the quality and value of seed, thereby encouraging them to purchase the best product, not merely the cheapest. With nature and market stabilized and coordinated under the Pure Seed Act, Idaho state officials and farmers hoped to establish, and eventually preserve and enhance, Idaho's position as a seed-growing region.[20]

As it turned out, the task proved more difficult than anticipated. Despite the Pure Seed Act and the fact that high-quality seed garnered the best prices, nature and the market eventually eluded the farmers' control. With demand so great, especially during and after the world war, farmers lowered quality in a rush to produce quantity. In some instances they cleaned their seed through a wire mesh, sold the pure seed, and then replanted the leftover, inferior screenings. Some irrigation farmers illegally purchased untested, weed-infested seed from neighbors because it sold below market price. Many growers were not diligent about keeping weeds out of their fields. As a result, the proportion of weed seeds gradually increased until it threatened the overall purity of the harvest. In particular, dodder, an especially pernicious weed seed that closely resembled the alfalfa seed in size and shape, polluted alfalfa seed, passing through screens along with the alfalfa. In 1916, the Idaho seed commissioner rejected about 21 percent of tested samples; in 1920, he found that 58 percent of the seed entering his laboratory fell below pure seed requirements. Dodder was responsible for 95 percent of the rejected samples. Economic and ecological factors—the rush to grow alfalfa to meet market demand and the consequent spread of dodder—had formed a dangerous partnership that threatened stable, predictable alfalfa seed production.[21]

George Clothier's striking metaphor of alfalfa-as-alchemist could not contain such a messy reality, for dirty, polluted, wrinkled seeds contradicted the notion of nature easily transformed into coin. And that reality became messier still when farmers confronted the problem of seed varieties. Identifying special seed varieties posed a major challenge to those farmers who remained devoted to profiting from the production and sale of high quality seed. In the form of seed, nature did not visually distinguish between different alfalfa varieties—to the naked eye, all seeds looked the same. Because it looked the same, unscrupulous dealers sometimes purchased seed from weak common alfalfa—generic alfalfa of no special variety—and then sold it as winter-hardy Grimm or Cossack seed. Dealers outside the state also found ways to cheat Idaho seed farmers. They sometimes raised the quality of mottled, cracked, wrinkled, inferior seed by mixing it with plump, smooth, yellow Idaho seed and then advertised the whole lot as "Idaho seed." In a sense, the dealers turned nature, the physical characteristics of seeds, to their own ends. When Earl Irwin said in 1918 that farmers were gamblers who bet their personal resources against nature and "the profiteers," he probably had in mind people such as alfalfa seed dealers.[22]

In response to the continuing problems of weeds and dishonest dealers, Idahoans redoubled their efforts to impose order on the cultivation and sale of alfalfa seed. They turned on weeds, especially dodder, with renewed fury; government and Farm Bureau officials informed the public of the weed threat and in conjunction with the farmers organized weed eradication campaigns that for the first time made use of herbicides. In 1920, Idaho started a seed certification program to counter unscrupulous dealers. Under this system, government agents, in cooperation with farmers, approved and tagged seed batches, from origin in a field to final sale. "Pedigreed" or "certified" seed, as it was called, commanded a high market value. "This system," reported officials in 1927, "assures the ultimate consumer that he receives the genuine hardy alfalfa. There is no guessing; sealed seed is certain seed."[23]

But Idahoans discovered that even their intensified efforts had limits: the certification program, like the Pure Seed Act overall, had no authority beyond the state's borders. Some Idaho seed growers argued that they should advertise and educate out-of-state farmers to recognize and demand Idaho seed. They contended that consumers could know the genuine Idaho commodity by its unique color. Idahoans desirous of promoting their seed envied and hoped to emulate the national reputation then developing for Idaho potatoes, Washington apples, and California's Sunkist oranges. But they overlooked an important point: people did not always recognize a particular agricultural commodity by its appearance alone. The Sunkist stamp, not just the orange's physical characteristics, revealed the fruit's identity to consumers. Idahoans could not stamp their tiny seeds the way California farmers stamped oranges. For

Idaho seed growers, only a label on a sack distinguished their product; a dealer could easily duplicate the label and affix it to a sack of inferior, non-Idaho seed.[24]

Farmers in the irrigated landscape probably achieved their greatest success in systematizing the production and sale of alfalfa seed through a popular agricultural organization, the cooperative. Like other farm cooperatives, the seed association operated, in effect, at the juncture of the natural world and the market. The association required farmers to keep their fields clean of weeds and to grow certified seed varieties. The organization graded the harvested seed, stored it in a warehouse, and then circumvented the dealers and marketed the crop directly to buyers. The Idaho Seed Growers' Association (founded c. 1911) was probably the first such organization in the Snake River valley, although it apparently was mostly concerned with wheat seed. More specialized associations appeared after World War I, the time when Idaho farmers were stepping up their overall efforts to rescue and stabilize alfalfa seed production. In 1921, farmers on the Aberdeen-Springfield irrigation project formed the Bingham County Grimm Alfalfa Seed Growers' Association. Member farmers had to grow pedigreed Grimm alfalfa, seed biologically descended from the Grimm strain that Henry K. Wiley had first cultivated near Springfield in 1907. In a warehouse in Blackfoot, the association cleaned the seed in a special mill and then stored it prior to sale. Because of the high quality of its product, the Bingham County Grimm Alfalfa Seed Growers' Association attracted buyers and garnered a large share of the national market. Wiley and the other members quickly became America's leading source of Grimm alfalfa seed. Major national seed companies—Dickinson, Northrup-King, Scott—and the U.S. Department of Agriculture purchased large quantities of their seed.[25]

Alfalfa thus became a money-maker for the farmers, but the process had not been simple, linear, or even miraculous, and it had generated a full measure of losses. At various moments the farmers had fortune on their side: a depleted range and starving animals, a growing national demand for milk, environmental conditions that suited the Aberdeen-Springfield area to seed production. But they also had to contend with a series of problems, including overproduction, insect and weed infestations, shady seed dealers, and the physical characteristics of seed itself. Through trial-and-error responses to these opportunities and constraints, farmers and government officials gradually developed specialized production and marketing arrangements to guide the crop from field to market. These systems worked well enough, but never perfectly, for alfalfa existed in economic and ecological relationships that were always undergoing potentially disruptive change. The story of alfalfa in the irrigated landscape is not, as A. G. Seiler and George Clothier suggested, simply one of farmers working hard, selling their crop, and then pocketing gold coin or picking up piles of money at the cashier's window. Rather, it is a story of farmers constantly struggling to profit from an environment and an economy in motion.[26]

That story is similar for other irrigated crops and for livestock as well. Yet each crop or animal involved particular challenges and required unique production and marketing methods. Each organism had its own biological and agronomic characteristics; each occupied a separate position in the shifting ecological relationships of the irrigated landscape. And the market operated differently for each. The demand for hay was not the same as the demand for a crop that people consumed; livestock and humans, after all, had different nutritional needs and tastes. Thus while the raising and selling of alfalfa and potatoes had much in common, it was one thing to grow and market the legume but quite another to produce and sell the tuber.

Potato production in Idaho began during the nineteenth century, when farmers discovered that the Snake River valley and the irrigated landscape were ideally suited to cultivation of the crop. Long summer days, cool nights, loam soils augmented by alfalfa and irrigation water—these promoted extraordinary potato growth. In 1890, Frank Drake, a rancher near Hailey, won a prize for the nation's third-heaviest one-acre yield. Drake's achievement prompted a local newspaper to dub his potato variety "Drake's Idaho." But Drake preferred to call it simply "the Idaho," a name that in the future would connote Idaho tubers of outstanding size and quality.[27]

While producers such as Drake were discovering that the irrigated landscape yielded exceptional potatoes, they were also becoming aware of the market that existed for their crop. Farmers first grew potatoes for the local mining camps. But with the advent of the Oregon Short Line (OSL) in the early 1880s, they furnished tubers in response to the American "beef and potatoes syndrome." In 1885, shortly after the OSL's completion, a Boise newspaper called attention to "the possible advantages of raising potatoes for the eastern market." By 1891, the Union Pacific Railroad (UP), of which the OSL was a subsidiary, reported large shipments of potatoes out of Idaho. "Many thousand car loads of potatoes," the UP stated, "were shipped from Idaho points . . . in 1887, 1888 and 1890, to eastern markets, where they are in great demand." Eventually, Idaho spuds went to terminal markets in all parts of the United States, but especially in the western half of the nation: Chicago, Kansas City, Los Angeles, and other cities. With markets opened by rail, Idaho farmers gradually expanded their potato acreage. In 1882, the U.S. Department of Agriculture reported approximately 2,000 acres of potatoes in Idaho. By 1890, the figure had risen to 7,300; in 1900, it stood at 9,300; and in 1915, Idaho had roughly 33,000 acres of potato fields.[28]

Thus by the early twentieth century, an Idaho potato bonanza was under way, and it generated much excitement in the state's irrigated regions. At harvest, the small towns along the Snake River were the scene of intense commercial activity. In each town's "spud alley," where storehouses lined the railroad tracks, farmers brought in wagonload after wagonload of potatoes. Special trains of insulated cars then hurried the cargo to urban destinations, often with great fanfare. In November 1912, for example, the Boyle Commission Company of Idaho Falls organized a thirty-car trainload, "big banners on each car announcing . . . that the shipment was made up of over 1,000,000 pounds of high-grade Idaho potatoes."[29]

While the trains were carrying the crop to America, Idaho potato farmers received regional and national acclaim for the quality of their product. In the Snake River valley, successful farmers such as Joe Taylor, who by 1908 worked 300 acres of spuds near Idaho Falls, became known as "potato kings." Following in the footsteps of Frank Drake, some growers won honors for their potatoes at national agricultural exhibitions. In autumn 1912, Leopold Pancheri of Idaho Falls pulled a magnificent yield from the ground, 606 bushels in one acre alone, and he displayed a sample of these at the New York Land Show. There his crop attracted the attention of the U.S. Department of Agriculture and national potato experts. It won third prize overall, and because he exhibited the fifty largest spuds at the show, he took home an additional award, one that symbolized the farmers' quest to transform nature into wealth: $100 in gold coin.[30]

It seemed so simple and wonderful: hard-working farmers, a highly productive environment, and growing consumer demand had come together to create champion potatoes. Yet problems accompanied the profits and accolades, the potato king title and gold coins, the simplistic representations of the farm-to-market process, like A. G. Seiler's "Story of An Irrigated Farm." Harnessed to potato production, the Snake River valley environment yielded generously, but the market did not necessarily reward farmers for the resulting bounty. Idaho's potato crop exacerbated a national glut that each year drove prices down. Most American farmers, including Idaho growers, harvested potatoes in late summer and autumn. Flooding the market at once, this massive late harvest sometimes resulted in low prices that brought farmers little but frustration. J. R. Simplot, who founded Idaho's greatest agricultural corporation, recalled that his father accompanied a potato shipment to Kansas City one year and there sold his crop, but the meager profit paid only for the freight cost and his fare back to Idaho.[31]

To ensure the highest prices for late potatoes, Idaho farmers tried to counteract the autumnal glut. Some pushed or delayed the harvest. Farmers who dug their potatoes slightly ahead of other growers avoided the eventual price decline, but if harvested too soon the crop was green and less valuable. Growers who delayed the harvest likewise might earn higher prices, although the potatoes might freeze or rot if left in the ground too long. It was a gamble either way.[32]

Farmers, however, had the best chance of countering the glut by storing the potatoes in special sheds or warehouses (also called houses and cellars). These were long, rectangular buildings set several feet in the ground and partly or completely covered with earth. These sheds, in effect, mediated between the environment and the economic system: they preserved the crop's physical condition while keeping it off the market. With the potatoes safely ensconced in temperate, dry sheds, farmers waited through the winter until prices rose to an acceptable level. As William Stuart of the U.S. Department of Agriculture (USDA) explained in 1917, "Good storage not only serves to hold perishable crops in a salable condition but insures . . . a more uniform market supply throughout the season." Wise farmers or farmers with enough capital built sheds early. Joe Taylor's two stone warehouses, built by 1908, probably helped him to achieve financial success and his potato king reputation.[33]

Economic constraints eventually led more Idaho farmers to build potato cellars. Until after World War I, the national potato market generally was strong, and farmers could profit without storage. But during the war a railroad car shortage forced growers to build more storage houses. The postwar price collapse and then huge yields, especially the 1922 harvest, created a vast excess of potatoes. This surplus compelled construction of still more facilities in order to keep the potatoes off the market. In August 1922, a Blackfoot newspaper reported dozens of farmers heading for the nearby Rocky Mountain foothills to cut pine for building material. "Practically all . . . the communities of this county," stated the paper, "are represented in a great exodus of teams and wagons going to the distant canyons for logs and poles to be used in building potato cellars . . . thirty-two teams and wagons started from Thomas alone on Monday." By the 1920s, such structures were a ubiquitous feature of southern Idaho's irrigated landscape. Yet they did not completely solve the marketing problem—as some observers noted, they only allowed growers to be better gamblers.[34]

Seasonal glut challenged early Idaho potato farmers; so too did crop diseases. In pursuit of profits, growers tended to delay rotation and plant the crop in the same fields year after year. This practice made the potatoes vulnerable to various bacterial, viral, and fungal diseases. Idaho's desert soils, which seemed to the first settlers to be so clean and free of disease, in fact contained indigenous fungi (*Fusarium* spp., *Rhizoctonia* spp.) that parasitized the tubers. Potatoes planted in Idaho soil immediately provided excellent habitat for the fungi; prolonged cultivation without rotation eventually allowed the fungi to overwhelm the crop. Poor seed potatoes compounded the problem of intense cultivation. The best potatoes went to market, and the worst—often the diseased leftover culls—went back into the ground as seed. By 1910, diseases endangered Idaho's potato fields. In 1913, the nationally renowned potato expert Eugene Grubb called attention to the worsening condition of Idaho spuds. Farmers, said Grubb, "have been making money so easily on potatoes that they neglected vigilance in keeping good seed and keeping out diseases, with the result that the earning capacity of their lands has fallen off rapidly and they are in much trouble." By 1914 some growers sustained losses as high as 50 percent.[35]

Farmers and agricultural scientists searched for ways to eliminate the diseases that threatened production. The potato boom had led to disease-ridden fields; in turn those infected acreages gave rise to a specialized market in clean seed potatoes. As early as 1920, farmers in the far reaches of the upper Snake River valley, around the towns of Ashton and Driggs, began to produce the crop. At more than a mile in elevation, the area had a relatively cold climate and short growing season, environmental conditions that inhibited the growth of diseases and made the land ideal for raising seed potatoes. Ashton-Driggs farmers became well known for their crop, which they sold to growers in lower elevations in Idaho and in other parts of the American West.[36]

Adoption of a new potato variety also allowed farmers to combat the effects of disease. During the early 1900s, Idahoans grew an array of potatoes—Rural New Yorker, Cobbler, Bliss Triumph, Delmany Challenge, Bull Moose, Idaho Rurals, Peachblow, Red Peachblow, and others. Among these, the Russet Burbank proved most resistant to scab fungi. The renowned plant breeder Luther Burbank had selected the variety in 1872 from a patch of Early Rose potatoes growing in his New England garden. Later, Lon D. Sweet of Denver, Colorado, isolated a strain of Burbank's spud with a crinkled, reticulated skin. It was this skin that made the Russet Burbank resistant to disease. Because of its hardiness, Idaho farmers increasingly preferred to raise the variety. According to Idaho's official potato history, "It was, in fact, potato scab which made the russet Burbank popular in Idaho and in the opinions of some people the savior of the Idaho potato industry."[37]

Fortunately for Idaho farmers, a combination of natural and market factors further contributed to the success of the variety. The Snake River valley proved better suited to Russet Burbank cultivation than any other region in America. Because of the local environment, the Russet Burbank exhibited qualities—color, firmness, texture, size—not found in potatoes of the same variety grown elsewhere. Idahoans had known about the Russet Burbank's high quality as early as 1907, from tests conducted by Alexander McPherson, director of the Twin Falls Land and Water Company's experiment farms. As more and more Idaho farmers grew the Russet Burbank, they drew the attention of wholesalers, grocers, and consumers to the potato. In turn, as people responded and bought increasing quantities, farmers expanded their cultivation of the variety. In this fashion, the Russet Burbank, also called Netted Gem, Idaho Russet, and eventually just the Idaho, came to dominate the potato fields of the Snake River valley. By 1929, it constituted roughly 60 to 70 percent of all the potatoes grown in the state. This percentage rose to 90 percent by 1939. According to a government report issued that year, "The consumer associates the characteristics of the Netted Gem with Idaho potatoes and is usually unwilling to accept other types in its place."[38]

The Russet Burbank's popularity helped achieve a goal of many Idaho farmers, businessmen, and government officials: potato standardization. During the early 1900s, these Idahoans advocated uniformity in crop varieties and quality because standardized produce commanded the highest prices from wholesalers and consumers. In 1910, newspaper editor and farmer Byrd Trego urged boosters "to encourage Idaho potato growers to plant only one or two staple varieties, instead of every man growing a different kind of spud, so that in a carload there [are] often as many varieties as are known." Trego and other like-minded Idahoans also contended that farmers should sort each variety into grades, because wholesalers and consumers paid more for graded than for field run tubers.[39]

Varietal uniformity and grading eventually influenced agricultural methods and regulated the entry of potatoes into the market. In order to grow a standardized crop that sold well, farmers had to make nature—that is, potatoes and, indirectly, the fields—more precise and systematic. Uniform tubers required varietal specialization and careful tillage, irrigation, and rotation. These exacting techniques, however, did not automatically produce potatoes of even size and quality. Through sorting and grading, farmers thus attempted to compensate for any remaining irregularity.

Grading structured the production and marketing of potatoes, but until World War I, standards were voluntary and informal, and they varied according to fluctuations in the market. Farmers and wholesalers, moreover, offered conflicting interpretations of the standards. Grading first involved passing the potatoes through sieves or machines that sorted them according to size. Then workers removed the damaged, diseased, or misshapen tubers by hand and grouped the remaining ones according to quality. Farmers exercised wide latitude in grading standards. They were reluctant to cull the smaller or irregular-shaped spuds when demand was high and prices were good. Potato wholesalers, however, decried lax grading habits. They wanted only the best potatoes, because these would easily meet the standards of bulk buyers in the cities. Consequently, when prices fell, wholesalers required farmers to meet higher standards. To ensure quality, some wholesalers began regrading the spuds before purchase. The wholesaler took the crop into a spud alley warehouse, graded and bagged the tubers, and paid the farmer for the weight of the top grade. Predictably, farmers protested. One dealer recalled the wrangling that sometimes ensued between wholesalers and farmers and between wholesale shippers and receivers:

> If the market went up the [grower's] tendency was to crowd in most everything he had in his cellar. If the market went down the dealer wanted them polished, perfect, and grown in a mold. There was no recognized criterion by which disputes could be settled and rights defined. The confusion and difficulty between the shipper here and the receiver at the other end was even greater. Every shipper and receiver had his own ideas of what "choice" potatoes were, and these ideas were in many instances influenced considerably by market conditions . . . prevailing when cars arrived at destination.[40]

The dealer's recollection illuminated some of the difficulties inherent in transferring potatoes from the fields through the various levels of the market. But his account also revealed a crucial connection between the conflicting metaphors of disorder and order that farmers, scientists, journalists, and other observers often used to describe irrigated agriculture in Idaho. Among farmers, a poor market typically evoked images of disorder—falling prices led them to believe that their work was a great risk, like gambling. Yet in the case of potato grading, flagging prices inspired a counter image that, in its emphasis on order, was compatible with A. G. Seiler's "Story of An Irrigated Farm." As the dealer remarked, the desired potatoes were "polished, perfect, and grown in a mold," like high-quality, mass-produced industrial goods. Thus, in Idaho potato culture, the opposing images of gambling and manufacturing, the loss of control and controlled production, sometimes derived from the same economic circumstances.

During World War I, the state of Idaho and the federal government established official grading standards and attempted to mediate conflicts between farmers and wholesalers. In August 1917, the director of the Idaho Bureau of Farm Markets, Harvey Allred, established three standard potato grades, to be administered by state inspectors. Allred stipulated that a grower or dealer who disagreed with an inspection could appeal to the director for reconsideration. The following month, the USDA and the wartime Food Administration recommended national grading standards, which Allred adopted for Idaho. Potatoes graded and stored according to USDA rules could serve as collateral for loans under the 1916 Farm Loan Act. In 1918, the Food Administration mandated the standards for its licensees.[41]

The new, more rigid standards compelled farmers to further refine their potato production techniques. A more organized and structured market required still more organized and structured cultivation. Farmers had to produce the right sort of potato if they wanted to get a federal loan, to profit from wartime economic planning, or to compete more effectively against other farmers.

Some farmers, however, could not make their farms so efficient. They could not exercise such precise environmental control and thus meet the new grading standards. These growers began to believe that the standards were unfair, a hindrance, or the imposition of a domineering government. In 1918, Earl Irwin told of disgruntled farmers on the Minidoka project:

> There has been considerable complaint in this section about the extraordinarily strict rules for grading potatoes. . . . The grades are so strict that most of the farmers find themselves with two-thirds of their crop left on their hands as culls. . . . It may be possible to grade potatoes so that they look as if they had been made by machinery, but it can not be very profitable to the grower.[42]

The collapse of commodity prices after World War I made the grading standards seem even more onerous than before. The best potatoes no longer brought high enough prices to offset the loss from culling out lower-quality ones. Farmers stepped up their complaints. Even Joe Taylor, the vaunted potato king, spoke out against the standards. In January 1921, he voiced his opinions in a speech to a group of about four hundred farmers at Idaho Falls. At the meeting, the assembled farmers "voted in favor of abolishing the present method of potato inspection, claiming it has been a damage to the growers rather than a help." In contrast to farmers such as Taylor, many growers—and, more important, wholesalers, other businessmen, and government officials—insisted on retaining the system. They believed that Idaho's position in the national potato market, and the high prices that the state's best tubers commanded, depended on maintenance of the standards.[43]

Indeed, the falling prices of the early 1920s reemphasized the need for intensely controlled production methods. A 1922 description of F. A. Starkweather's potato field near Blackfoot indicated the cultivation techniques that farmers hoped would produce high-quality potatoes:

> He has been grading his fields for several years to get them so they water perfectly, and part of the tract is laid out in rows half a mile long and he waters the full length of them from one ditch. The land is so even that only in a few places does [the water] get to the top of a ridge. The rows are as straight as gun barrals [*sic*] throughout the field, and it is one of the finest potato tracts in the country.

Here again, the underlying assumption of industrial precision and control sharply contrasted with depictions of farming as a gamble. Ironically, the fact that production for a constricted market entailed increased risk encouraged farmers to adopt practices that suggested factory-like regulation.[44]

Through sophisticated production methods and grading standards, Idahoans sought to create potatoes that brought high prices on the market. They also sought to augment the desirability and value of their product by the way they packaged it. Bulk buyers and consumers demanded and paid higher prices for graded, standardized tubers; they also tended to choose and pay more for potatoes packaged attractively.

A sack heightened the visual appeal and thus the worth of potatoes. In 1912, T. I. Hubbell, manager of the Western Milling and Elevator Company in Idaho Falls, described the almost intangible value that sacks imparted to spuds. "Good, clean sacks add nothing to the quality of potatoes," Hubbell stated, "but they do add much to their appearance. Potatoes are sold as much on their looks as on their eating qualities—

perhaps more." With sacks, farmers and wholesalers improved the market for the crop.[45]

Potato-farmers and wholesalers enhanced their produce through grading and sacking, but these same techniques gave them the means to unload poor-quality spuds on unsuspecting bulk buyers and consumers. A shipper filled the bottom of a sack with graded potatoes, then inserted a stovepipe into the sack. The shipper packed more graded spuds around the pipe, filled the pipe itself with culls, then withdrew the pipe and topped off the sack with good spuds. A wary buyer who opened the sack before purchase saw only high-quality tubers; the grocer or shopper eventually discovered the wretched produce within. Perhaps the wholesalers and farmers who practiced "stovepiping" appreciated its irony. A competitive market had compelled the adoption of grading standards that effectively excluded tons of small, diseased, or irregular potatoes from sale. That market also had encouraged Idahoans to use sacks to enhance the appeal of high-grade spuds. The grading and sacking meant to elevate potato quality, however, allowed farmers and wholesalers to hide and sell otherwise worthless culls. Grades and sacks improved the market in more ways than one.[46]

Sacks enticed buyers and consumers; advertising provided Idahoans with another means to influence the market for their spuds. During the late nineteenth century and through the early 1900s, farmers used informal methods of publicity in an effort to find markets for their burgeoning harvests. Wholesalers from outside the state, especially Colorado, initially had purchased and then shipped most Idaho potatoes, but Snake River valley production eventually outstripped the capacity of these dealers. Denver and Greeley wholesalers, moreover, often marketed Idaho spuds as a Colorado product, probably because Colorado potatoes had an established reputation and the Snake River valley tubers did not. Idaho farmers and produce dealers now wanted to sell their own crop, but first they had to attract the attention of bulk buyers. In 1908, the *American Falls Press* reported the efforts of the Idaho Falls farmer and businessman C. M. Grissom to find a buyer for his potatoes. Grissom, stated the paper,

> is doing some missionary work for Idaho by rustling a market for Idaho potatoes. . . . Last year Mr. Grissom had a mighty hard time disposing of a few cars of "Idaho potatoes" because, he said, the buyers knew nothing of them

and were decidedly skeptical about buying them but this year he thinks he can dispose of all he can get and they will go as Idaho products too.

To promote his spuds, Grissom probably just went to terminal markets and told potential buyers that Idaho produced good-quality tubers worthy of purchase. Once he won their confidence, urban produce dealers and grocers bought more and more Idaho potatoes. The fact that through the 1910s and 1920s Idaho produced increasing quantities of excellent Russet Burbanks no doubt helped Grissom and others to publicize and sell the crop.[47]

After prying open the market, Idaho farmers and wholesalers attempted to further expand their sales by strengthening the association of high-quality potatoes and "the Idaho" name. Through increasingly sophisticated advertising, they sought to make the name synonymous not just with good spuds but with outstanding, exceptional, wondrous ones. Agricultural shows and trade fairs gave Idahoans a venue for promotion. At the 1915 Panama-Pacific International Exposition in San Francisco, the Boise booster Robert Limbert topped a potato display with an eleven-foot model of a Russet Burbank.[48]

Ironically, the falling market of the 1920s as much as the actual quality and sales of the tubers compelled Idahoans to intensify their publicity. Indeed, risky markets and even failure provided much of the impetus for construction of "the Idaho" image. In response to the 1921–1922 price crash, the Nampa Chamber of Commerce exhorted people to "Buy A Sack of Potatoes" as an act of civic duty. At the urging of the Nampa chamber, Governor David Davis issued a proclamation asking Idahoans and people in other states to buy and consume more spuds. Paradoxically, the poor market soon intensified the use of advertising that connoted prosperity. In 1924, the Nampa Chamber of Commerce shipped twelve hundred potatoes to Philadelphia for a banquet of newspaper writers; each spud, neatly wrapped in tissue paper, was "perfect in form and size." The same year, the Idaho Falls Chamber of Commerce sent twenty-five "big, perfect, Idaho potatoes," each weighing one pound and wrapped in red and green paper, to President and Mrs. Coolidge for their Christmas supper. This promotional drive continued when, in 1926, the state first emblazoned the words "Famous Potatoes" on automobile license plates. Historians often associate modern advertising with 1920s prosperity, but Idaho's most enduring agricultural image owed much to economic depression. "Famous potatoes" initially expressed hopefulness more than triumphant fact.[49]

Gradually, however, the majority of the nation's buyers and consumers did come to associate "the Idaho" with superior quality. And as Snake River valley spuds achieved an enviable place in the market, wholesalers outside the state began to appropriate the potent "Idaho" cachet. These dealers sold inferior, non-Idaho potatoes as the real thing. Some even shipped tubers into Idaho and repackaged the imposters as Idaho potatoes. Much as Idahoans sometimes used grades and sacks to mask culls, the outside dealers now used "the Idaho" name to misrepresent lesser produce.[50]

Idahoans attempted to reassert control over "the Idaho." In the 1950s, the Idaho Potato Commission officially registered a "Grown in Idaho" trademark. Idahoans then placed a portion of the crop in labeled sacks that sold directly to grocery store patrons. The problem persisted in bulk sales, however, so the commission tried to develop a stamping machine that would identify each potato as an Idaho. In a sense, the stamp itself was an attempt to stabilize and join nature and the market. The stamp would give the perfect potato (nature) an identity that would signal consumers (the market) to buy it. The experiment failed, in part because the stamping process damaged nature: it hastened the potato's deterioration. Another cause of failure was that the market—consumers—found the stamp unappealing ("housewives really did not want stamped potatoes").[51]

Advertising had become an important part of potato production and marketing. As potato production expanded during the early twentieth century, farmers and businessmen linked all growing and marketing functions in cooperatives and integrated corporations. By manipulating nature and market more systematically and completely than any individual farm or wholesale operation, these agricultural organizations sought to make potato farming as efficient and profitable as possible.

The South Idaho Potato Producers' Association, the Blackfoot Potato Growers' Association (both founded about 1913), and similar groups performed numerous agricultural and economic functions. They promoted varietal specialization, crop rotation, and better seed; operated warehouses in which they sorted, graded, and stored potatoes; and sold farmers' spuds jointly in an effort to secure the highest prices. By combining growing, storage, and sales, the cooperatives allowed farmers to deal with the wholesale merchants with maximum power—or to circumvent them entirely. Farmers avoided haggling with wholesale merchants over grading standards. And they no longer had to worry that a dishonest wholesaler might steal a portion of the crop after it disappeared into the warehouse and underwent the sorting process. Still, the wholesalers whom the farmers sought to avoid often undermined the cooperatives by luring away their members with better prices. Potato associations attempted to control the market, but the market often subverted them.[52]

Like the cooperatives, large-scale agribusinesses tried to join potato production and sales, nature and market, in one stable unit. Much as the farmers had banded together to influence the market, some wholesalers tried to regulate production by raising their own crop. At the same time, a few individual farmers moved into sorting and shipping. In the early 1900s, young Jack Simplot labored on his father's potato farm near Declo, on the Minidoka irrigation project. As an adult, Simplot took up potato farming himself. Then, in what eventually became the J. R. Simplot Produce Company, he began operating a custom potato-sorting machine. During the 1930s, he built warehouses and started shipping tubers, many of them from farmers with whom he had contracted for the crop. In the Second World War, he continued to advance his agricultural empire. He moved into processing, dehydrating potatoes for the armed forces. He opened a phosphate mine and turned the mineral into phosphate fertilizer that he spread on his own fields and sold to other farmers. After the war, Simplot expanded horizontally into other types of agricultural production. His corporation represented the most thorough and successful effort in Idaho to systematize potato production and sales.[53]

Simplot's operation exemplified the efficiency that Idaho farmers in general relentlessly tried to attain. Yet despite achievements like Simplot's, the farmers' quest to order and control nature and the market was not a simple story of progress. It was not a story that could be easily represented by images of orderly farms, alchemy and gold coins, potato kings, and perfect, machine-made spuds. Rather, it was a story of struggle, of human beings constantly grappling with a changing world. Under the influence of nature and the market, the irrigated landscape and its people existed in a state of perpetual redefinition and refinement.

Contingency and irony were central to this process. Many of the Snake River valley's plants and animals were the unplanned consequences of unanticipated environmental and economic change, even disaster. Deterioration of the open range initially boosted the alfalfa hay market; that strong market encouraged prolonged cultivation that favored the alfalfa weevil; the weevil led to quarantines that intersected with a postwar price crash; farmers then moved into dairy production. Most Idaho farmers probably would not have recognized the alfalfa weevil and the dairy cow as distant ecological-economic cousins, but indeed they were. The expanding range of the weevil encouraged the proliferation of the cow; the two animals were inadvertent partners in the formation and evolution of the irrigated landscape.

Contingency, disruption, and even failure led to success in Idaho potato farming as well. Several factors contributed to the predominance of the Russet Burbank, but its triumph resulted in part from the diseases and fungi that destroyed other varieties. Disease and declining yields in the lower Snake River valley also helped to create the wealth and reputation of seed potato farmers around Ashton and Driggs. Finally, the failure of growers elsewhere in America contributed to the eventual success of Idaho potato farmers; contingency operated on a national, not just a regional, scale. From the 1930s through the 1950s, the unstable markets and eroding soils that hurt and ruined potato farmers in Maine's Aroostook Valley indirectly enhanced the market position of Idaho producers. By the mid-1970s, while Snake River valley farmers shipped their potatoes to cities around America, some Aroostook growers transported theirs only as far as the dump.[54]

Although disruption and disaster redounded to the benefit of Idahoans, the methods of making farming orderly and predictable inadvertently created new problems. Solving these problems necessitated still more systems for ordering nature and the market. Potato production exemplified this trend. The potato efficiently transformed the Snake River valley's sunlight, soil nutrients, and water into a profitable form. Yet potatoes (and other thirsty cash crops) required more water; consequently, new dam and reservoir construction projects paralleled expanding potato acreage. Simultaneously, Idahoans produced so many potatoes that the market could not easily absorb them all; to mitigate overproduction and regulate the supply of spuds, Idahoans built potato sheds. Potato storage and water storage thus mirrored each other—it was no mere accident that intensified shed construction in the early 1920s coincided with campaigns to build American Falls (1927) and other dams. Furthermore, as Idahoans erected dams and sheds, they established grading, sacking, and advertising to improve potato quality and give the spuds an edge in the market. Each of these devices in turn contained hidden ironies. Sacking and grading permitted stovepiping. Outside wholesalers stole the reputation of "the Idaho" to sell their inferior spuds. To counteract the sale of fraudulent potatoes, Idahoans initiated yet another ordering system, the stamping machine.

Today, the reputation of the Idaho potato belies the greatest irony of agricultural production in the irrigated landscape. The quest to harness nature and to influence the market, the struggle to make the farm efficient, at once evoked two sets of contradictory metaphors. The greater the ecological and economic instability, the greater the need for factory-like fields and machine-perfect spuds. Similarly, the more competitive and unstable the market, the stronger the need for a product image—"the Idaho"—that connoted prosperity and success. The Idaho came polished and perfect from a mold, from rows as straight as gun barrels. A. G. Seiler's farmer perhaps carried a load of Idahos in his wagon as he made his way to market.[55]

The need to control unstable conditions engendered images of orderly industrial production; the risk of failure inherent in farming under those conditions also evoked the metaphor of gambling. Indeed, order and gambling, control and the lack of it, stood in tense relation to one another. Eugene Grubb, who portrayed Idaho farmers as gamblers, overlooked the fact that more efficient and precise methods did not necessarily make a farm more stable. When Earl Irwin likened farming to a gambling game, in almost the same breath he reported the anger of farmers constrained by a grading system that demanded machine-perfect potatoes. Nature's instability and an erratic market made irrigators feel as if they had no control over their lives; the grading standards that purported to stabilize production made them feel no more secure. Even the cooperatives failed to ensure stability when members succumbed to the lure of the wholesalers' higher prices; farmers often proved their own worst enemy. Annie Pike Greenwood would have understood this tension between order and risk: her husband, Charles, was a model progressive agriculturist, an expert at systematizing the farm. The failure of his methods and his defeat filled her with a bitterness that she later vented in her image of farming as gambling.[56]

Modern technical systems, irrigated agriculture included, can seem the inevitable products of reason and the ability to rearrange nature according to plan. The landscape, as portrayed in Seiler's "Story of An Irrigated Farm," indeed can seem to be an instrument that smoothly transforms nature into money. Yet agricultural production in Idaho's irrigated landscape must be understood in terms of the dynamic, interacting environmental and economic forces that generated it. Contingency, not just planning and orderly progress, created the landscape and its crops. The wealth that alfalfa brought, the fame that eventually accrued to "the Idaho" potato—these resulted as much from accident and failure as from human design. Only history reveals what booster imagery will never disclose.

6/ **Industrial Eden**

Myth, Metaphor, and the Irrigated Landscape

Idaho's irrigated landscape was a world made from many things: water, earth, and concrete, plants and animals, laws, policies, and social institutions, sweat, and economic relationships. But it was also a world that was inseparable from the activity of the mind, a world made from the imagination. It was here, in thought, that the inhabitants of the irrigated landscape made sense of the material reality that surrounded them; it was here that they envisioned the future and its possibilities, struggled to comprehend the present, and dreamed of the past.

To these people, the irrigated landscape frequently, although not always, objectified a deeply compelling story—the garden myth. It was an ancient tale, as old as Western civilization. Through the centuries, its tellers had modified it to match the changing circumstances of history. In America, the garden myth became an epic of personal and national regeneration. In this version, westward-moving pioneers conquered the howling wilderness and transformed it into beautiful, productive fields and farms. Triumphing over chaotic wildness, re-creating the lost Eden, the pioneers redeemed themselves and the land, restored the agricultural base of the Republic, and realized God's plan for the earth. It was a story that became central to American identity. But in the arid West, in places such as Idaho, it perhaps took on its greatest significance and power. Here, as in the original promised land, farmers reclaimed an arid wasteland and made it "blossom as the rose."

The garden myth offered a compelling explanation of Idaho's irrigated landscape, but it was not the only means by which people understood this place. By the early twentieth century, the Snake River valley also manifested a modern secular objective: the uniting of human activity and nature in an efficient industrial order. Engineers, government agencies, and corporations shaped the valley as much as the agrarian pioneers. Indeed, irrigated agriculture by the early twentieth century reflected the highest engineering ideals. As members of a society that sought an advanced technological control of nature, the people of the irrigated landscape resorted to another interpretation of their world, one in which the irrigated landscape was a sort of factory or a component in a larger industrial process.

But when they sought to find meaning in their land and lives, the irrigated landscape's inhabitants did not always resort to one set of cultural images or the other. Rather, they often blended the garden and the factory, creating complex, sometimes contradictory images that rested on combinations of organic and mechanical, female and male, secular and divine metaphors. The garden myth served as the archetype for discussing and defining irrigated agriculture, but Idahoans adapted it to new circumstances by assimilating into it elements of modern technology and industry. By bringing the factory into the garden, Idahoans evinced a desire to reconstitute an agrarian existence in a modern context. Poised between an agricultural past and a technological future, they imagined a mythic, metaphorical landscape that might be called the industrial Eden.[1]

Each rendition of the industrial Eden offered a variation on the basic story of humans imposing order on nature. Each person emphasized particular themes, narrated from the standpoint of a selected moment, or rearranged the story into a chosen sequence of events. Many versions contrasted the desert waste with the garden landscape that superseded it. In "The Romance of A River" (1920), E. B. Darlington spoke of the contribution that engineers—agents of a new industrial order—had made to the transformation: "Those of us who have dwelt for months in the sagebrush, who have worked long hours into the night with the weird wail of the coyote in our ears . . . can now take keenest delight in the fair fields of alfalfa, the waving grain, the splendid buildings, the thriving stock, and the canals flowing full where once our lines of [survey] stakes were driven."[2]

Other representations of the industrial Eden, in contrast, adopted a forward-looking perspective. Around 1910, the Electric Studio of Emmett, Idaho, produced a photograph of an orchard—a stock feature in many representations of the garden—in which technology (road, fence, poles, wires) and nature (orchard trees) appeared as parallel lines that converged at a vanishing point on the horizon (fig. 27). The entire scene suggested a systematized and controlled landscape in which the orchard was a component in a progressive technological order. Here the transition from desert to garden was at most only implied; the photograph announced a movement not from past to present, but from present to future.[3]

These are but two examples of how the people of the irrigated landscape mixed the garden myth and technology to create the industrial Eden. There were many more versions of the myth and its modern variant, almost as many, perhaps, as the people who settled and worked in the Snake River valley. In part because of this, there is no simple or obvious way to describe the garden and its industrial restatement. No version dominated any given period, nor did the basic story evolve in a neat chronological fashion between the late nineteenth and early twentieth centuries. An image evoked by someone in the 1890s might make just as much sense to another person in the 1920s. And even in the twentieth century, versions of the older garden myth, the garden without the industry, still appeared. To partake of the archetype was to mix and match its particular elements to fit circumstances that could be quite personal or idiosyncratic.

One way to describe the many versions is to assess them in a loose order that is consistent with one of their major themes: the transition from desert to garden. One group emphasized, in part, the beginning of this narrative trajectory—the conquest. In these stories the pioneers, predominantly men, subdued the desert or nature (often represented as female) and created a garden landscape that was often, although not necessarily, industrial. A second set of images and stories concentrated not on the conquest but on the creation of the garden by men who, like Johnny Appleseed, were nurturers or nurserymen. A third group moved beyond stories of conquest or creation by focusing on the garden itself. These renditions of the myth depicted the garden as metaphorically female, with men and women living in a soft, harmonious landscape in which technology was subsumed, or pastoralized, within a larger organic metaphor. A fourth group, further expanding on the basic desert-to-garden transformation, focused on what grew in the garden: crops, livestock, and children. A final array of stories offered a tale of disaster, the antipastoral, in which the Edenic dream threatened to become a nightmare.

Conquest constituted the essential first step that irrigators took in their efforts to reclaim what they saw as a dreary, arid waste. In myth, men subjugated or even destroyed wild nature, then resurrected it—redeemed it—in a productive form. Conquest was a necessary, moral act, but to male Idahoans it was also an emotional redemption, a regeneration of the masculine self. Violence defined masculinity as nothing else could. As the Idaho poet Vardis Fisher expressed it in "Joe Hunter" (1928),

> Time built a pioneer and set him down
> Upon the grayest waste of Idaho.
> He clubbed the desert and he made it grow
> In broad and undulating fields of brown.
> He laid his might upon it, stripped its frown
> Of drought and thistles; till by sweat and glow
> He left the aged and barren hills aglow
> With color—and its flame was his renown. . . .[4]

The rabbit drive was perhaps Idahoans' most masculine and overt demonstration of the desert conquest. In part, the drive was a means by which farmers controlled the jackrabbit hordes that periodically came in from the desert to eat irrigated crops. Men and boys armed themselves with sticks and clubs and gathered at a prearranged site, usually open land or a farm on the edge of an irrigation project. Then, advancing in a semicircle, they drove the "jacks" into a wire enclosure and there beat them to death. As important as the drive's functional objective, however, were its social and cultural purposes. It was an opportunity for neighborly socializing—men and boys conversed and joked; women, girls, and small children mingled and looked on; and afterward everyone came together for refreshments. But the rabbit drive also served as a ritual in which farmers reenacted the subjugation of the desert. A primitive quality permeated the event: men and boys reveled in the slaughter, in the satisfying crunch of club on furry skull, and in their blood lust they often whooped and hollered in a fashion that some observers thought was crude and barbaric. Like Vardis Fisher's Joe Hunter, men and boys clubbed the desert, beat it into submission so that they could create and maintain the garden.[5]

The writer and settler Annie Pike Greenwood described a rabbit drive in *We Sagebrush Folks* (1934), her memoir of life on an Idaho irrigated farm during the 1910s and 1920s. Staged at dusk at a nearby homestead, the drive drew approximately 200 men and boys (Greenwood's husband and sons included) and an unspecified number of women and girls. In typical fashion, the drive itself was predominantly a male event in which participants experienced a sense of primitivism. As men and boys rounded up the rabbits and clubbed them, they broke into yells that, to Greenwood, sounded like "Indians on the warpath." She could not stand the slaughter and the sickening smell of carcasses left over from previous drives, and she was horrified when she observed her young daughter, Rhoda, momentarily caught up in the thrill of killing. So Greenwood fell back from the moving line of drovers and turned her attention toward the farmhouse where other women were preparing the postdrive refreshment, ice cream.[6]

In describing the socializing and relaxation that followed the drive, Greenwood created an image of the garden that, in the great mythic cycle, came after the conquest—in this case, after the slaughter of jackrabbits. As the drive concluded, she sat beneath a tree and observed the antics of her children as they ate ice cream cones. To her, the tree, along with the other farmyard trees, created a mood of peace and contentment that contrasted with the tension and disorder of the drive and the harshness of the desert. "You cannot appreciate what a tree means until you have lived in this treeless desert country," she wrote.

> Trees always seem to me so compassionate. As I lean my head back against the trunk of this tree, which with a few of its fellows makes an oasis in this part of our desert, I gain a sense of calm detachment which is heightened by the obscurity of the night. Out of my cave of darkness, in which I have become a dryad part of the poplar's trunk, I can see the jovial group around the lantern that is set on a table by the ice-cream freezers.

In this placid setting, the rabbit drive at last fulfilled its purpose; what began with the slaughter of screaming animals ended in a pleasant ice cream social beneath the trees. Dead rabbits and ice cream cones—no two images could better represent the transformation of desert into garden that constituted the central drama of the irrigators' mythos.

Of course, not everyone experienced or interpreted the rabbit drives in exactly the same way. In some cases, participants and observers recognized a great deal of humor in the event and even used the farmers' primitive experience as an occasion for self-mockery. The comedy derived from ironic, symbolic inversions: civilized men going wild or "native"; violent, brutal conquest of pacific animals. A 1921 account from the upper Snake River valley illustrated some of these satirical meanings:

> About 150 of the braves went out for a rabbit drive west of Rockford Sunday morning and there in the grand and solemn solitudes of the sagebrush, they formed a phalanx to attack the bunnies. . . . Most of them were armed with rudimentary weapons, sticks and stones and one thing or another such as the cave men used and they did rejoice much that they were getting that close to nature.

But even when it seemed funny, the drive's primitiveness revealed a paradox in the mythic experience of conquering nature: to overcome wildness, to progress, men first had to become wild men themselves.[7]

In the rabbit drive, Idahoans reenacted the desert conquest in a form that gave no sense of the modern, technological control of nature characteristic of the industrial Eden. In other versions of the garden myth, violent conquest served as an archetype, a template, for depictions of the industrial domination of the land. In 1906, the *Idaho Register* reported the activities of railroad official Thomas W. Lee (also known as Commodore Lee), who had purchased land near Idaho Falls and was then busily turning it into a modern farming operation, an exemplary industrial Eden. In this version of the myth, wild land was female, and Lee was subduing and domesticating it.

> The last large piece of sage in the country is now figuratively "under the axe." The twelve hundred acre ranch owned by T. W. Lee and the largest in the valley under one fence is rapidly losing its virginity under the vigorous and efficient managership of its owner. . . . Amid [the] busy scene [of construction] looms the tall form of Mr. Lee personally superintending the work

which goes on with as precise a systemization as if it were a piece of railroad construction.

In this account, clearing the land, laying out irrigation ditches—in short, establishing a farm—was a masculine act in which the farmer forcefully made the female (virgin) land productive. Typical of such conquest images, the transformation of the land entailed a violent, rather primitive subjugation (in this case by ax, not club). Tellingly, though, this act also involved modern industrial construction techniques.[8]

E. B. Darlington, chief engineer for the Twin Falls North Side Land and Water Company, similarly depicted the subjugation of a female nature and the creation of an industrialized garden landscape. In his 1920 essay on irrigation in southern Idaho, he recapitulated the mythic passage from desert to completed garden. The "untamed Snake" violently rebuked the "white men" who first attempted to travel it by canoe, he wrote, but with irrigation development (in essence, the conquest) "the great river began to yield her beneficence to man." Darlington interpreted developers and engineers in divine terms, as "understudies of the Creator," men who implemented God's great plan for turning the Snake River valley into not only a garden but an industrial Eden. Irrigation, in his estimation, would one day produce more than just crops:

> There will be better transportation facilities and therefore better markets; great industrial plants, such as sugar factories, cheese factories, creameries, ice plants, dehydrating plants, flour mills, alfalfa meal mills, canning factories, great elevators and warehouses. Numerous cities and towns will spring up, with attendant business facilities. Train loads of cattle, sheep, wool and farm produce will roll out, and returns will roll in.

In a sense, Darlington depicted a reconstituted garden in which structures and machines were the dominant features. Cities, towns, and businesses would "spring up," almost like sprouts after irrigation. Industrial plants would rise from the fertile, moist soil; the rhythm of trains rolling out and profits rolling in, like the passing of seasons, would mark society's progressive movement toward a perfect future.[9]

Much like Darlington, U.S. Reclamation Service official C. J. Blanchard portrayed the engineer as a lordly creator who miraculously transformed the desert waste into a garden. His essay, "The Minidoka Project, Idaho" (1915), recalled his 1904 meeting with a visionary engineer at work there. This story featured no overt, violent conquest; its meaning rested on the image of the omnipotent, controlling engineer.

> At that time the desert held full sway. About our tent on a wind-swept knoll the coyotes howled dismally all night long. The blazing sun sprang up in the morning from a waste of sagebrush and sank at night below a horizon equally monotonous. It was truly a desolate landscape, out of which it seemed impossible to visualize the dream of the engineer who planned the great work of reclamation. Beside the smelly campfire of sagebrush and by the light of a candle, briefly he told of the transformation which was to follow the completion of his ditches. Together we discussed the location of the villages which were to be laid out, the path of the iron horse which was to transport the tons of hay and grain, the fruitage of orchard and garden, the fat stock, the luxuries and necessities of the thousands who were to flock to this new land of promise and fulfillment.

Under the guiding influence of this powerful engineer, the irrigation ditches—"his ditches"—would bring into being a redeemed, fruitful, productive landscape, indeed, an industrial Eden.[10]

A similar depiction of omnipotence appeared in Fannie G. Brunt's historical pageant, "The Saga of the Snake River, or Magic Valley's Yesterdays" (date unknown). Brunt portrayed Twin Falls irrigation developer Ira Perrine as a visionary, commanding patriarch. She borrowed the verse of Irene Welch Grissom (appointed poet laureate of Idaho in 1923) to characterize Perrine:

> A dreamer comes as dreamers will—
> To watch the swirling torrents spill
> Between the steep black lava walls,
> And thunder into foaming falls.
>
> He sees the desert disappear,
> A fair and smiling land appear,
> With streams that sparkle here and there,
> And new life springing everywhere.

Grissom's representation differed from Blanchard's in the final landscape; in Grissom's poem, the land was less an industrial Eden than a conventional garden, with even the canals turned into pastoral streams. Still, the image projected was one of masculine control.[11]

Perhaps more than any other feature in the irrigated landscape, dams—massive, angular, towering structures—signified man's ability to exert his will over wildness. Artwork (c. 1927) commemorating American Falls Dam illustrated man's triumph over the unruly Snake River: "Conservation Giant of the Snake River Valley" (fig. 6) proclaimed the optimism and vaulting ambition of the engineers, government officials, farmers, and businessmen who saw themselves engaged in a titanic struggle against the Snake. In one way, the dam was the structural counterpart to the "tall form" of Commodore Lee looming over the subjugated sagebrush land. In another sense, the muscular Conservation Giant resembled a club-wielding man at a rabbit drive; in the conquest of nature, a man had to strip off the garments of civilization and become more natural himself. The control of raw nature, the picture seemed to say, required an equally raw, even primitive power. There was irony in this organic man-versus-nature theme: one of the supreme achievements of modern engineering, the dam was here defined by brawn and sweat.[12]

Images of rivers and water similarly represented the theme of conquest that was central to the industrial Eden. Idahoans often portrayed a river or stream as an animate force that men subjugated and put to work. In one popular expression, men "harnessed" the river much as they harnessed wild horses. In 1894, engineer D. W. Ross portrayed the Snake River as "a giant force, . . . terrible when uncontrolled, but [when] harnessed [it] can be made to do the work of half a million horses." Annie Pike Greenwood characterized the Snake as an impoverished worker; she wrote of "the spillway where the Jerome Canal begins to voice the protest of its father, the Snake River, at being forced by man into the degradation of common labor." In Fannie Brunt's historical pageant, the river took on an almost biblical meaning: it was a serpent that man conquered and transformed into docile irrigation water.[13]

If conquest in one sense defined the relationship of dams to nature, in other images dams seemed to rise from the landscape itself, much like the factories, machines, and cities of Darlington's industrial Eden. H. C. Wones expressed this notion in a poem (1923) that celebrated the future American Falls Dam:

> To-day across the mountain peaks
> A glorious light appears,
> And men with gladdened hearts arise,
> To greet their dreams of years.
>
> Deserts remote, and sage-clad plains,
> In symphonies sublime,
> Shall herald to the ends of earth
> This monument of time.

In Wones's verse, the power and significance of the dam derived less from its inherent masculinity than from its relationship to the surrounding topography. To conquer the awful power of nature, to fill a deep canyon and control a mighty river, a dam had to be of such massive proportions that it acquired its own sublime characteristics. In short, the dam became a force of nature.[14]

Ultimately, sublime dams figuratively embodied the modern society that inhabited the industrial Eden. On 14 October 1915, thousands of Idahoans gathered to dedicate Arrowrock Dam on the Boise River. During the ceremony, businessman, land developer, and farm owner J. H. Lowell spoke of Idahoans as "pioneers" attempting to "realize . . . progress" by using "natural resources . . . to lay the foundations of a great Commonwealth," a land and society of modern farms, communities, and industries. He reviewed the history of the Arrowrock project and in conclusion asked that the "structure . . . stand through the years, not as a pyramid of old, a monument to a dead king, but as the symbol of an efficient American Government and as the sign of protection to the American home." To Moses Alexander, governor of Idaho, Arrowrock Dam encapsulated the essential qualities of the society that had built it:

> He remarked that the concreted unity of the giant structure, built to stand for the ages, should be the emblem of unity and strength of the people

through proper organization, permitting them to stand as one, working . . . for their own development and the development of their country.

The huge crowd then intoned, "My country! 'tis of thee / Sweet land of liberty / Of thee I sing . . . ," thereby making the dam, in a sense, the monumental objectification of America itself. Yet it was, as Lowell and Alexander made clear, a decidedly industrial America: efficient, properly organized (a "concreted unity"), and progressive.[15]

Like Lowell and Alexander, the Blackfoot newspaper editor Byrd Trego recognized giant dams as monumental symbols of progress. In a 1923 discussion of the future American Falls Dam, Trego invoked a popular American icon—the noble, defeated Indian—in an effort to create a vivid image of the advancement that the structure would represent. Trego first rhapsodized about the dam, calling it a "new type of pyramid" that would make the Snake "a new valley of the Nile." He then took a more somber tone, providing a verbal portrait of the reservoir covering the Snake River bottoms, the home of the Shoshone-Bannock Indians. "Go out on the bluffs overlooking the scene," Trego wrote, "and stand side by side with the Red man in contemplation of the change and picture, if you can, his grief at losing his own, his nature, so dear to him, and taking this artificial thing in its place." Here was a common, although remarkably versatile, feature of the imagined West: the stoic Indian surveying the land while a symbol of progress superseded him. The image emphasized the inevitable transformation from savagery to civilization (or wilderness to garden) that European-American settlement—and the construction of dams—purportedly represented.[16]

Stories and other representations of conquest and regeneration followed from a narrative perspective that placed powerful men (and their society) in a commanding position relative to nature and the landscape. Whether wielding a club at a jackrabbit, implementing the grand design of a patriarchal God, pushing back a river, or shouldering Indians aside, the conqueror was always potently masculine.

If one were to climb down, so to speak, from the crest of the towering dam (or from Commodore Lee's shoulders), one could find images of the industrial Eden in which men played less domineering, but no less masculine, roles. Creating the garden and farming it, after all, required men to nurture crops and livestock. In describing men who figured prominently in the creation of the garden—even the industrial Eden—Idahoans made use of metaphors and mythic archetypes appropriate to a soft, organic image of farming. In some instances they drew on the Johnny Appleseed myth. Scholars of western or frontier myth have overlooked the Appleseed archetype in their efforts to understand figures such as Boone or Crockett. Yet in many ways, Appleseed was the perfect image for men whose purpose was the planting of trees and crops. Appleseed's story was about planting and cultivating a landscape, not conquering wild nature. In the Appleseed myth, farmers achieved regeneration through horticulture, not violence.[17]

Idahoans cast Lafayette Cartee in the role of Appleseed. Cartee, an engineer, settled in Boise in 1863, and for fourteen years he served as territorial surveyor. More importantly, he devoted himself to horticulture, importing all manner of fruit trees, berries, and other plants and then propagating them in a nursery. After his death, residents of the Boise Valley remembered him much as other Americans recalled the mythical Appleseed: he was the pioneer who planted trees and then passed on, leaving a legacy for future generations. Cartee's obituary, while noting his engineering achievements, credited him with "placing within reach of his neighbors and fellow-pioneers" the plants they needed for their orchards, groves, and arbors. "He may justly be called the planter of all the orchards in the Boise and neighboring valleys," read the eulogy; "the first orchards all owed their origin to his enterprise in rearing and furnishing the first fruit trees."[18]

Depictions of Appleseed also rested on an overt mixture of organic and modern technological metaphors. In 1922, the *Idaho Republican,* a Blackfoot newspaper, announced the upcoming lecture of C. L. Smith, also known as Farmer Smith. A railroad official, Smith gave inspirational lectures on farming to rural and small-town people. He blended two realms, agriculture and industry, a quality that the *Idaho Republican* attempted to capture in its description of him: "'Farmer' Smith is the 'Jim Hill' of the O.W.R. & N. company; he is the Luther Burbank of horticulturists; he is the 'Johnny Appleseed' of Oregon and Washington." The *Republican,* furthermore, recalled a previous lecture Smith had given in Blackfoot and remarked on his ability to gain the audience's attention by avoiding technical terms and concepts. "He didn't say a word," reported the newspaper, "about balanced rations, carbo-hydrates, vitamines or anything that puzzles ordinary people." In sum, he was at once a symbol of industrialism, an advocate of progressive agriculture, and a pioneer of mythical stature who spread the gospel of farming in the plain speech of ordinary folk. It is revealing that the three people used to describe Smith formed a sort of iconic trinity of the industrial Eden: Hill, the railroad magnate and captain of industry; Appleseed, the agrarian; and Burbank, the plant breeder whose work represented a union of the technical and the organic.[19]

The Appleseed myth provided a peaceful counterpoint to the conquest theme evident in so many versions of both the garden and industrial Eden. A third, even more tranquil variation on the archetypal story depicted a harmonious union of male farmer and female garden. These tales described a transformation of desert into garden that resembled a wedding, in which the farmer and the land appeared as mates. In transforming the desert into fruitful garden, the farmer in a sense found a wife for himself. Implicit in the union was a marriage pact: the farmer tended the garden and made her productive, and she in turn lived up to her wifely role by serving and nurturing him. Irene Welch Grissom's "Desert Reclaimed" (1920) spoke of how "The land gives man the promise fair / That he shall reap from what he sows / Reward abundant, rich and rare." Or, in another author's poem, "The Man With the Shovel" (c. 1907), a feminine garden enveloped the farmer in tranquillity, abundance, and bliss:

> 'Tis the man with the shovel who turns the stream,
> From its mountain source to a paradise dream;
> Where the orchard bloom perfumes the air,
> Where plenty and beauty are everywhere.
>
> He is not at all like the man with the hoe,
> This man with the shovel, in Idaho;
> For the future to him, if he does his best,
> Is bright with the promise of plenty of rest.[20]

Such depictions of male farmer and female garden constituted a softened form of the industrial Eden. They pastoralized the technological, instrumentalist, masculine image of modern irrigation by subsuming it within an organic, feminine metaphor. In this mythic vision, man was not a "Conservation Giant of the Snake River Valley," a conqueror of a female river. Nor was he an industrialist who efficiently defoliated (or deflowered) the virgin land and turned it into a machine. Instead, he was a farmer who, with his shovel, gently turned the stream and created a beautiful, perfumed paradise.

In certain versions of the garden myth, this image of a female landscape in which work was not hard merged with an idea that many irrigation proponents espoused—that irrigation technology eased the burdens of agriculture. This meaning was evident in a song that the Reverend H. N. Ruddock composed in 1905 for a celebration held by Boise Valley settlers from Illinois and Minnesota. "A Song of Idaho" went to the tune of "Beulah Land" (land of Israel):

> The farmer here may be at ease,
> And work amid fruit laden trees,
> What'er he plants is sure to grow,
> And make a crop in Idaho.
>
> For genial showers you need not wait,
> You only have to hoist the gate,
> And let the waters overflow,
> Our valleys rich in Idaho.

In merely lifting the headgate, the farmer might release ample water into a fecund garden or orchard.[21]

Similarly, a feminine industrial Eden perhaps informed the meaning of the modern farm home. Idahoans sometimes gave their farm residences pastoral names (such as Sylvandale, Hillcrest, and Pleasant View Farm) indicating the rest and ease typically associated with the female garden. In addition, they often surrounded their houses with trees, shrubs, lawns, and flowers, which in their softness and beauty carried feminine connotations. At the same time, a farm home might also have a technological component associated with women's work. On the federal Minidoka irrigation project, for example, Benjamin Spittler (a Standard Oil Company employee) and his wife owned a small farm, End-of-the-Rainbow Ranch, that featured a "vine-covered bungalow" surrounded by a "bluegrass lawn shaded by ornamental trees." Like other settlers, the Spittlers had come to Idaho because they sought a simple farm life, but the Minidoka project also offered cheap electricity, allowing Mrs. Spittler to operate an array of modern appliances to ease her domestic work.[22]

Although representations of the union between man and nature might subsume technology within an organic, feminine metaphor, not all of them did so. Charles J. Lisle's semiautobiographical story, "The Book Farm" (1916), featured a wedding of farmer and land, but it also included a vision of a masculine industrial Eden. Originally an urbanite, Lisle suffered financial ruin on the stock market. Humiliated, his confidence destroyed, he watched passively as his wife, Mabel, stepped in as head of the family, bought an Idaho farm, and took him back to the land to make a man of him again. At first, Lisle quailed at the sight of their undeveloped homestead. In his words, "I had heard the Call of the Wild—mostly the stock-ticker and the anguished bleat of the shorn sheep in the stock market—but I wasn't prepared for the wild land that Mabel led us to and told us was to be our farm home." With her help and encouragement, however, he turned the raw homestead into a productive farm. (His was no account of conquest, only of struggle for personal redemption.) In the process, he recovered his self-assurance and stature as family head. And, most revealingly, he regained his potency: Mabel—now returned to her wifely role—bore two more children, "the little Rose that bloomed in our family garden that year, and the little Billy who came to us a year later." Thus in Lisle's masculine regeneration, wife and female landscape merged in the role of mate; Lisle both made the land productive and impregnated Mabel.[23]

While an organic metaphor underlay Lisle's tale of redemption, industry and technology—the basis of a manly industrial Eden—also permeated his story. Lisle depicted his farm as a virtual extension of the state agricultural college and its modern methods. "Engineering and irrigation experts" from the institution helped him lay out his ditches and provided other valuable advice. He learned more about progressive, profitable farming from technical books, his neighbors thus referring to his place as the Book Farm or the College. Knowledge of electricity gained from his study earned him winter employment at a nearby hydroelectric plant, his wages in turn paying for the electrification of the farm. Ultimately, Lisle merged organic, technological, and religious metaphors in his own vision of the industrial Eden. "The marvelous divine plan of plant and animal growth and development," he wrote, "and man's agency to grasp and to guide these marvels, are boon companions in our farm-university-home."

Other Idahoans, particularly women, who made the male farmer–female land relation their theme reversed the typical relationship. In these versions of the industrial Eden, men did not have a central place; rather, they served the garden and its female inhabitants. Mary Hallock Foote, artist, writer, and wife of the engineer Arthur Foote, in "The Irrigating Ditch" (1889) (fig. 19) portrayed the farmer as the servant of a feminine garden. Her essay represented male irrigators as patriarchal pioneers, but the accompanying illustration told a somewhat different story. In the foreground of an overall feminine and organic landscape stood a mother holding a baby, an image of Madonna and child. She gazed downward at an irrigation ditch and a small wooden headgate. The ditch looked not so much like an engineered structure as it did a natural creek, its grassy edges irregularly defined, its course circuitous, its origins lying somewhere in the hazy distance. A grove of trees directed the eye to the background, as did the ditch. There, a farmer—a husbandman—tended the ditch with a shovel, and the water slowly wended its way toward the woman and child, finally flowing to them through the headgate. In Foote's picture, the ultimate purpose of the ditch was the creation and nurturing of a family.[24]

The popularity and persistence of this theme of man serving the garden is illustrated by Agnes Just Reid's poem "The Man in the Rubber Boots" (1947), written decades after Foote composed her scene. Reid, a native of Blackfoot and daughter of a farm family, depicted a chivalric hero watering, tilling, and fertilizing a female nature:

> In the land of irrigation
> Where the desert blossoms as the rose,
> There dwells a knight in armor
> Whom everyone loves that knows.
> He guides the little streamlets
> To the famishing stems and roots,
> He carries life in his shovel—
> The man in rubber boots. . . .

The river out in the valley
Where man has scarcely trod,
Keeps calling, calling to him
To till her virgin sod;
And the song of the river is music
To him as she cries for recruits,
So he hurries away to her service,
Shod in his rubber boots.

Reid's verse was especially revealing because she transformed a heroic figure into a servant of the land. Nature called for recruits, and the knight responded—not with a weapon, but with a shovel, and not to do battle, but to perform a gentle, procreative act.[25]

Annie Pike Greenwood, writing in the early 1930s, similarly portrayed the farmer as a servant of the garden. But in her imagery, masculinity dissolved almost completely within the feminine landscape. To Greenwood, farming was a means for men to vicariously experience the female act of giving birth.

Farming is giving birth. That is why nearly all men yearn for the soil. In spite of pain, giving birth is the most ecstatic experience possible. . . . Man unconsciously envies woman this privilege. His surrogate is cultivation of the soil. He thrills for the earth at the receipt of the seeds. He feels her gestation, with anticipatory bliss, anxious fear, faithful preoccupation. The harvest is for him the birth.

This assessment of farming was no doubt unusual, but it revealed the extent to which inhabitants of the irrigated landscape, women especially, could imagine male farmers as supporters and servants, not dominators, of the female garden.[26]

Whether viewed as female or male, irrigated farming ultimately was about procreation, growth, and increase, as Greenwood so vividly expressed. Thus many versions of the garden myth emphasized the products of the farm: crops, livestock, or children. As might be expected, these were usually represented in organic terms. But not always. Rather, the same organic-technological interplay that defined the industrial Eden sometimes defined images of garden produce.

That interplay informed the aesthetic standards by which Idahoans judged fruit and vegetable exhibits at agricultural fairs. Idahoans celebrated the fecundity of the land by glorying in crops of freakish size—a pumpkin more than six feet in circumference, a radish twenty-three inches around, a Gloria Mundi apple weighing some twenty-seven ounces. In 1887, Ephraim Smith of Boise told a U.S. Senate committee on irrigation about a corn stalk twenty-two feet high, bearing six ears of corn, that "we had to chop down with an axe." Exhibited at a New Orleans fair, the stalk "was a pride to look at." However, just as Idahoans delighted in the giants, they also subscribed to another aesthetic, one that made size secondary to the industrial ideal of standardization. Large produce was still desired, but it must be uniform and not detract from other qualities. In 1910, J. R. Shinn of the Idaho State Horticultural Association offered advice to exhibitors at agricultural fairs:

> Size is important and counts for much in the judge's eye. However, there is a danger always accompanying fruits of too large size, not only because they may develop irregularity but also that they may become water-cored and be inferior in quality. . . . Correlated with size and color is . . . uniformity. Not only is it necessary that the fruit be large size but every specimen in the entry should be the same size as every other specimen. The same is true of color.

This union of size and standardization was similarly apparent in the "big, perfect, Idaho potatoes," each weighing one pound, that the Idaho Falls Chamber of Commerce delivered to President and Mrs. Coolidge for their 1924 Christmas supper.[27]

Like representations of crops, images of children celebrated the organic productivity of the garden. Idaho farmers often metaphorically depicted their offspring as crops or livestock. In "The Book Farm," Lisle wrote of the new "Rose . . . in our family garden." Similarly, popular agricultural photography juxtaposed children against rows of corn, orchard trees, or other crops, visually associating humans and plants. The meaning was clear: Idaho's fertile (and female) environment—the garden—produced not only fine crops but also healthy children. And there were lots of them; Idahoans took pride in their large families much as they did in their immense fruits and vegetables. In 1916, the *Idaho Farmer* reported that "some of our Idaho contemporaries are engaged in a debate on the subject of big families and as to who can show the biggest one."

> Monday's [*Idaho Statesman*] threw out a challenge to produce a larger family than that of Mr. and Mrs. Adoph Eldredge of Fruitland. There are 14 children in that family. Emmett [Idaho] can go that record two better. The family of Mr. and Mrs. C. T. Moulton consists of 16 living children and one that passed away. All are children of one father and mother and consist of 11 girls and five boys.[28]

Clearly, the family garden symbolized the fertility and organic productivity of the irrigated landscape. In certain cases, however, Idahoans also began to discuss children in technical terms that resembled the fruit, vegetable, and livestock standards of the industrial Eden. At the 1924 Twin Falls county fair, parents of 554 infants and young children brought their offspring to a baby clinic, where public health officials apparently evaluated them according to eugenic standards. The four children of Mr. and Mrs. Grover Beem of Filer were judged to be "perfect," like winning entries in the fair's crop and livestock competitions. Indeed, the *Idaho Farmer* reported the story under the headline, "Four Perfect Idaho Products." When May Nessly, author of the article, asked Mrs. Beem to explain how her children could be perfect, Beem outlined their feeding and training program, which resembled procedures for raising livestock: "Plenty of good milk, very few eggs, whole wheat and graham bread and from 12 to 15 hours of sleep form the basis of their care." Her young ones were unfamiliar with "the taste of white bread and hot cakes," she added, noting that they "are put to bed at 8 o'clock and are trained to go to sleep immediately." Although Nessly and Beem did not intend to portray the children as specially bred and nurtured animals, the parallels were obvious.[29]

The comments of L. S. Otto, a hog breeder from Filer, resembled the better-babies discourse but with a masculine emphasis. Otto, the "father of the Duroc [swine] industry in Idaho," in 1921 commented on the Farm Bureau's program of sponsoring "pig clubs" in which boys learned to raise purebred hogs. In Otto's opinion, the purpose of the clubs was not just to produce high-quality swine, but to rear boys who would one day become expert hog breeders like himself. "What we are doing," he said, "is making men. . . . What will it mean to have 50 such breeders when these boys grow up? Such breeders are hard to find now." On a metaphorical plane, raising hogs and boys blurred almost imperceptibly together.[30]

Of course, the figurative equation of livestock and children also had its own uniquely female variant. In 1910, W. O. Paisley, general manager of the Western Land Products Exhibition in Omaha, visited Idaho and lauded the state's crops, especially potatoes. Paisley's praise, however, went beyond just marketable produce. "I do not mean to create the impression that Idaho is just a potato state," he said. "Many other good things are raised there, including beautiful women, healthy babies, and fine horses." This association of women and livestock reached its apotheosis in an event that became a typical feature of agricultural fairs: the beauty pageant. In 1927, a fair in Boise featured a statewide competition in which judges selected the "queen of beauty" from among a group of young women. Not unlike entries in the livestock competition, the contestants (clad in bathing suits) paraded before a crowded grandstand. The judges then announced the winner: Miss Nampa, Irene Aguer. Clearly, the pageant revealed a combination of organic and technical ideals of the sort that informed the better-babies clinics and the boys' pig clubs. The girls were products of Idaho agriculture whose beauty conformed to standards of appearance and behavior.[31]

As products of the family garden, children were not merely the passive objects of their parents' dreams and ambitions. Offspring of the Snake River valley irrigated lands composed their own versions of the garden myth. In 1917, a young girl with the nickname "Spud" (again, the plant-child metaphor) wrote a poem that expressed her feelings about farm life. "My Rambles Around the Farm" featured the tranquil, pastoral landscape so typical of the mythic garden, but from a child's perspective.

> First thru the barn-yard gate I sped,
> The wonders catching my eye ahead;
> Then through the meadows fresh and clear,
> And all the sights that met me here.
> The cows a feeding on fresh green hay,

> The ducks in the pond swimming away,
> The tall green trees in their stately form
> Protecting the birds from various harm. . . .
>
> Then to the fields in the scorching sun;
> Work is a pleasure and lots of fun—
> Then comes the corn and the golden grain;
> Been thru the fields and back again. . . .
>
> The golden sun is sinking beneath the western hills,
> The birds good-night song, each tone with melody thrills,
> I hope that nature's children may always be kept from harm,
> And now I will say good-bye to my rambles around the farm.

Spud had imbibed and restated in her own words the elements of the pastoral form; indeed, so classic was her imagery that it suggested an English countryside rather than the irrigated landscape of the upper Snake River valley. Perhaps only in reference to "the scorching sun" did she reveal something of Idaho's desert environment. In other ways, children rendered a more indigenously Idaho pastoral. Many children, for example, dramatized a life close to nature by using ditches and canals for play. "One of the main pastimes of the children on the old Malm homestead" near Blackfoot, recalled Amanda Clark, "was swimming in the irrigation ditches." In this fashion, young people did more than just imaginatively transform hydraulic technology into an idyllic setting befitting the mythic garden. To the children, the garden myth was also lived experience.[32]

As an end result of the mythic narrative in which pioneers redeemed nature in a fruitful form, children provide a good place to begin analyzing the garden myth more deeply, to go beyond an overview of the metaphors and narrative structure that usually constituted the story. Both the garden and its modern variant, the industrial Eden, after all, were idealizations of nature and society. They were cultural forms, modes of expression, that allowed Idahoans to resolve imaginatively the problems or contradictions of daily life by positing a world where everything came out all right in the end. Happy images of perfect children, beautiful girls, and young boys who carried on their fathers' agricultural traditions masked discomfiting realities. Baby clinics alleviated anxieties about the loss of rural population and the declining importance of farm life. Otto's vision of a new generation of hog breeders and a vigorous and competent agriculture hid the turmoil of the agricultural economy in the 1920s. Beauty pageants likewise obscured the harsh existence of many farm women.[33]

Other scenarios similarly masked natural and social conflicts in the irrigated landscape. Exhibits of huge or perfectly uniform produce provided Idahoans not only with an objectified agricultural ideal but with the comforting illusion that the Idaho environment did not produce undersize or misshapen fruits and vegetables. Ruddock's "Song of Idaho," in which the farmer merely lifted the headgate to release the magic of irrigation, hid any evidence of grueling labor, falling prices, ruined crops, and competition for scarce water. Even Spud's "Rambles Around the Farm" seemed to put a pleasant face on aspects of farm life that might have been hard on a child. "Work is a pleasure and lots of fun," she wrote, but it might often have been otherwise. Indeed, her concern for the safety of "nature's children"—an obvious reference to human children as well as songbirds—perhaps indicated that her image of the farm hid a troublesome reality.

Some stories of the garden, however, explicitly acknowledged and discussed social and environmental conflicts within irrigated agriculture. Myth and metaphor did not only disguise problems; sometimes they were a means of expressing concerns and warning about conditions perhaps too disturbing to address openly. Weeds, for example, provided a particularly vivid image for discussing social disruption. In the irrigated landscape, weeds and undesirable people were metaphorical equivalents. In 1895, the University of Idaho president F. B. Gault, in an address to the Idaho Horticultural Association, conjured an image of a human garden—including its children—gone to weed.

> Is it not remarkable that we give such infinite pains and fostering to weeds? No family can afford to rear boys and girls for infamy when love, wholesome discipline, and wise example will make honorable men and lovely women. . . . Let us wage a war of extermination upon weeds, the unprofitable weeds, in orchard, home, community, and society, in the body politic and in the national life. . . . Everywhere we are sowing this lack of confidence, spreading mistrust,

> enervating public virtue, and traducing righteousness by the devotion to the luxuriant growth of weeds which we will not suppress. . . . We must . . . quit raising weeds and cultivate the best in life.

In the aftermath of the economic depression, political turmoil, and social unrest that characterized the mid-1890s, Gault found a fitting metaphor to help him admonish an audience of orchardists.[34]

Conversely, Idahoans sometimes represented actual threats from weeds or insects in social terms. In 1898, for example, botanist Louis Henderson noted the spread of a weed, prickly lettuce, through the irrigated farmlands of southern Idaho. This plant, Henderson said, "was generally known as the 'Chinese Lettuce,' reflecting probably the too prevalent idea that all of the ills on the Pacific Coast come from China and the Chinese, and that everything that comes from China must be bad!"[35]

When describing their efforts to eradicate weeds, Idahoans spoke of war against a hated foe. In this metaphorical warfare, they returned in a sense to the theme of conquest that defined the garden myth's narrative beginning. Farmers had defeated the desert and created the garden; now they must conquer, in military fashion, garden invaders. Henderson, discussing the Russian thistle, invoked the image of a race war, in which Western civilization defended itself against an invasion of eastern barbarians. He wrote that "all should rise against this pest as did the whole Roman world against the threatened invasion of Attila the Hun." Much of the warfare imagery, including the struggle against Huns, borrowed from the First World War. Even the use of chemical poisons evoked parallels between fighting pests and battling the World War I enemy. In winter 1923, the entomologist Don B. Whelan warned Idaho farmers: "The time to begin fighting next season's crop of grasshoppers is right now. . . . Order your poison and other ingredients early, so as to be ready when the 'zero hour' arrives." Whelan's metaphor of chemical warfare-provided an interesting variant of the industrial Eden. Technology had helped to create the garden; chemical technology would assist in its defense.[36]

Some Idahoans used an appropriately modified rendition of the garden to explain another problem: the economic failure of irrigated farms. In the standard narrative, the pioneer victory over the desert signaled the redemption of humans and nature in the garden. When the struggle ended in defeat for some farmers, a revision of the myth was in order. In 1923, Idaho state engineer and commissioner of reclamation Warren Swendsen tried to account for farm failure in the upper Snake River valley. "The early settlers . . . experienced many hardships and difficulties," he said in reference to the beginning phase of the garden narrative. "Because of these conditions, a natural process of elimination was brought about, the weak, less thrifty and unenduring settlers leaving, and none but the most thrifty, energetic and enduring remaining." The survivors, according to Swendsen, constituted a "cosmopolitan class" of articulate, knowledgeable farmers, superior agricultural "personnel." Thus in his story, pioneers struggled not just against the desert wilderness but against each other in a Darwinian competition ("a natural process of elimination"). When the dust finally settled, only the competent remained—farmers whose designation as "personnel" indicated their place in a redeemed garden, an efficient, rigorous, industrial Eden.[37]

Others subscribed to Swendsen's explanation of farm failures. During the early 1920s, when failure and tenantry on federal reclamation projects officially came to light, self-appointed moralists placed new emphasis on the business and industrial qualities of irrigation farming. Noting congressional revisions of the National Reclamation Act that enabled the Bureau of Reclamation to select project settlers, the *Salt Lake Tribune* in 1926 said that the government would now "hand pick" the farmers. This phrase corresponded with the notion of carefully choosing only the best seeds for next year's crops or the most uniform, standardized fruits for display at agricultural fairs. The newspaper also characterized the selection process as a solution to the painful struggle of the sort described by Swendsen. "'Handpicking' settlers may sound cruel," commented the *Tribune,* "but it will eliminate in the long run a great deal of the suffering and sorrow that come when the law of the survival of the fittest begins to operate."[38]

If opinion mongers and government officials used Darwinian struggle to dismiss farm failures as natural and inevitable, other people associated with Idaho irrigation—ruined farmers included—just as easily employed the garden myth to express their discontents. In a version of what the geographer John Rennie Short calls the "anti-pastoral," the redemption promised in the garden myth turned into disaster and tragedy. Thus, a triumphal, celebratory narrative became a form of protest.[39]

Mary Hallock Foote exposed the folly of irrigation development in *The Chosen Valley* (1892), a novel that centered on the construction and failure of a dam. She portrayed the structure as an extension of masculine, capitalist ambition, but rather than competence and triumph, it stood for hubris, greed, and ruin. Here, irrigation engineers and developers did not implement God's plan, they foolishly meddled in his mysterious ways. Margaret, wife of the project foreman and housekeeper to Dunsmuir, the dam's engineer, viewed the nearly completed dam with a deep sense of foreboding. Dunsmuir and the developers had arrogated divine powers, she believed, and trouble loomed ahead:

> There's a mony mair warnin's than promises in the Scripture set over against that word water. . . . He that presumes to know the ordinances of Heaven; that brak's the seal o' the auld, ancient, fearsome waters, to turn them from their given course—he'll need to mind!

Her words were prophecy. Dunsmuir, an otherwise scrupulous engineer, had capitulated to the demand of Norrisson, the financier, to hurry the dam's completion. At the end of the story, a laughing, mocking river undermined the poorly built structure, taking Dunsmuir with it and wiping out the pastoral valley lying downstream. Norrisson then erected a new, stronger dam, a monument to the fallen engineer, but to Margaret and Dolly, Dunsmuir's daughter, it was morally hollow, a symbol of arrogance and duplicity.[40]

Foote's antipastoral novel can be understood, perhaps, in light of her relationship with her husband, the engineer Arthur Foote. The book expressed frustrations that arose from the failure of Arthur Foote's Boise Valley irrigation scheme, and it revealed her doubts about "the cruel expansion of our country's material progress," as she wrote in the conclusion. On one level, the story hinged on the conflict between Norrisson and Dunsmuir; the capitalist greed of the financier overcame the principles of the engineer, with tragic consequences. On another level, attributes of the male characters—the ambition of Norrisson and the stubborn integrity of Dunsmuir—held other characters, particularly women, hostage. It was the aged, eccentric, superstitious Margaret who recognized that Dunsmuir had wrongly appropriated God's power. In a sense, the crumbling dam provided a female counterpoint to triumphal, masculine images such as the "Conservation Giant of the Snake River Valley."[41]

Perhaps the greatest Idaho antipastoral was Annie Pike Greenwood's memoir, *We Sagebrush Folks*. The book recounted the years between 1912 and 1924 that she and her family spent on an Idaho irrigated farm. In many ways, it celebrated the spiritual liberation and happiness that the garden myth promised. Ultimately, however, it was a lament for failed dreams, a protest for exploited victims, a tale of the industrial Eden gone awry. Arriving in Idaho a reluctant but hopeful pioneer, Greenwood gradually came to know hardship and failure, not a blissful pastoral existence.[42]

Near the beginning of the book, she recounted an experience that established a pattern or metaphor for the rest of the memoir. On the way to their new farm, along the Jerome Canal, she and her husband "passed the wreck of an old steam shovel, lying among the boulders. It had settled there two years before, when it had blown up, killing two men." Around the machine lay discarded wire, piles of empty tin cans, and the bleached bones of slaughtered animals. Snagged in the sagebrush were tufts of wool from passing flocks of sheep. "The wilderness," Greenwood wrote, "bore evidence that it had been desecrated by man; his justification was not yet apparent." The justification, of course, was supposed to be the garden, the mythic landscape that Greenwood and her neighbors never realized. Her attention to the remnants of construction set her story in opposition to the industrial Eden; the wreck of the steam shovel prefigured the human wreckage that appeared again and again as her story unfolded.[43]

To a large degree, Greenwood's memoir was an indictment of patriarchy, a tirade against male farmers who treated their women much as they treated their livestock. For the wife—"the Great Beast of Burden of the American nation"—farming consisted of drudgery and unremitting toil, emotional and physical exhaustion, pain, sexual exploitation, and mental illness. Greenwood herself suffered bodily ailments and, around 1920, a nervous breakdown. Indeed, Greenwood's depiction of farm females as beasts of burden sharply contrasted with the beauty pageants, in which women also resembled livestock, but of a different sort: pampered purebreds. In Greenwood's estimation, women were captives of the farm patriarch and his dreams. "The father of a sagebrush family is its god or its demon," she wrote. "There is no escape for the wife or children." Her anger spilled over in a parody of "America," the mythic song that celebrated freedom and patriarchy:

> My country 'tis of thee,
> Sweet land for two or three,
> Oh, ting-a-ling!
> Land where the farmer bragged
> While his poor wife was fagged
> And agriculture sagged
> Like everything.

In Greenwood's view, farming was essentially a male pursuit that women ultimately rejected. "Adam, I am sure, regretted the lost Garden of Eden," she concluded, "but I suspect that Eve fed him the apple in order to get off the farm." In the end, Greenwood and her husband lost their homestead and then separated from one another, ending their attempt to create a rural idyll but perhaps fulfilling her desire for relief.

Although Greenwood decried the abuse of farm women at their husbands' hands, she did not blame farmers entirely for the difficulty of farm life. Following a long American tradition, her memoir ultimately cast the farmers and their families as heroes and victims, and capitalists—particularly the "middlemen"—as the villains. This group, abetted by an amoral economic system and an incompetent government, held final responsibility for the hardship and failure of the Greenwoods and other Idaho farmers. Describing her brief stint as a teacher, Greenwood's heart broke at the memory of her students' backs and hands made sore from picking potatoes that only brought a low price. (Here Greenwood's depiction of agriculture differed markedly from the pastoral dreamland of the child-poet Spud.) She railed at the "lordly," cigar-smoking buyers who came to the farmers and said, "'I'll give you forty-five cents for your potatoes, and you'll take it, or be damned.'"

In contrast to Foote or Greenwood, some versions of the antipastoral continued to hold out the promise of the garden. In 1923, a farmer wrote to an agricultural newspaper to express his thoughts about the family farm. His letter combined a jumble of ideology and myth into a reactionary protest—so typical of 1920s America—against immigrants, the wealthy, and the large, impersonal organizations that seemed to be taking over the country. He described an agriculture sullied by "the peasant of the European type" and by "wealthy land holders" who employed "Chinese, Japanese, and Mexicans" to till their great farms. He complained that America's economy "and, in fact, just about everything that is the work of man" had been "specialized, incorporated and syndicated, until they are controlled by a small number of men in comparison to our population." Nonetheless, the farmer still believed that real Americans could regenerate the garden:

> But we do want every acre of American soil over which floats the Stars and Stripes to be utilized for the establishment of real American homes, with modern conveniences and standards of living[,] and to furnish employment to educated, patriotic and ambitious citizens of our own country. . . . Government propaganda should be directed toward the growing of and protection of good working citizens. . . . Let us save, if we can, the Creator's gifts of land, rain, and sunshine for the benefit of the masses, especially the boys and girls.[44]

In the garden myth, the people of Idaho's irrigated landscape found a supple, mutable narrative form that allowed them to impart meaning to the varied circumstances and conditions of their lives. The archetypal passage from desert waste to redeemed and regenerated land provided an explanatory structure that contained and organized complex realities.

Perhaps the most striking feature of the garden myth was the manner in which Idahoans modified it with technological metaphors to create the industrial Eden. On one level, the industrial Eden seems contradictory; plants and animals on the one hand and artifacts of modern technology on the other have virtually nothing in common. But on another level, Idahoans often found parallels, resonances, between the two categories. In the broadest sense, they linked the garden and technology within a larger theme of divine providence. God's will lay behind all things; just as he meant for humans to turn the desert into a garden, so he intended the engineers—understudies of the Creator—to build the dams and canals that made the garden possible. The theme of human control over the material world provided another means of reconciling the garden and technology; whether in an orchard or a factory, humans had mastered their environment.[45]

Most important, implicit in the industrial Eden was the notion of growth, increase, and production. Orchard trees bore fruit; factories produced goods. A 1912 Idaho government report described a leaf from a crop plant as a "chemical laboratory" and a "foliage laboratory," a place where nature transformed sunlight and nutrients into plant growth. In this metaphor—the plant as laboratory or factory—the industrial Eden began to take shape. From here it was a small step to Charles Lisle's "farm-university-home," where it was man's task to "grasp and to guide" biotic growth. Another small step led to E. B. Darlington's industrial landscape, where factories rose from the irrigated soil. And embodying the qualities and skills of the people who made the garden thrive was "Farmer" Smith, a metaphorical composite of railroad magnate, renowned plant breeder, and Johnny Appleseed.[46]

In creating the industrial Eden, Idahoans turned the old conventional garden myth into a modern account of progress. Engineers, Mary Hallock Foote's fictional Dunsmuir or the real-life E. B. Darlington, may have justified their work with references to the Bible or divine providence, but they were, after all, modern men, vanguards of a new technological and industrial order. C. J. Blanchard's visionary engineer gazed through the dim light of a candle and saw a landscape with a railroad locomotive in it. Darlington's "Romance of A River" marked the passage from desert to garden to a future industrial Eden where factories and machines droned on in a sort of perpetual motion. In the Electric Studio's photographic portrait of the orchard, the past disappeared in a representation of linear movement from present to future that suggested the continual, progressive expansion of capitalist production.

Indeed, most formulations of the industrial Eden involved some notion of spatial, temporal, or technological progress. The Spittler bungalow on the Minidoka project merged the comfort and leisure of the garden with the supposed ease and convenience of modern consumer technology. Even Warren Swendsen's view of irrigated agriculture as evolutionary struggle implied progress: competition among the pioneers resulted in a new bureaucratic, industrial order. The march of civilization cast aside the weak, unfit specimens; now the irrigation projects in the Snake River valley could get on with the real business of farming.

The people of the irrigated landscape also infused their economic values into the garden and industrial Eden. Versions of the myth, especially pictorial ones, contained motifs that evoked a sense of wealth and richness: fat livestock, stacks of hay, sheaves of grain, boxes of fruit, green and gold colors. Most important, the myth itself became a commodity hawked by Babbit-like town boosters and, finally, sold by corporations. Some of the most striking representations of the garden and industrial Eden were advertisements. It is important to keep in mind, however, that these were not simply a means for businessmen and developers to lure unwary settlers. In large part, the people who chose to take up irrigated farming in the Snake River valley helped invent and embellish the myth, investing it with their desires and thus abetting capitalist constructions of it.[47]

In certain instances, the garden myth masked the relationship of irrigated agriculture to the disruptive and exploitative aspects of the larger economic system that contained it. The myth implicitly promised that irrigated farming offered a refuge from the nation's economic vagaries. With important exceptions, it denied farm failure, and it also virtually erased any sign of Indians or the wageworkers and migrants whom the farmers needed to create and sustain their garden in the desert. If these people appeared at all, it was only as dramatic props that provided a standard against which to measure progress or decline.

Ultimately, however, the security and success promised by irrigated agriculture rested on its modern capitalist and industrial qualities. The presence of factories and railroads in the industrial Eden indicated the wealth and economic vigor that irrigated agriculture would generate. At the same time, industrial images could be symbols of the very conditions from which farmers sought escape in the first place. This ambiguity perhaps explains why Idahoans sometimes represented the irrigated landscape as a softened industrial Eden or even as a garden landscape that had no sign of industry at all. Irrigators may not have wanted Darlington's factories and all the economic and social ills they implied, but these people did want an easier, wealthier, more fulfilling way of life. They wanted the benefits of industry, urbanism, and a modern capitalist economy, only without the problems. They were like the angry farmer who wanted the "modern conveniences and standards of living" but who resented the immigrants, ethnic minorities, laborers, bureaucrats, and big businessmen who made possible those amenities. They could not have it both ways, but in "The Man With the Shovel," Ruddock's "Song of Idaho," and the Spittlers' electric bungalow they indicated the depth of, and contradictions in, their desire.

Whatever form the garden myth took, gender was often central to it. The relationship of male and female portrayed in words and images was complicated, even ambiguous. Although the masculine, industrial quest to conquer and dominate a female nature was a frequent theme, it was not the only one. There were varieties of masculinity: brawny conquerors, husbands, Appleseeds, servants, midwives envious of the female soil's capacity to give birth. And there were various representations of femininity. Mary Hallock Foote's "Irrigating Ditch" illustrated the classic pastoral, feminine landscape. But just as men were not always conquerors, neither were women always passive, gentle nurturers. Mrs. Beem, the disciplinarian, boasted of the efficient principles she used to bring up her children. Mabel Lisle assumed the leadership of her family and brought her shattered husband to Idaho so that he could restore his masculinity and build an industrial Eden.

In fiction and in life, however, the irrigated landscape was often a male dominion, and a farm wife was often just another beast of burden. In a way, this fact emphasized the importance of the feminine perspective of Mary Hallock Foote, Annie Pike Greenwood, and other female writers and artists. Their versions of the garden envisioned a peaceful, harmonious world in which men were not conquerors and dominators but servants and tenders of a feminine social and natural order. Through their representations of the garden, they questioned and opposed the hard, exploitative, industrial Eden.

Indeed, in Greenwood's antipastoral memoir we can see traces of a version of the garden myth that is now widely told: the environmentalist's parable of decline. Throughout much of her book, Greenwood expressed an affinity for the wild sagebrush lands that the farmers vanquished through irrigation. The "wilderness" to which she and her husband came in 1912 was, in her view, the real Eden. Tragically, in turning the wilds into a farm, they and other irrigators reenacted the fall from grace. Their effort at redemption was really a decline; farming marked their transition from innocence to knowledge. The basic form of this story is now quite popular among many Americans: in the progressive construction of farms, canals, dams, and factories, in the expansion of our capitalist system, we have not improved nature and turned it into an Eden—we have destroyed it.[48]

In certain respects, this interpretation is no more valid or realistic than the triumphal visions of the people who have celebrated irrigated agriculture and other forms of natural resource development. It does, however, underscore an important fact of how Americans, then and now, understand themselves in relation to the wider world in which they live. The irrigated landscape, indeed any landscape, is more than just a set of physical things, because those things are always entangled in webs of meaning that, often as not, take the form of myth. A landscape, in sum, is not just a place, it is a story. For the people of the irrigated landscape, life could make sense only in terms of such a story, as progress toward or regression from one of the most versatile, durable, and powerful images Western civilization has ever created.

Conclusion

A World in the Making

When I began my exploration of the irrigated landscape and its history—when I first stood and looked at Lateral L—I did not realize that I would eventually travel so far into mythic terrain. I had anticipated that my journey would keep me squarely in the realm of the physical landscape and the laws, policies, and institutions founded on it. But once I had glimpsed the myth I could not turn back. Whether you believe it or merely study it, myth has a nearly irresistible allure. Wherever I went, I found more and more evidence of it, until at last, not just the irrigated landscape but all of the American West, and all of America, stood revealed to me in mythic form.

I discovered one of my favorite examples of western American myth in a poem that appeared in the early twentieth century. If you travel through the West you will likely see this poem in common places. I have found it encased in a picture frame hanging in a lonely tavern; printed on a place mat in a restaurant; and embossed in copper plate and nailed to the wall of a shabby motel room. Somewhere, in Twin Falls or Rock Springs, I don't recall where, I purchased a copy of it printed on a postcard. "Out Where the West Begins" (1916), so the caption on the back of my card informs me, "was written by Arthur Chapman, then a reporter on a Denver newspaper, on the occasion of a dispute as to the real boundary of the West." It is an astute work, for Arthur Chapman recognized that the West's boundary is not simply physical, but mythical. More than just a region, the West is an imaginary place of fresh starts, a place where friendship is stronger, the sky is bluer, the snow whiter, the sun brighter, and people are less greedy, more generous, and more happy than anywhere else:

> Out where a fresher breeze is blowing,
> Where there's laughter in every streamlet flowing,
> Where there's more of reaping and less of sowing,
> That's where the West begins.
>
> Out where the world is in the making,
> Where fewer hearts in despair are aching,
> That's where the West begins. . . .[1]

"Out Where the West Begins" would have appealed to the people of Idaho's irrigated landscape, for it spoke to their mythic quest to achieve redemption through the garden. In the American West, in Idaho, farmers, engineers, journalists, poets, and others propounded a tale in which people renewed themselves by turning wilderness into a beneficent, productive, beautiful land. Through irrigation, they transformed a desert waste into a fresh, modern, industrial-agricultural landscape, an environment in which they reaped more and sowed less, and where water flowing through ditches was laughter to their ears. In Idaho's Snake River valley the world was, indeed, in the making.

We can never escape our myths, nor perhaps should we want to, but we need to recognize that whatever form they take, they are masks that obscure a complex, sometimes baffling world. Myths serve a purpose—they allow us to resolve imaginatively our contradictions, conflicts, and fears, and by doing so they can give us hope and summon us to action. But myths cannot fully explain our lives, especially our lives in relation to nature. So leave the mask as it is, but every now and then gently lift it up and look behind it. Understand the garden myth and its proponents, but look also at the living landscape, at the unadorned face of the world.

In Idaho's irrigated landscape, farmers never just tamed the wilderness and turned it into a garden. Engineers did not transform rivers and land into likenesses of industrial systems. In their efforts to make a new world, these people had to contend with a tough, resilient land that they could not shape exactly as they wished. Nature often responded to dams, canals, and farms in unpredictable, uncontrollable ways. Confronted with its changeability and with evolving environmental conditions, irrigators tried various technological, economic, and institutional means to shape and systematize land, water, and organisms. This reciprocal, human-nature interplay brought into existence a complex, hybrid landscape, a compromise between human design and natural processes.

In the hybrid landscape, clear distinctions between technology and natural systems dissolved. Nowhere was this more evident than in hydraulic technology, in the dams, reservoirs, canals, and ditches that provided the basis for irrigated agriculture. Hydraulic technology existed in complicated relation to geological formations, soils, creeks and rivers, and wild flora and fauna. Dams did not dominate rivers as Idahoans thought and as many people today still believe. Dams indeed created reservoirs, but as the controversies surrounding the early operation of Jackson Lake dam showed, these structures altered but did not "conquer" the complex, erratic hydrology of streams. Rather than turn the Snake into plumbing (a favorite metaphor of environmentalists and developers alike), dams instead became part of the river system itself. Moreover, although reservoirs destroyed substantial amounts of riparian land that harbored many plants and animals, it is undeniable that they in turn created habitat suitable for certain other species.

Canals and ditches similarly reflected a blurring of technological aims and natural processes. Made from the very ground from which irrigators excavated them, made of permeable soil and rock, canals and ditches lost water to aquifers. They also attracted plants, and thus insects, birds, and rodents that in turn threatened to destroy the canals themselves. Drainage ditches, largely a technological response to seepage from canals, also drew wild flora and fauna.

Even the irrigated landscape's crops and livestock represented the convergence of technology and nature. The Russet Burbank potato, for example, was a technological innovation through which Idaho farmers transformed sunlight, soil nutrients, and water into caloric energy. This potato, however, was not the same as a machine; although an artifact of human ingenuity, it was also the product of a contingent, dynamic natural world. Environmental factors beyond the control of humans helped to shape the variety and make it popular. An unplanned hybrid from Luther Burbank's garden, it serendipitously resisted the soil fungi that ravaged other Idaho potatoes. Thus the spread of the Russet Burbank through Idaho's irrigated fields did not represent only human choice and artifice; it reflected the reciprocal interplay of culture and nature that created the irrigated landscape as a whole.

Just as an intermingling of nature and technology defined the irrigated landscape, so did a dynamic relationship between nature and the market. The irrigated landscape was a shifting, often indeterminate zone where the economic system and climate, soil, water, plants, and animals uneasily converged. The law of prior appropriation attempted to divide rivers into many privatized parts, but it did so imperfectly; irrigators could not reconcile legal and real rivers smoothly. Capitalism turned land into individual parcels, but land took on productive power and value only with the addition of natural things: irrigation water and legumes. In addition, farms did not function as so many individual units—larger hydrological and ecological relationships linked them in ways that challenged the notion of pure private property. Russet Burbank potatoes proved immensely popular with American consumers, yet farmers had to construct commodity associations, potato sheds, and grading standards to negotiate the passage of the crop from fields to buyers.

A network of social organization, a dense human ecology, accompanied the creation of the irrigated landscape. To understand the history of irrigation, it is necessary to analyze the land and its human occupants as a social and ecological whole. On a basic level, farmers created many institutions to manage water. These included irrigation districts and companies, each characterized by varying degrees of family-based labor, voluntary cooperation, bureaucracy, industrial organization, and affiliation with the federal government. Then, in response to ecological changes induced or influenced by irrigation development, Idahoans established other types of organization: drainage and pest districts, insect and rodent drives and fences, and water associations such as the Committee of Nine. Like irrigation companies and districts, these multifarious environmental organizations often exhibited elements of voluntarism, cooperation, bureaucracy, and attachments to the federal government. In response to spatial aspects of the irrigated landscape and characteristics of particular crops, Idahoans established an array of economic associations and labor systems, each of which combined various forms of corporate, cooperative, family, and community organization. In a sense, the human network that rested on the irrigated landscape mirrored the texture of the land itself. In attempting to change and control a dynamic environment, irrigators themselves changed. Culture and nature, social system and natural system, shaped each other: the result was a hybrid landscape and a hybrid social order.

One of the most striking aspects of this changing social order was the extent to which the landscape encouraged irrigators to act cooperatively, in contradistinction to the individualism and private property boundaries that otherwise divided them. There were many social and cultural factors that contributed to this cooperation, of course. But it is clear that irrigators found it expedient, at the very least, to organize themselves and behave in ways that were consistent with the ecological commons. There is a lesson here for the American West and its periodic obsession with private property. Nature—water, soil, or organisms—will seldom if ever follow the boundaries that we try to place upon it. It will always draw us out of our individual plots and, whether we choose to recognize it or not, transform us into groups in which we, as individuals, have standing only in relation to the community.

In the irrigated landscape, ecological change often led to social change; alterations in the land also brought about chastened attitudes toward nature. Idaho irrigators realized quite early that flowing water was not like bottled milk, rivers were not pipelines, rodents and weeds would probably never disappear, and that farming was akin to gambling. Such acknowledgements provided a subtext to the grand mythic, metaphorical pronouncements that Idahoans had conquered the desert and turned it into a garden or industrial Eden. Even though they idealized a tame, controlled land, irrigators recognized that raising crops and livestock involved a daily struggle to counter and control an array of dynamic environmental conditions. It is intriguing that a belief in the garden myth and acknowledgment of environmental problems existed side by side. Perhaps the appeal of the myth increased in proportion to the inability of Idahoans to ever attain it.

Frank admissions of environmental resilience and the difficulty of farming demonstrate the problems inherent in equating beliefs with environmental actions. Yi-Fu Tuan, Jeanne Kay, and Steven Simms have shown that a belief in conservation or nature-human harmony does not necessarily mean that people behave with restraint in their use of natural resources. Ideals and actions do not automatically coincide. Given the history of Idaho's irrigated landscape, a corollary might be added to this basic observation. A belief that humans should conquer and exploit the environment does not necessarily mean that they will actually achieve their objectives. As the irrigated landscape showed, nature often eluded ideals: a conquest myth did not produce a conquered land.[2]

The implications of this corollary should stand as a warning to those of us who view developed landscapes as declensions from, as opposed to progress toward, the mythic garden. Too often, in this age of environmentalism, any human activity becomes just another story of ecological degradation. North America, we are told, once was pure and humans lived in harmony with each other and the land; Europeans then arrived and destroyed the primeval Eden. It is true that Americans have often severely disrupted the environment, but we need to acknowledge that the story of the ruined Eden is another of the many masks that prevent us from realistically viewing the world as it is.

As Idaho's irrigated landscape demonstrates, our activities are not always opposed to nature and do not necessarily dominate or wreck it. Leaky reservoirs, muskrat-infested canals, weedy fields full of magpies and crows, rivers teeming with carp, seemingly natural "springs," wild bees fortuitously pollinating alfalfa fields, Lateral L—all illustrate a complex interrelation between human and natural systems. It may seem an exaggeration to say that the irrigated landscape had (and still has) wildness in it, but this is a far more realistic assessment than one that posits a pristine ecosystem that humans reduced to a tame, artificial state. Indeed, all landscapes that we inhabit and use—whether a wilderness area, a potato field, or a parking lot—are variations of the same dialectic, the same tension between human design and natural process. All landscapes are hybrid landscapes, fusions of artifice and nature, offspring of our dreams and a natural world that we can neither fully comprehend nor totally control.

In depicting nature as resilient and protean, I have not intended to present an apology for unbridled efforts to control and exploit the environment. In the irrigated landscape of southern Idaho, alteration of the environment at times did have adverse consequences for people and nature alike. Yet no matter what humans do to nature, the moral of the story remains the same: whether grand and imperialistic, or small and apparently benign, attempts to shape or control the environment do not necessarily have predictable outcomes. People indeed make their landscapes, but they do not make them exactly as they please.

The people who established irrigation systems in Idaho's Snake River valley sought to recapture an old American ideal in a new historical and environmental context. They found that Idaho, like other parts of the American West, really was a place of fresh starts, a world in the making. They soon learned, however, that this land was not a blank slate waiting the inscription of a mythic dream. It was a dynamic environment with great capacity to limit, circumvent, confound, and in turn shape human systems. The interaction of irrigators with the land wedded artifice and nature in a hybrid landscape whose complexity and irony we have only begun to appreciate.

Notes

INTRODUCTION

1. By *metaphor,* I mean the figures of speech that people use to make sense of the world around them. An Idahoan might have said that a farm manufactured potatoes or that the land, like a factory, manufactured potatoes; both phrases suggest a factory or machine metaphor. I also use *image* in ways similar to and sometimes synonymous with metaphor. Idahoans who thought of irrigated farming in terms of machines or factories wanted to promote an image of Idaho farms as mechanistic or industrial. By *myth,* I mean a story that cuts through the disorder of lived experience to provide people with a simple, timeless, viscerally compelling explanation of how the world works. Idahoans, for example, told a myth when they said that they came to the desert as pioneers and transformed a wasteland into a garden. Like *metaphor* and *image,* I sometimes use *myth* and *image* (and *story*) synonymously. My use of all these terms is conventional and is grounded in a large body of scholarship; see Chapter 6, note 1, for relevant sources.

2. My work is part of a broader dialogue among scholars, writers, and the public in general concerning the human relationship to nature. Some thinkers posit a sharp division between Americans and nature and interpret modern irrigated agriculture as environmental decline. See Tim Palmer, *The Snake River: Window to the West* (Washington, D.C., and Covello, Calif.: Island Press, 1991); William Preston, *Vanishing Landscapes: Land and Life in the Tulare Lake Basin* (Berkeley: University of California Press, 1981); Marc Reisner, *Cadillac Desert: The American West and Its Disappearing Water* (New York: Penguin Books, 1987); Wallace Stegner, *The American West as Living Space* (Ann Arbor: University of Michigan Press, 1987); and especially the brilliant and controversial work by Donald Worster, *Rivers of Empire: Water, Aridity, and the Growth of the American West* (New York: Pantheon Books, 1985). Of course, I am not engaged in dialogue with these people only. My intent is to examine a range of assumptions about humans and nature that most of us hold.

3. Numerous historical works, especially those that deal with the historical ecology of landscapes, have influenced my thinking. Most important have been the books and articles that emphasize complex interconnections between human societies and nature. These include, among others, William Cronon, *Changes in the Land: Indians, Colonists, and the Ecology of New England* (New York: Hill and Wang, 1983); idem, *Nature's Metropolis: Chicago and the Great West* (New York: W. W. Norton, 1991); idem, ed., *Uncommon Ground: Toward Reinventing Nature* (New York: W. W. Norton, 1995); David Igler, "When Is a River Not a River? Reclaiming Nature's Disorder in *Lux v. Haggin*," *Environmental History* 1 (April 1996): 52–69; Nancy Langston, *Forest Dreams, Forest Nightmares: The Paradox of Old Growth in the Inland West* (Seattle: University of Washington Press, 1995); Stephen Pyne, *Fire in America: A Cultural History of Wildland and Rural Fire* (1982; rpt., Seattle: University of Washington Press, 1997); idem, *World Fire: The Culture of Fire on Earth* (1995; rpt., Seattle: University of Washington Press, 1997); Mart A. Stewart, *"What Nature Suffers to Groe": Life, Labor, and Landscape on the Georgia Coast, 1680–1920* (Athens: University of Georgia Press, 1996); idem, "Rice, Water, and Power: Landscapes of Domination and Resistance in the Lowcountry, 1790–1880," *Environmental History Review* 15 (Fall 1991): 47–64; Richard White, *Land Use, Environment, and Social Change: The Shaping of Island County, Washington* (Seattle: University of Washington Press, 1980); idem, *The Organic Machine: The Remaking of the Columbia River* (New York: Hill and Wang, 1995). Several works on ecology and natural history have also informed my thought. These emphasize ecological change and the role of human activity in shaping landscapes. See Daniel B. Botkin, *Discordant Harmonies: A New Ecology for the Twenty-first Century* (New York: Oxford University Press, 1990); Stephen Budiansky, *The Covenant of the Wild: Why Animals Chose Domestication* (New York: William Morrow and Company, 1992); Peter Goin, *Humanature* (Austin: University of Texas Press, 1996); Andrew Goudie, *The Human Impact on the Natural Environment*, 4th ed. (Cambridge, Mass.: MIT Press, 1994); Michael Pollan, *Second Nature: A Gardener's Education* (New York: Dell Publishing, 1991); Hugh Miller Raup, *Forests in the Here and Now* (Missoula: Montana Forest and Conservation Experiment Station, School of Forestry, University of Montana, 1981).

1/ GENESIS

1. D. W. Ross, *Biennial Report of the State Engineer to the Governor of Idaho for the Years 1899–1900* (Boise: Capital Printing Office, n.d.), 7; *Twin Falls News*, 2 December 1904.

2. *Idaho Republican* (Blackfoot), 22 February 1907; W. F. G. Thacher, "Buhl, Idaho—the Land That Beckons," *Pacific Monthly* 24 (December 1911): 718. See also H. A. Thomas, "Carey Act Lands Surrounding Shoshone, Idaho,"

Irrigation Age 24 (March 1909): 151–52. "A Farmer in the irrigated section does not depend on the weather for a successful crop," wrote Thomas. "He regulates the rainfall and is the manufacturer of farm products."

3. Hamlin Garland, *A Son of the Middle Border* (New York: Macmillan, 1935), 434–435.

4. For overviews of Snake River valley geological history, see Bill Hackett and Bill Bonichsen, "Volcanic Crescent," in *Snake: The Plain and Its People,* ed. Todd Shallat (Boise: Boise State University, 1994), 25–59; David Alt and Donald W. Hyndman, *Northwest Exposures: A Geologic Story of the Northwest* (Missoula: Mountain Press Publishing Company, 1995), 267–276, 333–343; David D. Alt and Donald W. Hyndman, *Roadside Geology of Idaho* (Missoula: Mountain Press Publishing Company, 1989); Eugene H. Walker, "The Geologic History of the Snake River Valley," *Idaho Yesterdays* 7 (Summer 1963): 18–31.

5. On the concept of the hydrologic cycle, see Luna B. Leopold, *Water: A Primer* (San Francisco: W. H. Freeman and Company, 1974); U.S. Department of the Interior, Geological Survey, *The Hydrologic Cycle* (Washington, D.C.: Government Printing Office, 1989); U.S. Department of the Interior, Geological Survey, *Ground Water* (Washington, D.C.: Government Printing Office, 1986). On the hydrologic cycle within the Snake River valley, see Israel C. Russell, *Geology and Water Resources of the Snake River Plains of Idaho,* U.S. Geological Survey, Bulletin 199 (Washington, D.C.: Government Printing Office, 1902); Harold T. Stearns et al., *Geology and Ground Water Resources of the Snake River Plain in Southeastern Idaho,* U.S. Geological Survey, Water Supply Paper 774 (Washington, D.C.: Government Printing Office, 1938); Alt and Hyndman, *Northwest Exposures,* 333–343.

6. For information on Idaho settlement and economic development, see Leonard J. Arrington, *History of Idaho,* 2 vols. (Moscow: University of Idaho Press; and Boise: Idaho State Historical Society, 1994), I:183–206, 231–283, 313–343, 345–365, 471–492.

7. There is no single source entirely devoted to the social history of the people who came to inhabit the irrigated landscape. Some information can be gleaned from general sources. See, for example, Arrington, *History of Idaho,* I:259–283, II:255–291; and especially Laurie Mercier and Carole Simon-Smolinski, *Idaho's Ethnic Heritage: Historical Overviews,* 3 vols. (n.p.: Idaho Centennial Commission and National Park Service, U.S. Department of the Interior, 1990). Another useful source is Davis Bitton, "Peopling the Upper Snake: The Second Wave of Mormon Settlement in Idaho," *Idaho Yesterdays* 23 (Summer 1979): 47–52. See also the many excellent local histories compiled by dedicated amateur historians, some of whom are natives of the irrigated regions of southern Idaho. Among my favorites are Kate B. Carter, ed., *Pioneer Irrigation: Upper Snake River Valley* (Salt Lake City: Daughters of Utah

Pioneers, 1955), and Alvin C. Holmes, *Swedish Homesteaders in Idaho on the Minidoka Irrigation Project* (n.p.: n.p., 1976). Perhaps the single best social history of Idaho irrigation is Annie Pike Greenwood's memoir, *We Sagebrush Folks* (New York: D. Appleton-Century, 1934; rpt. Moscow: University of Idaho Press, 1988). Two agricultural journals, *Gem State Rural* (Caldwell) and *Idaho Farmer* (Caldwell), frequently carried news that indicated the social and economic background of the irrigated landscape's people. Much of the rest of the story is scattered through various sources too numerous to mention here. For a good overview of the Carey Act and National Reclamation Act in Idaho, see Arrington, *History of Idaho,* I:476–485. According to the terms of the Carey Act, settlers on an irrigation project acquired 40-, 80-, or 160-acre tracts for fifty cents per acre; the irrigation company that built the canal system then sold shares in the canal system equivalent to the number of acres that each settler claimed. After the settler established residence on the land and brought it under irrigation, he or she could claim title to it. Under the National Reclamation Act, settlers on an irrigation project could claim up to 160 acres of land, but with the proviso that they repay the federal government for the cost of the irrigation works. In Idaho, the Carey Act and the National Reclamation Act opened about 1 million acres of land to development and, according to Arrington (I:485), "were the principal factors in causing the population of Idaho to jump by 164,000 between 1900 and 1910."

8. There is no single source that assesses the reasons people moved to Idaho's irrigated landscape. But one can find examples of all of the reasons listed here. A farmer in Italy, Texas, sought an Idaho farm because his own locale had "either too much or too little rain"; T. J. Tingle to Twin Falls Land and Water Company, 25 June 1906, Box 5, Twin Falls Land and Water Company Papers, Idaho State Historical Society and Archives, Boise (hereafter cited as IHS). During the early 1900s, a Salt Lake City businessman, Glen Bothwell, moved to southern Idaho and became involved in real estate transactions and land speculation on an irrigation project; Mark Fiege, "'A Higher Type of Personnel': The Bingham and American Falls Canal and the Creation of a Modern Irrigation Order, 1895–1923" (1989), unpublished manuscript in author's possession. Because land was so expensive in California, a farmer decided to move to a place where it was cheaper; N. J. Granbeck to Twin Falls Land and Water Company, 16 June 1906, Box 2, Twin Falls Land and Water Company Papers, IHS. A conductor for the Denver and Rio Grande Railroad acquired a forty-acre irrigated farm in an effort to regain the agricultural life that he had known as a boy; Homer McClure to A. H. Thompson, [5 October?] 1916, Box 1, American Falls Canal and Power Company Papers, Merrill Library, Utah State University, Logan. In 1911, a man purchased a small tract of irrigated land, hoping that the "open, invigorating,

Idaho air" would restore his daughter's health; John Scott to D. W. Davis, 3 May 1922, Box 16, Davis Papers, IHS.

9. Conquest was a popular metaphor for irrigation development in the arid American West. For the classic statement of this idea, see William E. Smythe, *The Conquest of Arid America* (New York: Harper and Brothers, 1899; rpt. Seattle: University of Washington Press, 1969). For an Idaho example, see *Reclamation Record* 12 (May 1921): 206, which describes a two-reel film, "Taming the Snake," that celebrated irrigation development in the Snake River valley. On the desire to both conquer and work with nature, see Carol Sheriff, *The Artificial River: The Erie Canal and the Paradox of Progress, 1817–1862* (New York: Hill and Wang, 1996), 32–35.

10. Russell, *Geology and Water Resources of the Snake River Plains,* 28–30; F. H. Newell, *Report on Agriculture by Irrigation in the Western Part of the United States at the Eleventh Census: 1890* (Washington, D.C.: Government Printing Office, 1894), 143–147; J. W. Powell, *Eleventh Annual Report of the United States Geological Survey to the Secretary of the Interior, 1889–'90,* Part. 2, *Irrigation* (Washington, D.C.: Government Printing Office, 1891), 78–83; U.S. Department of the Interior, General Land Office, *Territory of Idaho,* 1876, 1879 (maps); F. H. Newell, *Survey of Reservoir Sites,* 55th Congress, 2d Session, 1899, Senate Document 116, pp. 6–8. See also Carter, *Pioneer Irrigation,* for numerous accounts of early dam and ditch construction.

11. *Twin Falls News,* 2 December 1904, 24 March 1905. W. D. Trego of Blackfoot, Idaho, similarly wrote of "great natural reservoir sites on the headwaters" of the Snake River's tributaries. See *Farmer's Friend* (Blackfoot), 27 March 1906.

12. J. W. Powell, *Eleventh Annual Report of the Geological Survey, 1889–'90, Part 2, Irrigation,* 190–191; J. W. Powell, *Tenth Annual Report of the United States Geological Survey to the Secretary of the Interior, 1888–1889,* Part 2, *Irrigation* (Washington, D.C.: Government Printing Office, 1890), 106–108; D. G. Martin, "Delivery of Stored Water and Regulation of River Flow and Canal Diversions From Rivers Where Stored Water is Carried Together With Natural Flow," in "Report of Boise Conference of Operating Engineers for Irrigation Canal Systems Located in Idaho, Oregon, and Washington, November 1911," p. 84, Roll 9, Microcopy M96, Project Histories and Reports of Reclamation Bureau Projects, 1905–1925, Record Group 115, Records of the Bureau of Reclamation, National Archives, Washington, D.C. (hereafter cited as RG 115).

13. E. B. Darlington "Construction of the Milner-Gooding Canal," *New Reclamation Era* 22 (November 1931): 242–243, 245.

14. On the design and construction of irrigation systems, see Newell, *Report on Agriculture by Irrigation in the Western United States,* 19; U.S. Department of the Interior, Bureau of the Census, *Twelfth Census of the United States,* 1900, Vol. 6: *Agriculture,* Part 2, *Crops and Irrigation* (Washington, D.C.: U.S. Cen-

sus Office, 1902), 807–810; C. R. Burky, "Maintenance Problems—Earthwork, Wooden Structures, Metal Flumes, etc.," in "Report of Third Annual Conference of Operating Engineers, Held in Boise, Idaho, February 3, 4 and 5, 1914," Roll 23, Microcopy M96, Project Histories and Reports of Reclamation Bureau Projects, 1905–1925, RG 115. See also P. J. Flynn, *Irrigation Canals and Other Irrigation Works* (San Francisco: George Spaulding and Company, 1892), and Frederick Haynes Newell and Daniel William Murphy, *Principles of Irrigation Engineering* (New York: McGraw-Hill, 1913).

15. On the early survey of canals in the Snake River valley, see Powell, *Eleventh Annual Report of the Geological Survey,* Part 2, *Irrigation,* 190–200; and D. W. Ross, *Biennial Report of the State Engineer, 1899–1900,* 64–70.

16. For descriptions and histories of Boise Valley systems, see H. H. Caldwell and Merle Wells, *Economic and Ecological History Support Study: A Case Study of Federal Expenditures on a Water and Related Land Resources Project: Boise Project, Idaho and Oregon* (Moscow: Idaho Water Resources Research Institute, 1974), 135–157; *Idaho Statesman* (Boise), 17, 22, 29, 31 May 1877. For descriptions and histories of upper Snake River valley irrigation systems, see Carter, *Pioneer Irrigation,* 6, 8, 11, 13, 52, 71, 81–83, 92, 98, 161, 177–188, 228–229. On the Farmers' Progress Canal: *Idaho Register* quoted in Carter, *Pioneer Irrigation,* 82.

17. *Twin Falls News,* 2 December 1904, 7 April, 19 May 1905, 12 April, 12, 26 July 1907; James Stephenson, Jr., *Irrigation in Idaho,* U.S. Department of Agriculture, Office of Experiment Stations, Bulletin 216 (Washington, D.C.: Government Printing Office, 1909), 42–44; H. G. Raschbacher, *Irrigation From Snake River, Idaho,* U.S. Department of Agriculture, Office of Experiment Stations, Circular 65 (Washington, D.C.: Government Printing Office, 1906), 3–5; W. G. Hoyt, *Water Utilization in the Snake River Basin,* U.S. Geological Survey, Water-Supply Paper 657 (Washington, D.C.: Government Printing Office, 1935), 125–126; William J. Powell, "I'll Take It If It's Legal," *Pacific Northwesterner* 13 (Summer 1969): 33–40, (Fall 1969): 55–64; Lamont Johnson, "Canals of Idaho Carry Living Water," *Idaho Farmer* (24 November 1932): 3, 14, in Irrigation vertical file, IHS; Ross, *Biennial Report of the State Engineer, 1899–1900,* 66–68; Twin Falls Canal Company, "System Map," August 1986, Twin Falls Canal Company Records, Twin Falls, Idaho (hereafter cited as TFCC). Bickel's system was not the only project to use natural draws and waterways to carry water. So did the Twin Falls North Side project, on the north side of the Snake from the Twin Falls Land and Water Company project. See Crandall, "Use of Water on Twin Falls North Side Project, Idaho," [1923], Box 2, George N. Carter Papers, American Heritage Center, University of Wyoming, Laramie. Irrigation systems on the Boise project similarly made use of natural waterways to carry water to farms. See C. C. Fisher, "General Report on Lateral System of Boise Division, Boise Project, Idaho," March 1912,

pp. 9–10, Roll 18, Microcopy M96, Project Histories and Reports of Reclamation Bureau Projects, 1905–1925, RG 115.

18. For examples of field irrigation techniques, see Raschbacher, *Irrigation From Snake River, Idaho,* 9–11; "Prize Winning Irrigator and His Methods," *Gem State Rural* 11 (8 November 1906): 2; *Twin Falls News,* 25 November 1904, 18, 25 May, 7 September, 26 October 1906.

19. Richard J. Hinton, *Irrigation in the United States,* 49th Congress, 2d Session 1887, Senate Misc. Document 15, p. 100.

20. For typical statements on harnessing the Snake and making it do duty for man, see D. W. Ross, "Home Making in Idaho Under the Carey Law," *Irrigation Age* 6 (December 1894): 255; U.S. Congress, *Report of the Special Committee of the United States Senate on the Irrigation and Reclamation of Arid Lands,* Vol. 1: *The Northwest* (Washington, D.C.: Government Printing Office, 1890), 357; C. J. Blanchard, "Taming the Snake: Controlling a Mighty River to Reclaim a Million Acres of Idaho's Desert," *Reclamation Record* 10 (December 1919): 563–565. During the years around 1920, boosters circulated a two-reel film, "Taming the Snake," that advertised the development of the Snake River valley. See *Reclamation Record* 9 (November 1918): 543, and 12 (May 1921): 206. For an example of a comment about making water run uphill, see Greenwood, *We Sagebrush Folks,* 453.

21. *Twin Falls News,* 2 December 1904; E. B. Darlington, "The Romance of a River," *Reclamation Record* 11 (March 1920): 122–123. For insight into how European peoples have imagined the design of nature, see Daniel Botkin, *Discordant Harmonies: A New Ecology for the Twenty-first Century* (New York: Oxford University Press, 1990), and Raymond Williams, "Ideas of Nature," in *Problems in Materialism and Culture: Selected Essays* (London: Verso, 1980).

22. On dams in the Snake River valley, see Hoyt, *Water Utilization in the Snake River Basin.* Estimates on the length of canals and ditches, the number of irrigated acres, and numbers of farms were derived from figures in U.S. Department of Commerce, Bureau of the Census, *Fourteenth Census of the United States,* 1920, Vol. 7: *Irrigation and Drainage* (Washington, D.C.: Government Printing Office, 1922), 163–176. And see the excellent map and accompanying text by G. F. Lindholm and S. A. Goodell, *Irrigated Acreage and Land Uses on the Snake River Plain, Idaho and Eastern Oregon,* Department of the Interior, U.S. Geological Survey, Atlas HA-691 (n.p.: U.S. Geological Survey, 1986).

23. *Gem State Rural* 11 (1 February 1906): 6. For the history of drought and irrigation in the irrigated landscape of southern Idaho, see Lynn Crandall, "History of Irrigation Development in the Snake River Valley" [1946?], and Crandall, "Irrigation History Given," undated newspaper clipping (1934?), both in Idaho–Reclamation of Land vertical file, Idaho Room, Idaho Falls Public Libarary, Idaho Falls (hereafter cited as Idaho Room). See also

Gooding Leader, 11 April 1935; *Idaho Falls Post-Register,* 22 July 1952; *Aberdeen Times,* 23 January 1936; *Idaho Statesman* (Boise), 3 October 1960; and *Idaho State Journal* (Pocatello), 11 March 1977. L. Carl Brandhorst, "The Panacea of Irrigation: Fact or Fancy," *Journal of the West* 7 (October 1968): 491–508, analyzes the influence of drought, hail, and other conditions on irrigated agriculture in Scotts Bluff and Morrill counties, Nebraska.

24. Lindholm and Goodell, *Irrigated Acreage on the Snake River Plain;* Crandall, "History of Irrigation Development in the Snake River Valley"; Hoyt, *Water Utilization in the Snake River Basin.*

25. Hugh Lovin, "Footnote to History: 'The Reservoir . . . Would Not Hold Water,'" *Idaho Yesterdays* 24 (Spring 1980): 12–19; idem, "How Not to Run a Carey Act Project: The Twin Falls–Salmon Falls Creek Tract, 1904–1922," *Idaho Yesterdays* 30 (Fall 1986): 14; Hoyt, *Water Utilization in the Snake River Basin,* 45, 51; A. P. Davis, "Why Some Irrigation Canals and Reservoirs Leak," *Engineering News-Record* 80 (4 April 1918): 663–665; Crandall, "Use of Water on Twin Falls North Side Project, Idaho." On extremely rare occasions, seeping water and unstable earth caused dams to fail. For the failure of the Orchard Farmers Company dam near Caldwell, see *Caldwell Tribune,* 17 March 1894.

26. Carter, *Pioneer Irrigation,* 177–196; Gary Wayne Gneiting, "An Economic History and Analysis of the Great Feeder Canal of Southeastern Idaho," M.S. thesis (Utah State University, 1972), 18–30. For herding the water, George N. Carter, *Sixth Biennial Report of the Department of Reclamation, State of Idaho, 1929–1930* (Boise: n.p., n.d.), 36–37.

27. See Chapter 3 for a history and analysis of water rights in Idaho and especially in the upper Snake River valley.

28. *Twelfth Census of the United States,* 1900, Vol. 6: *Agriculture,* Part 2, *Crops and Irrigation,* 807. See also the comments by J. B. Camp, manager of the King Hill irrigation project, on distribution of water through the project's canals: "Operation [of irrigation systems] is far from an exact science, and none of our systems is perfect." See *Idaho Farmer* 27 (7 April 1921): 377.

29. The Stewart Decree is covered in Paul Lloyd Murphy, "Irrigation in the Boise Valley, 1863–1903: A Study in Pre-federal Irrigation," M.A. thesis (University of California, 1947), 104–114; see also Murphy's "Early Irrigation in the Boise Valley," *Pacific Northwest Quarterly* 44 (October 1953): 177–184. On the Danskin Ditch, see *Idaho Republican,* 1 August 1919. See Chapter 3 for a detailed history and analysis of the Snake River Committee of Nine.

30. "Twin Falls Canal Company Operations Policy, Floods due to Natural Runoff," 27 February 1989, General Information on the Canal System file, TFCC; U.S. Department of the Interior, Reclamation Service, "History of the Boise Project, Idaho, From Beginning to 1912," Vol. 1: 126, Roll 8, Microcopy M96, Project Histories and Reports of Reclamation Bureau Proj-

ects, 1905–1925, RG 115; Carter, *Pioneer Irrigation,* 109, 119; D. W. Cole, "Patrol of Canals on Irrigation Projects," *Reclamation Record* 9 (January 1918): 24–25.

31. Davis, "Why Some Irrigation Canals and Reservoirs Leak"; F. N. Cronholm, "Drainage System for the North Side Minidoka Irrigation Project," *Engineering News-Record* 69 (25 April 1914): 468; Don H. Bark, "Duty of Water Investigation," in *Tenth Biennial Report of the State Engineer to the Governor of Idaho, 1913–1914* (n.p.: n.p., n.d.), 159–163. Warmer climates with narrower temperature ranges than Idaho's allowed engineers and farmers to build and maintain concrete canals. Indeed, a mild climate permitted California irrigators to make concrete linings thinner, and thus less expensive, than those in colder environments. For discussions of seepage, canal linings, and the weathering of concrete, see Samuel Fortier, "Design and Structural Details of Concrete Lining for Irrigation Canals," *Concrete* 18 (April 1916): 174–178; R. C. E. Weber, "Economy of Concrete for Irrigation Canal Linings," *Concrete* 32 (October 1923): 158–159; F. W. Hannah, "Losses in Canals and Methods of Prevention," *Journal of the Idaho Society of Engineers* 1 (May 1911): 29–36; Newell, *Report on Agriculture by Irrigation in the Western United States,* 19; Tim Palmer, *The Snake River: Window to the West* (Washington, D.C., and Covello, Calif.: Island Press, 1991), 106. When irrigators did line canals with concrete, the lengths were usually quite short. In 1918, the U.S. Reclamation Service lined 500 feet of the Knorr lateral in the Boise Valley with concrete; "Annual Project History of Boise Project for 1918," p. 145, Roll 10, Microcopy M96, Project Histories and Reports of Reclamation Bureau Projects, 1905–1925, RG 115.

32. Davis, "Why Some Irrigation Canals and Reservoirs Leak," 665.

33. *Twin Falls News,* 7 July, 4 August 1905; McCaffrey quoted in *Idaho Register* (Idaho Falls), 21 June 1910.

34. On the continuous flow method, the response of engineers and government officials to it, and alternate methods of distribution and application, see, for example, Ross, *Biennial Report of the State Engineer; 1899–1900,* 81–89; *Twin Falls News,* 25 May 1906; see also Cronholm, "Drainage System for the North Side Minidoka Irrigation Project."

35. On the general problem of seeped land, see James Thorp and C. S. Schofield, "Drainage in Arid Regions," in *Soils and Men: United States Department of Agriculture, Year Book of Agriculture, 1938* (Washington, D.C.: Government Printing Office, n.d.), 717–722; Milton Fireman and H. E. Hayward, "Irrigation Water and Saline and Alkaline Soils," in *Water: United States Department of Agriculture, Yearbook of Agriculture, 1955* (Washington, D.C.: Government Printing Office, n.d.), 321–327.

36. On the problem of seepage and salt, see Fireman and Hayward, "Irrigation Water and Saline and Alkaline Soils"; O. C. Magistad and J. E. Chris-

tiansen, *Saline Soils: Their Nature and Management,* U.S. Department of Agriculture, Circular 707 (Washington, D.C.: Government Printing Office, September 1944). Soil survey reports describe salinity conditions in Idaho soils; see, for example, F. O. Youngs et al., *Soil Survey: Minidoka Area, Idaho,* U.S. Department of Agriculture, Bureau of Chemistry and Soils, Series 1923, No. 27 (Washington, D.C.: Government Printing Office, 1928).

37. P. M. Fogg, "A History of the Minidoka Project, Idaho, to 1912 Inclusive," August 1915, Vol. 1: 21–25, 78–83, 131, Roll 96, Microcopy M96, Minidoka Project, Idaho, 1903–1919, RG 115.

38. Mrs. J. P. Butler to D. W. Cole, 24 September 1918, Boise Project Manager to Chief of Construction, 27 September, 24 October 1918, "Memorandum with reference to recommendation of Chief of Construction to Director, November 15, in regard to proposed irrigation district for the Arena Basin portion of the Boise project," 20 November 1918, all File 260D, General Administrative and Project Records, Boise Project, 1902–1919, RG 115; Project Manager to Chief Engineer, 28 December 1922, and to Director, 4 January 1923, Mrs. J. P. Butler to Secretary Work of the Interior, 29 March 1923, and to ex-Governor Davis, 29 March 1923, all File 330, General Administrative and Project Records, Boise Project, 1919–1929, RG 115. For a personal account of seepage on the Minidoka project, see Holmes, *Swedish Homesteaders on the Minidoka Irrigation Project,* 59–63, 93, 111–112.

39. On the figures for drainage in Idaho in 1920 and 1930, see *Fourteenth Census of the United States,* 1920, Vol. 7: *Irrigation and Drainage,* 433; and U.S. Department of Commerce, Bureau of the Census, *Fifteenth Census of the United States,* 1930, *Drainage of Agricultural Lands* (Washington, D.C.: Government Printing Office, 1932), 95, 96. On the design and construction of drains, see F. T. Crowe, "Draining the Pioneer Irrigation District," *Reclamation Record* 5 (October 1914): 373–374; Percival M. Fogg, "Drainage of Irrigated Lands on the Minidoka Project, Idaho," *Journal of the Idaho Society of Engineers* 2 (June 1912): 106–118; W. L. Powers, "The Drainage of Irrigated Lands," *Reclamation Record* 9 (September 1918): 424–426.

40. Stearns et al., *Geology and Ground-Water Resources of the Snake River Plain,* 124–135; Fogg, "Drainage of Irrigated Lands on the Minidoka Project, Idaho"; E. R. Richards, "Report of Construction and Operation, Minidoka Project, Year of 1911," pp. 29–30, Roll 96, and "Report of Construction and Operation (Annual Project History), Minidoka Project—Idaho, Vol. VIII, Year of 1914," Roll 98, Microcopy M96, Minidoka Project, Idaho, 1903–1919, RG 115.

41. For a partial review of Idaho laws pertaining to drainage districts, see George N. Carter, *Fourth Biennial Report of the Department of Reclamation, State of Idaho, 1925–1926* (Boise: n.p., n.d.), 165–188. The community emphasis in the drainage law was evident in Article 7, which stated that the law "shall be liberally construed to promote the public health and welfare by reclaim-

ing wet or overflowed lands, building embankments or levees and the preservation of any system of drainage. . . ." Ibid., 188. The law also authorized irrigation districts to levy charges against all members of the district to pay for drains that alleviated seepage or that protected land from becoming seeped. Embroiled in controversy for years, by 1925 it had become fixed in Idaho statutes. Ibid., 92–93. See also Board of Engineers to Director, 30 August 1913, District Counsel to Chief Counsel, 2 September 1915, Boise Project Manager to Chief of Construction, 24 October 1918, all file 260D, Boise Project, General Administrative and Project Records, 1902–1919, RG 115; B. E. Stoutemyer to Director, 22 September 1920, Memorandum for the Secretary, 18 January 1924, Ivan Hiler to Hubert Work, 8 January 1923, all file 330, Boise Project, General Administrative and Project Records, 1919–1945, RG 115; "Drainage for Irrigation Projects in Idaho," *Reclamation Record* 9 (February 1918): 68; "Interpretation of Idaho Drainage Act," *Reclamation Record* 10 (October 1919): 474; "Drainage on Federal Irrigation Projects," *Reclamation Record* 14 (July 1923): 250; and Powers, "The Drainage of Irrigated Lands."

42. An account of the Nampa-Meridian Irrigation District case is recorded in "Annual Project History of Boise Project for 1917," pp. 165–166, Roll 9, and "Annual Project History of Boise Project, Idaho, for 1918," pp. 162–164, Roll 10, both Microcopy M96, Project Histories and Reports of Reclamation Bureau Projects, 1905–1925, RG 115. See also "Drainage for Irrigation Projects in Idaho," *Reclamation Record* 9 (February 1918): 68; and "Interpretation of Idaho Drainage Act," *Reclamation Record* 10 (October 1919): 474.

43. Russell, *Geology and Water Resources of the Snake River Plains,* 161. On the history of the duty of water, see Hugh Lovin, "'Duty of Water' in Idaho: A 'New West' Irrigation Controversy, 1890–1920," *Arizona and the West* 23 (Spring 1981): 5–28.

44. I. D. O'Donnell, "Duty of Water," *Reclamation Record* 7 (September 1916): 406–407.

45. For discussions of beneficial seepage, see Stearns et al., *Geology and Ground-Water Resources of the Snake River Plain.* For an example of developed drains supplying water for irrigation, see F. T. Crowe, "Draining the Pioneer Irrigation District," *Reclamation Record* 5 (October 1914): 373–374. For an example of seepage flowing into creeks used to transport irrigation water, see Fisher, "General Report on the Lateral System of the Boise Division," 9–10.

46. "Annual Project History of Boise Project, Idaho, for 1915," p. 310, "Annual Project History for Boise Project, Idaho, for 1916," pp. 157–158, 269, both Roll 9, Microcopy M96, Project Histories and Reports of Reclamation Bureau Projects, 1905–1925, RG 115; A. P. Davis et al., *Fifteenth Annual Report of the Reclamation Service, 1915–1916* (Washington, D.C.: Government Printing Office, 1916), 592.

47. Davis, "Why Some Irrigation Canals and Reservoirs Leak," 664.

48. Ross, *Biennial Report of the State Engineer, 1899–1900,* 17–18; *Twin Falls News,* 3 February 1905; Carter, *Pioneer Irrigation,* 249; Verl King, *Egin Bench Sprinkler Irrigation Impact Study Report* (Boise: Idaho Department of Water Resources, 1988), 1, 6.

49. Julius Hornbein, "Seepage Water, Formerly Wasted, Now Pumped for Irrigation," *Engineering News-Record* 78 (26 April 1917): 192–193.

50. Ross, *Biennial Report of the State Engineer, 1899–1900,* 14.

51. Stearns et al., *Geology and Ground-Water Resources of the Snake River Plain,* 142–166; M. J. Mundorff et al., "Ground Water for Irrigation in the Snake River Basin in Idaho," in U.S. Bureau of Reclamation and United States Army Corps of Engineers, *Upper Snake River Basin,* Vol. 3, Part 2: *Coordination and Reports of Cooperating Agencies: U.S. Geological Survey—Ground Water Branch* (n.p.: n.p., 1961), 1–4, 10–12, 58–59, 63, 72, 131–134, 149, 151.

52. Lindholm and Goodell, *Irrigated Acreage on the Snake River Plain;* B. B. Bigelow et al., *Water Withdrawn for Irrigation in 1980 on the Snake River Plain, Idaho and Eastern Oregon,* Department of the Interior, U.S. Geological Survey, Atlas HA-690 (n.p.: U.S. Geological Survey, 1986); Lynn Crandall, "Ground Water Flows of the Snake River Plain," 31 March 1953, Idaho–Reclamation of Land file, Idaho Room.

53. George N. Carter, *Twenty-first Biennial Report of the Department of Reclamation, State of Idaho, 1959–1960* (n.p.: n.p., n.d.), 33–34; *Idaho Statesman,* 26 December 1963, 23 November 1960; Charles Brockway, Idaho Water Resources Research Institute, Kimberly Research and Extension Center, lecture at Thousand Springs Reserve, Idaho, 16 July 1992.

54. E. B. Darlington, "Greater Minidoka Project," *Burley Bulletin,* 6 September 1928, newspaper clipping in Irrigation vertical file, IHS. See also Darlington's earlier, although carefully qualified, characterization of an irrigation system as a machine: E. B. Darlington, "Use of Water on the Salmon River Tract," *Reclamation Record* 9 (May 1918): 228.

2/HABITAT

1. Jessie Warrington, "The Wonderful Redemption of a Desert: A Story of the Past, Present, and Future of the Famous Twin Falls Country," *See Idaho First* 5 (September 1914): 3–6; Irene Welch Grissom, "The Desert Reclaimed," *New West Magazine* 11 (December 1920): 15, also reprinted in *Idaho Falls Daily Post,* 16 June 1923. For a more complete analysis of the garden myth, see Chapter 6. Compare the recollection of Mrs. A. C. De Mary of Rupert, Idaho: "Every one of that early number [of pioneer women] had made a great sacrifice to come to this new land in order that she might help to create in the desert a new garden." See *Reclamation Record* 7 (February 1916): 60–61.

2. E. B. Bentley and Glenn Oakley, "A Climate of Change," in *Snake: The Plain and Its People,* ed. Todd Shallat (Boise: Boise State University, 1994), 63–73; Eugene H. Walker, "The Geologic History of the Snake River Country of Idaho," *Idaho Yesterdays* 7 (Summer 1963): 18–31.

3. Mark Plew, "Native Traditions," in *Snake: The Plain and Its People,* 99–121; Earl Swanson, Jr., "The Snake River Plain," *Idaho Yesterdays* 18 (Summer 1974): 2–11; Dean A. Shinn, "Historical Perspectives of Range Burning in the Inland Pacific Northwest," Journal of Range Management 33 (November 1980): 415–423. See also P. J. Mehringer, "Prehistoric Environments," and K. T. Harper, "Historic Environments," in *Handbook of North American Indians,* Vol. 2: *Great Basin,* ed. W. L. D'Azevedo (Washington, D.C.: Smithsonian Institution Press, 1986), 31–50, 51–63. Steven R. Simms, "Wilderness as a Human Landscape," in *Wilderness Tapestry: An Eclectic Approach to Preservation,* ed. Samuel I. Zeveloff et al. (Reno and Las Vegas: University of Nevada Press, 1992), 183–201.

4. Leonard Arrington, *History of Idaho,* 2 vols. (Moscow: University of Idaho Press; and Boise: Idaho State Historical Society, 1994), I:87–113, 141–160, 492–494; Bentley and Oakley, "A Climate of Change," 75–77; F. Ross Peterson, "Confronting the Desert," in *Snake: The Plain and Its People,* 124–135; U.S. Department of the Interior, Census Office, *Report on the Productions of Agriculture as Returned at the Tenth Census,* 1880 (Washington, D.C.: Government Printing Office, 1883), 140–146. See also James A. Young and B. Abbott Sparks, *Cattle in the Cold Desert* (Logan: Utah State University Press, 1985).

5. The phrase is in Tim Palmer, *The Snake River: Window to the West* (Washington, D.C., and Covello, Calif.: Island Press, 1991), 185.

6. T. S. Palmer, "The Deer Flat National Bird Reservation, Idaho," *Reclamation Record* 7 (March 1916): 221–223; F. M. Dille, "The Minidoka National Bird Reservation, Idaho," *Reclamation Record* 7 (December 1916): 564–565.

7. For the harmful effects of dams, reservoirs, and irrigation systems on fish, see U.S. Department of the Interior, Fish and Wildlife Service, "A Preliminary Survey of Fish and Wildlife Resources, Upper Snake River Basin," in U.S. Bureau of Reclamation and U.S. Army Corps of Engineers, *Upper Snake River Basin,* Vol. 3, Part 1: *Coordination and Reports of Cooperating Agencies* (n.p.: n.p., 1961); H. H. Caldwell and Merle Wells, *Economic and Ecological History Support Study: A Case Study of Federal Expenditures on a Water and Related Land Resources Project: Boise Project, Idaho and Oregon* (Moscow: Idaho Water Resources Research Institute, 1974), 14, 17; James Simpson and Richard Wallace, *Fishes of Idaho* (Moscow: University Press of Idaho, 1982). For some contemporary accounts of fish destruction, see, for example, *Idaho Statesman* (Boise), 29 May 1877; *Idaho Republican* (Blackfoot), 9 June 1916. The matter of fish entering irrigation canals engendered controversy between wildlife

officials and irrigators. Idaho law required irrigators to place screens over the entrances of the canals. The irrigators resisted, and the poorly funded Idaho Fish and Game Warden officials apparently lacked the resources to fully enforce the law. The biennial reports of the Idaho Fish and Game Warden record some of the early history of this controversy, as well as fish kills caused by irrigation systems. See, for example, Otto M. Jones, *Eighth Biennial Report of the Fish and Game Warden of the State of Idaho, 1919–1920* (n.p.: n.p., n.d.), 45–58, 63–64. For information on fish that thrived in the new world of Idaho rivers, see Allan D. Linder, "Idaho's Alien Fishes," *TEBIWA: The Journal of the Idaho State Museum* 6 (2) (1963): 12–15; Simpson and Wallace, *Fishes of Idaho,* 89, 113, 115, 117, 119, 143–153, 161; Walter R. Courtenay, Jr., et al., "Distribution of Exotic Fishes in the Continental United States," in *Distribution, Biology, and Management of Exotic Fishes,* ed. Walter R. Courtenay, Jr., and Jay R. Stouffer, Jr. (Baltimore: Johns Hopkins University Press, 1984), 42–49.

8. Dille, "The Minidoka National Bird Reservation, Idaho"; Palmer, "The Deer Flat National Bird Reservation"; Ira N. Gabrielson, *Wildlife Refuges* (New York: Macmillan, 1943), 188; "Deer Flat National Wildlife Refuge, Narrative Report, September 1 to December 31, 1951," pp. 7–8, Deer Flat National Wildlife Refuge Records, Deer Flat National Wildlife Refuge, Nampa, Idaho (hereafter cited as Deer Flat).

9. L. N. Huggins's activities were reported in *Reclamation Record* 9 (January 1918): 5. See also C. J. Blanchard, "Playgrounds on Reclamation Projects," *Reclamation Record* 8 (August 1917): 375. For another example of an individual who sought to plant many new species in the irrigated landscape, see Herb Pollard, "William H. Ridenbaugh, Pioneer Wildlife Entrepreneur," *Idaho Wildlife Review* 27 (May–June 1975): 8–9.

10. "Report of Construction and Operation, Minidoka Project, Year of 1912," p. 13, Roll 97, Microcopy M96, Minidoka Project, Idaho, 1903–1919, Record Group 115, Records of the Bureau of Reclamation, National Archives, Washington, D.C. (hereafter cited as RG 115); Blanchard, "Playgrounds on Reclamation Projects"; "Our Reservoirs As Pleasure Resorts," *Reclamation Record* 8 (November 1917): 526–527; Alvin C. Holmes, *Swedish Homesteaders in Idaho on the Minidoka Irrigation Project* (n.p.: n.p., 1976), 56–57. Inhabitants of the irrigated landscape frequented groves for picnics, religious observances, and political rallies. See Annie Pike Greenwood, *We Sagebrush Folks* (New York: D. Appleton-Century, 1934; rpt. Moscow: University of Idaho Press, 1988), 315, for an example of Grangers meeting in a grove near Eden, Idaho. See also the *Idaho Farmer* for occasional news about such meetings. For a discussion of the meaning of groves in American culture, see J. B. Jackson, *The Necessity for Ruins, and Other Topics* (Amherst: University of Massachusetts Press, 1980), 77–88.

11. *Caldwell Tribune,* 28 January 1888; Linder, "Idaho's Alien Fishes," 13; Simpson and Wallace, *Fishes of Idaho,* 113, 115. On attempts to exterminate the carp, see *Idaho Register* (Idaho Falls), 20 August 1912, and Rich Mauer, "Plan Abandoned to Kill Lake Lowell Carp," *Idaho Statesman,* 14 October 1978. My notion of the carp as the aquatic equivalent of the hog is drawn from Aldo Leopold, *A Sand County Almanac, With Essays on Conservation from Round River* (New York: Oxford University Press, 1966; reprint New York: Ballantine Books, 1970), 26. For a brief history of the ecology of the carp (*Cyprinus carpio*), see Christopher Lever, *Naturalized Fishes of the World* (San Diego: Academic Press, 1996), 92–111.

12. Early accounts of so-called moss tended to say very little if anything about specific plants and their biology. Rather, they emphasized removal techniques. The historical record makes it difficult to tell exactly when the moss began to spread through the canals. Certainly it was becoming a problem after 1910, if not before. For early accounts, see P. M. Fogg, "A History of the Minidoka Project, Idaho, to 1912 Inclusive," August 1915, Vol. 1: 85–86, Roll 96, Microcopy M96, Minidoka Project, Idaho, 1903–1919, RG 115; Barry Dibble and T. W. Parry, "Control of Moss, Weeds, and Willows on the Minidoka Project," *Reclamation Record* 8 (April 1917): 192–193; R. M. Adams, "Removal of Vegetation from Twin Falls Irrigation Canals," *Engineering News-Record* 68 (1920): 319–320; Hugh L. Crawford, "Report On Cleaning Canals With Modified Disk Harrows, Minidoka Irrigation Project, Idaho," *Reclamation Record* 13 (February 1922): 22–24; *Idaho Farmer* (Caldwell) 31 (26 July 1923): 60 (which mentions the problem of "frog moss" on the Minidoka project). See also "Keep the Ditches Clean," *Gem State Rural* (Caldwell) 13 (September 1908): 12; and "The Waste of Water," *Gem State Rural and Live Stock Journal* (September 1910): 41. Later sources were more specific. See, for example, Robert Balcom, *Preliminary Survey of the Aquatic Weed Problem in Irrigation Systems* (Denver: U.S. Department of the Interior, Bureau of Reclamation, Operation and Maintenance Division, March 1943). The best description of individual aquatic plants and their irrigation system habitats is Gary W. Hansen et al., *Herbicide Manual* (Denver: U.S. Department of the Interior, Bureau of Reclamation, 1983), 1–69. This volume is clearly illustrated with colored drawings.

13. On willows and grass, see Fogg, "A History of the Minidoka Project," 132; W. M. Wayman, "Methods of Combatting Weeds, Moss, and Burrowing Mammals in Irrigation Canals," in "Report of Boise Conference of Operating Engineers for Irrigation Canal Systems Located in Idaho, Oregon, and Washington, November, 1911," pp. 81–82, Roll 23, Microcopy M96, Project Histories and Reports of Reclamation Bureau Projects, 1905–1925, RG 115; Alfred Peters, "Early History of the Twin Falls Canal Company and My Adult Life With the Company," file 1511.01, History of the Twin Falls Canal Com-

pany, Twin Falls Canal Company Records, Twin Falls, Idaho (hereafter cited as TFCC); "Watermaster's Annual Report, Aberdeen-Springfield Canal Company," 15 February 1923, p. 15, Aberdeen-Springfield Canal Company Records, Aberdeen, Idaho (hereafter cited as ASCC). For clover, alfalfa, Russian thistle, and other ditch bank weeds, see *Caldwell Tribune,* 4 May 1894; L. F. Henderson, *Twelve of Idaho's Worst Weeds,* University of Idaho Agricultural Experiment Station, Bulletin 14 (Moscow: Moscow Mirror, 1898); R. K. Bonnett and H. W. Hulbert, *Sweet Clover,* University of Idaho Agricultural Experiment Station, Circular 22 (Moscow: University of Idaho, March 1922). In addition to these sources, Idaho's agricultural journals, primarily the *Gem State Rural* and the *Idaho Farmer,* regularly carried news about weeds and their spread through canals and other parts of the irrigated landscape.

14. For examples of these sorts of animals inhabiting irrigation canals, see "Wants Pests Exterminated," *Gem State Rural* 13 (October 1907): 25; "Report of Construction and Operation, Minidoka Project, Year of 1912," p. 102, Roll 97, Microcopy M96, Minidoka Project, Idaho, 1903–1919, RG 115; "Annual Project History of Boise, Project, Idaho, for 1915," p. 200, Project Histories and Reports of Reclamation Bureau Projects, 1905–1925, Roll 9, Microcopy M96, RG 115; *Idaho Farmer* 28 (21 July 1921): 44; Lester C. Walker to W. H. Crouch, 12 August 1921, H. G. Hilfiker to E. H. Neal, 16 June 1938, both file 4.07, Rodent Control, ASCC; "Watermaster's Annual Report, Aberdeen-Springfield Canal Company," 15 February 1923, p. 14, ASCC; "The Twin Falls Canal Company System," August 1929, p. 37, and Alfred Peters, "Early History and My Adult Life With the Company," both TFCC; Holmes, *Swedish Homesteaders on the Minidoka Irrigation Project,* 40, 73. The habits of these animals are discussed in Earl J. Larrison and Donald R. Johnson, *Mammals of Idaho* (Moscow: University Press of Idaho, 1981); Vernon Bailey, *Harmful and Beneficial Mammals of the Arid Interior,* U.S. Department of Agriculture, Farmers' Bulletin 335 (Washington, D.C.: Government Printing Office, 1908); Vernon Bailey, *Revision of the Pocket Gophers of the Genus Thomomys,* U.S. Department of Agriculture, Bureau of Biological Survey, North American Fauna 39 (n.p.: n.p., November 1915); William T. Hornaday, *The American Natural History,* Vol. 1: *Mammals* (New York: Charles Scribner's Sons, 1914), 195–221; Thomas B. Scheffer, *Habits and Economic Status of the Pocket Gopher,* U.S. Department of Agriculture, Technical Bulletin 224 (Washington, D.C.: Government Printing Office, January 1931). Edith May Fawcett Moore recalled that a mountain lion once came through her family's farm on Mud Creek, one of the coulees that carried water through the irrigated landscape near Buhl (see Introduction and Chapter 1): "Life and Memories of Amos William, Theresa Ellen Gengler Fawcett and children," MC 21-5-162, Special Collections Department, Eli M. Oboler Library, Idaho State University, Pocatello (hereafter cited as Oboler). Irrigation systems also became

havens for insects harmful to agriculture. The alfalfa weevil, for example, lived on sweet clover that grew on the banks of canals and ditches. See, for example, *Idaho Farmer* 27 (14 April 1921): 401.

15. "Are There Clams in Your Ditch?" *Twin Falls News,* 16 November 1906. The Snake River has forty-three native species of shellfish, including fifteen clams. For information on shellfish in the Snake River, see Peter Bowler, "Mollusks of the Middle Snake River," in *Snake: The Plain and Its People,* 84. On people fishing in the canals, see Alfred C. Pettybone et al. to Officers and Directors of the Twin Falls Canal Company, 17 November 1932, File P, TFCC.

16. For accounts of beaver and muskrat in the irrigated landscape, see *Twin Falls News,* 14, 28 April 1905; Twin Falls Canal Company, "Minutes," 14 November 1911, TFCC; Dibble and Parry, "Control of Moss, Weeds and Willows on the Minidoka Project"; "Annual Project History for Boise Project, Idaho, for 1916," p. 162, Roll 9, Microcopy M96, Project Histories and Reports of Reclamation Bureau Projects, 1905–1925, RG 115; "Report of Construction and Operation, Year of 1911," p. 93, Roll 96, Microcopy M96, Minidoka Project, Idaho, 1903–1919, RG 115; "1931 Annual Report, Aberdeen-Springfield Canal Company," ASCC; E. H. Neal to State Game Warden, 12 November 1931, file 4.07, Rodent Control, ASCC; *Idaho Farmer* 25 (3 June 1920): 885; 37 (7 January 1926): 2; 38 (25 November 1926): 476; E. L. Davis, "Memories of Early Aberdeen," MC 21-2-52, Oboler; Peters, "Early History and My Adult Life With the Company."

17. For examples of this romantic attitude toward irrigation canals, see Greenwood, *We Sagebrush Folks,* 14, 41–42, 170–171, and Holmes, *Swedish Homesteaders on the Minidoka Irrigation Project,* 78–79, 83, 112.

18. See, for example, "Report of Construction and Operation, Minidoka Project, Year of 1911," pp. 85–86; Weymouth, "Methods of Operation and Management"; and Wayman, "Methods of Combatting Weeds, Moss, and Burrowing Animals in Irrigation Canals," 9, 81; Fred C. Scobey, *The Flow of Water in Flumes,* U.S. Department of Agriculture, Technical Bulletin 393 (Washington, D.C.: Government Printing Office, December 1933). See also the sources listed in note 12.

19. "Report of Construction and Operation, Minidoka Project, Year of 1911," pp. 92–93; Weymouth, "Methods of Operation and Management," 9; Wayman, "Methods of Combatting Weeds, Moss, and Burrowing Animals in Irrigation Canals," 81; Dibble and Parry, "Control of Moss, Weeds, and Willows on the Minidoka Project"; "Keep the Ditches Clean," *Gem State Rural* 13 (September 1908): 12; "The Waste of Water," *Gem State Rural and Live Stock Journal* (September 1910): 41; Peters, "Early History and My Adult Life With the Company." The problems that developed with willows similarly developed with Reed canary grass, another plant that irrigators introduced to canals. In the 1930s, irrigators and agricultural scientists touted the grass as stabi-

lization for canals and as a productive forage crop in waterlogged areas. Advocates of the plant claimed that it was easy to eradicate and would not become a problem. See J. H. Christ, *Reed Canary Grass,* University of Idaho, Agricultural Experiment Station, Circular 71 (Moscow: University of Idaho, June 1933). Eventually, the grass did become a major problem in irrigation systems. See W. Dean Boyle, "Summary of Field Trials in Suppression of Reed Canary Grass," in Bureau of Reclamation, Region 1—Boise, Idaho, "Thirteenth Irrigation Operators' Conference: Management's Responsibility to the Water Users," January 21–22, 1964, document on file in Enterprise Irrigation District Records, Teton, Idaho.

20. *Idaho Republican,* 12, 19 July 1923. Burrowing mammals damaged canals throughout the irrigated landscape. See, for example, *Gem State Rural* 8 (17 September 1903): 20; 13 (August 1907): 22; *Twin Falls News,* 19 July 1907; "Report of Construction and Operation, Minidoka Project, Year of 1911," p. 93; "Report of Construction and Operation, Minidoka Project, Year of 1912," p. 102; "Annual Project History of Boise Project, Idaho, for 1915," p. 200; "Annual Project History of Boise Project Idaho[,] for 1918," pp. 145, 147–148, Roll 10, Microcopy M96, Project Histories and Reports of Reclamation Bureau Projects, 1905–1925, RG 115; *Idaho Republican,* 19 July 1923 (a break in the Peoples Canal caused by muskrats); *Idaho Farmer* 25 (3 June 1920): 885; 28 (24 July 1921): 44; 29 (26 January 1922): 71; 37 (7 January 1926): 4; 38 (25 November 1926): 476; Carter, *Pioneer Irrigation,* 38, 172, 262; and, for a more recent example, *Twin Falls Times-Register,* 12 May 1978.

21. Numerous sources describe the pioneer process of transforming the land into farms. See, for example, Carter, *Pioneer Irrigation;* Greenwood, *We Sagebrush Folks;* Holmes, *Swedish Homesteaders on the Minidoka Irrigation Project.* For more specific examples of settlers removing sagebrush as preparation for crop production, see Hugh Lovin, "Sage, Jacks, and Snake Plain Pioneers," *Idaho Yesterdays* 22 (Winter 1979): 15, 18; "Opening of Great Reclamation Project," *Gem State Rural* 14 (March 1909): 1; H. G. Raschbacher, *Irrigation From Snake River, Idaho,* U.S. Department of Agriculture, Office of Experiment Stations, Circular 65 (Washington, D.C.: Government Printing Office, 1906), 5–11; *Caldwell Tribune,* 6 September 1890; Carter, *Pioneer Irrigation,* 23, 25, 39, 174.

22. Israel C. Russell, *Geology and Water Resources of the Snake River Plains of Idaho,* U.S. Geological Survey, Bulletin 199 (Washington, D.C.: Government Printing Office, 1902), 14–22; Raschbacher, *Irrigation From Snake River, Idaho,* 11–15; James Stephenson, Jr., *Irrigation in Idaho,* U.S. Department of Agriculture, Office of Experiment Stations, Bulletin 216 (Washington, D.C.: Government Printing Office, 1909), 7–19; F. H. Newell, *Report on Agriculture by Irrigation in the Western Part of the United States at the Eleventh Census,* 1890

(Washington, D.C.: Government Printing Office, 1894), 136–156. For examples of Idaho irrigation farmers attempting to grow different kinds of crops, see *Idaho Statesman,* 15, 24 March, 2 October 1870, 1 April, 23 June 1874, 24 June 1875, 17 May 1877; "A Look at the Boise Valley," *Idaho Yesterdays* 28 (Winter 1985): 12–19; *Caldwell Tribune,* 14 February 1891, 5 August 1910; *Twin Falls News,* 28 April, 19 May 1905; U.S. Congress, *Report of the Special Committee of the United States Senate on the Irrigation and Reclamation of Arid Lands,* Vol. 1: *The Northwest* (Washington, D.C.: Government Printing Office, 1890) 377–379; *Idaho Farmer* 28 (14 July 1921): 22; Articles of Incorporation, Nampa Cactus Hedge Fence and Orchard Company Limited, 13 October 1892, File 101, Articles of Incorporations Files, Canyon County Courthouse, Caldwell, Idaho.

23. Lovin, "Sage, Jacks, and Snake Plain Pioneers," 15. The belief that sagebrush signaled soil fertility resembled the midwestern pioneers' belief that trees indicated the same. On the latter view, see John Mack Faragher, *Sugar Creek: Life on the Illinois Prairie* (New Haven: Yale University Press, 1983), 43, 62–63.

24. U.S. Department of Commerce and Labor, Bureau of the Census, *Thirteenth Census of the United States,* 1910, Vol. 6: *Agriculture* (Washington, D.C.: Government Printing Office, 1913), 405; Elias Nelson, *Alfalfa,* University of Idaho Agricultural Experiment Station, Bulletin 66 (Moscow: Idaho Post Print, 1909), 3. See also *Idaho Statesman,* 23 June 1874; *Caldwell Tribune,* 14 February 1891 ("Alfalfa is king of all Idaho grasses"); *Gem State Rural* 9 (16 March 1905): 4; *Idaho Farmer* 24 (3 July 1919): 7; 25 (4 March 1920): 362; 39 (31 March 1927): 351. The quotation is from Richard Lowitt and Judith Fabry, eds., *Henry A. Wallace's Irrigation Frontier: On the Trail of the Corn Belt Farmer, 1909* (Norman: University of Oklahoma Press, 1991), 125.

25. On the popularity of grain and hay among early irrigators, see, for example, D. W. Ross, *Biennial Report of the State Engineer to the Governor of Idaho, for the Years 1899–1900* (Boise: Capital Printing Office, n.d.), 11, 12, 31; and Byron Hunter and Samuel B. Nuckols, *An Economic Study of Irrigated Farming in Twin Falls County, Idaho,* U.S. Department of Agriculture, Bulletin 1421 (Washington, D.C.: Government Printing Office, 1926), 10, 54–59. See also Holmes, *Swedish Homesteaders on the Minidoka Irrigation Project,* 28, for an example of a homesteader planting a first crop of wheat. For wheat production during the 1919 drought, see *Idaho Farmer* 24 (28 August 1919): 5; 25 (30 September 1920): 1420; 26 (18 November 1920): 1658; 27 (3 February 1921): 119.

26. Leonard J. Arrington, *Beet Sugar in the West: A History of the Utah-Idaho Sugar Company, 1891–1966* (Seattle: University of Washington Press, 1966), 54–73; James W. Davis and Nikki Balch Stilwell, *Aristocrat in Burlap: A History of the Potato in Idaho* (n.p.: Idaho Potato Commission, 1975).

27. Peter J. Valora, "A Historical Geography of Agriculture in the Upper Snake River Valley, Idaho," Ph.D. dissertation (University of Colorado, 1986), 310–316; Davis and Stilwell, *Aristocrat in Burlap,* 26–32, 49–53.

28. R. A. Oakley, "The Seed Supply of the Nation," in U.S. Department of Agriculture, *Yearbook of the United States Department of Agriculture, 1917* (Washington, D.C.: Government Printing Office, 1918), 497–536, discusses seed conditions brought on by World War I and mentions Idaho's leadership in the production of legume seeds, particularly alfalfa, clover, and peas. See W. A. Wheeler and G. C. Edler, "Some Effects of the War Upon the Seed Industry of the United States," in *Yearbook of the United States Department of Agriculture, 1918* (Washington, D.C.: Government Printing Office, 1919), 195–214. See also E. J. Iddings, *Idaho Pure Seed Law,* University of Idaho Agricultural Experiment Station, Circular 65 (Moscow: University of Idaho, 1931), 1; *Idaho Farmer* 25 (1 January 1920): 17, (12 February 1920): 230; 26 (25 November 1920): 1699; 35 (19 March 1925): 270; *The Farming Business in Idaho,* University of Idaho Agricultural Experiment Station, Bulletin 151 (Moscow: University of Idaho, July 1927), 50–67. For the history of Grimm alfalfa: *Idaho Farmer* 25 (1 January 1920): 6; 26 (25 November 1920): 1699; 27 (10 February 1921): 169, (17 February 1921): 190; 29 (9 February 1922): 123, (27 April 1922): 394; *Idaho Republican,* 5 August 1919; Nelson, *Alfalfa,* 5–6.

29. In 1926, a representative of the University of Florida visited Idaho bean fields and announced that Florida farmers wanted bean seed free of disease, diseases such as anthracnose having infested Florida's fields. See *Idaho Farmer* 38 (19 August 1926): 140; 25 (29 April 1920): 699. For pea seed history, see T. A. Brindley and Joseph C. Chamberlain, "The Pea Weevil," in U.S. Department of Agriculture, *Insects: The Yearbook of Agriculture, 1952* (Washington, D.C.: Government Printing Office, n.d.), 530; A. O. Larson et al., *Biology of the Pea Weevil in the Pacific Northwest with Suggestions for its Control on Seed Peas,* U.S. Department of Agriculture, Technical Bulletin 599 (Washington, D.C.: Government Printing Office, April 1938); *Idaho Farmer* 26 (4 November 1920): 1699, (25 November 1920): 1699; 29 (9 February 1922): 120; 30 (31 August 1922): 147, (28 December 1922): 504; 39 (24 February 1927): 6; *Idaho Falls Post-Register,* 10 September 1934.

30. Agriculture in Idaho and other parts of the American West initially lacked many of the plant and animal species that troubled farming in older regions. Evidence suggests that weed, insect, and disease infestations impelled the movement of agriculture, especially monocropping systems, from settled areas to the newly developed lands in the Far West. Certainly the weevil infestation drove the pea industry into the upper Snake River valley. On the westward flight of agriculture from weeds, see H. R. Cates, "The Weed Problem in American Agriculture," in *Yearbook of the United States Department of Agriculture, 1917* (Washington, D.C.: Government Printing

Office, 1918), 212. Many farmers took their clean landscape for granted. "When the land was new and irrigation projects were in their infancy," L. C. Aicher of the University of Idaho observed in 1917, "comparatively little attention was given to the control of weed growth"; see Aicher, *The Production of Clover Seed Under Irrigation in Southern Idaho,* University of Idaho Agricultural Experiment Station, Bulletin 100 (Moscow: University of Idaho, June 1917), 14.

31. Alfred Crosby, *Ecological Imperialism: The Biological Expansion of Europe, 900–1900* (New York: Cambridge University Press, 1986), 145–170, discusses the ecological characteristics of weeds and their historical relationship to Euro-American agriculture.

32. Henderson, *Twelve of Idaho's Worst Weeds,* 97–99, 110.

33. Henderson, *Twelve of Idaho's Worst Weeds,* 94, 109–116, 131–132; O. M. Osborne, *Weed Pests of Idaho and Methods of Eradication,* University of Idaho Agricultural Experiment Station, Bulletin 71 (n.p.: n.p., July 1911): 10–13.

34. *Idaho Statesman,* 12 August 1875; *Idaho Farmer* 25 (5 February 1920): 195, (23 September 1920): 1383; 26 (4 November 1920): 1586, (18 November 1920): 1662; 27 (5 May 1921): 458, (24 March 1921): 322, (31 March 1921): 346; and George Stewart, *Alfalfa Growing in the United States and Canada* (New York: Macmillan, 1926), 1–6.

35. *Caldwell Tribune,* 4 May 1894; Henderson, *Twelve of Idaho's Worst Weeds,* 95; Jessie C. Ayres et al., *Idaho Weeds: How to Know and Control Them,* University of Idaho, College of Agriculture, Extension Bulletin 65 (Boise: n.p., April 1926), 9; *Idaho Farmer* 26 (14 October 1920): 1485, (4 November 1920): 1586; Lester Vance Benjamin, "Experimental and Historical Development of Weed Control in Idaho," M.S. thesis (University of Idaho, 1932), 18–19.

36. Bonnett and Hulbert, *Sweet Clover;* Henderson, *Twelve of Idaho's Worst Weeds,* 91 ("great curse"); *Twin Falls News,* 10 August 1906 (McPherson). "A weed is a plant out of place," wrote Alfred Atkinson in 1907; see Atkinson, "Eradicating Weeds," *Gem State Rural* 12 (2 May 1907): 1. Atkinson also distinguished between "relative" and "absolute" weeds. Rye growing in a wheat field was a relative weed; a dandelion growing in a wheat field was an absolute weed.

37. Stewart, *Alfalfa Growing in the United States and Canada,* 361; for the history of the alfalfa weevil in Idaho, see the multipart article by Claude Wakeland, "The Alfalfa Weevil," *Idaho Farmer* 27 (31 March 1921): 349, (7 April 1921): 375; (14 April 1921): 401; (21 April 1921): 425; (28 April 1921): 438.

38. For examples of grasshopper and cricket infestations, see *Idaho Statesman,* 2, 30 May 1868, 25 May, 22 June 1869; *Rigby Star,* 1 June, 6 July 1906; *Idaho Farmer* 25 (29 July 1920): 1129; 31 (8 March 1923): 219. The 1923 irruption in Bingham County destroyed 25 percent of the alfalfa seed crop. See also J. M. Aldrich, *Grasshopper and Cricket Outbreaks,* University of Idaho Agri-

cultural Experiment Station, Bulletin 41 (Moscow: Moscow Weekly Mirror Print, March 1904); Claude Wakeland and W. E. Shull, *The Mormon Cricket, With Suggestions For Its Control,* University of Idaho College of Agriculture, Extension Bulletin 100 (Moscow: n.p., February 1936); James A. Young, "Mormon Crickets," *Rangeman's Journal* 5 (December 1978): 193–196.

39. Rowland W. Haegele, *The Beet Leaf-Hopper (Eutettix tennellus Baker): A Survey in Idaho,* University of Idaho Agricultural Experiment Station, Bulletin 156 (Moscow: University of Idaho, July 1927); Walter Carter, *Ecological Studies of the Beet Leafhopper,* U.S. Department of Agriculture, Technical Bulletin 206 (Washington, D.C.: Government Printing Office, November 1930); Rowland W. Haegele, *Field Studies of the Beet Leafhopper,* University of Idaho Agricultural Experiment Station, Bulletin 182 (Moscow: University of Idaho, February 1932); R. L. Piemeisel, *Weedy Abandoned Lands and the Weed Hosts of the Beet Leaf Hopper,* U.S. Department of Agriculture, Circular 229 (Washington, D.C.: Government Printing Office, July 1932); Arrington, *Beet Sugar in the West,* 101–122.

40. For descriptions of the effect of irrigated agriculture on native birds, see, for example, U.S. Department of the Interior, Fish and Wildlife Service, "A Preliminary Survey of Fish and Wildlife Resources, Upper Snake River Valley." On the magpie, killdeer, and sparrow, see *Idaho Farmer* 24 (9 January 1919): 5, (16 January 1919): 4; 39 (14 April 1927): 408; Holmes, *Swedish Homesteaders on the Minidoka Irrigation Project,* 60. On pocket gophers, ground squirrels, and jackrabbits, see Lovin, "Sage, Jacks, and Snake Plain Pioneers"; Harold Forbush, "A Trilogy of Affliction upon Our Valley Pioneers," *Snake River Echoes* 11 (1982): 100–102; J. Kent McAdoo and James A. Young, "Jackrabbits," *Rangelands* 2(4) (August 1980): 135–138; *Idaho Statesman,* 7 July 1874; 31 May 1877; *Gem State Rural* 8 (17 September 1903): 20; 13 (August 1907): 25; *Idaho Farmer* 24 (19 June 1919): 2; 29 (9 March 1922): 222; 35 (1 January 1925): 4; 38 (26 August 1926): 162; General Manager, ASCC to W. E. Crouch, 21 January 1926, file 4.07, Rodent Control, ASCC; Ezra J. Poulsen, "Too Many Squirrels," *Harper's Magazine* 178 (February 1939): 302–305; Alice Fern Ravenscraft Crandall, "From Dawn to Dusk: An Autobiography," MC 21-2-45, Oboler. C. Clifford Robinson also recollected rabbits eating hay in the midst of the livestock. "Rabbits used to be so thick they would eat our crops and hay in the winter. We had to keep them fenced out or they would eat more than our stock." C. Clifford Robinson Oral History, MC 21-6-186, Oboler. Delbert Fowler recalled a "farm where the ground squirrels were so plentiful, they ate more grass in the large pasture than the cattle did." See "Personal Journal of Delbert Fowler," MC 21-1-21, Oboler. See also the sources listed in note 14.

41. C. Clifford Robinson, "The Life Story of Charles Clifford Robinson," MC 21-6-186, Oboler. For other sources that mention the presence of coy-

otes in the irrigated landscape, see *Gem State Rural* 13 (August 1907): 22; John Theodore Montgomery and James Warren Barber, "An Economic Study of the History, Present Situation, and Outlook of Agriculture on the Minidoka Irrigation Project," M.S. thesis (University of Idaho, 1927), 20.

42. My analysis here owes much to Hildegard Binder Johnson's observation that the rectangular survey, upon which private property boundaries are based, cannot fully accommodate landforms and drainage patterns. See *Order upon the Land: The U.S. Rectangular Land Survey and the Upper Mississippi Country* (New York: Oxford University Press, 1976), and "Rational and Ecological Aspects of the Quarter Section: An Example from Minnesota," *Geographical Review* 47 (July 1957): 330–348. My analysis also owes much to John Brinckendoff Jackson, *Discovering the Vernacular Landscape* (New Haven: Yale University Press, 1984), and to Mart A. Stewart, *"What Nature Suffers to Groe": Life, Labor, and Landscape on the Georgia Coast, 1680–1920* (Athens: University of Georgia Press, 1996).

43. Annie Pike Greenwood's family had fourteen cats (as well as five dogs and a pet magpie). Greenwood, *We Sagebrush Folks,* 269. On cats as predators, see Scheffer, *Habits and Economic Status of the Pocket Gophers,* 17; "Deer Flat Narrative Report, May 1 to August 31, 1950," p. 6, and "Deer Flat Report, September 1 to December 31, 1951," p. 5, both Deer Flat; Holmes, *Swedish Homesteaders on the Minidoka Irrigation Project,* 70. Herbert E. Salinger, "Factors Affecting Pheasant Production on the Irrigated Lands of Southern Idaho," M.S. thesis (University of Idaho, 1950), also briefly mentions cats (as well as dogs) as predators. For a summary of research on the importance of cats as predators, see Peter B. Churcher and John H. Lawton, "Beware of Well-fed Felines," *Natural History* 98 (July 1989): 40–46. On wildlife refuges as bird feeders, see Dille, "The Minidoka Bird Reservation, Idaho"; "Deer Flat Report, September 1 to December 31, 1951," pp. 7–8, Deer Flat. On the presence and problems of pheasant and other game birds in the irrigated landscape, see U.S. Department of the Interior, Fish and Wildlife Service, "A Preliminary Survey of Fish and Wildlife Resources, Upper Snake River Valley"; Salinger, "Factors Affecting Pheasant Production in Southern Idaho"; J. Burton Lauckhart and John W. McLean, "Chinese Pheasants in the Northwest," in *Pheasants in North America,* ed. Durward L. Allen (Harrisburg, Pa.: Stackpole Company; and Washington, D.C.: Wildlife Management Institute, 1956); Donna M. Gleisner, *Ring-necked Pheasant,* Idaho Department of Fish and Game, Wildlife Leaflet 3 (n.p.: n.p., 1983); *Idaho Farmer* 21 (21 September 1916): 3; 24 (6 November 1919): 5; 29 (4 May 1922): 416; 32 (3 April 1924): 340; 37 (7 January 1926): 2; 39 (5 May 1927): 488. On farmers caring for pheasants, see *Idaho Republican,* 26 January, 2 February 1922. Consider, also, the Idaho Fish and Game Warden concerning the quail: "They are a half-domestic bird and, unless hunted too closely, like to be around farm and

ranch buildings, corrals, and feed lots. The farmers and ranchmen report them unusually plentiful the season of 1908." See *Second Biennial Report [1907–1908] of the Fish and Game Warden of the State of Idaho* (n.p.: n.p., n.d.), 14.

44. On bees in the irrigated landscape, see L. C. Aicher, *The Production of Alfalfa Seed in Southern Idaho,* University of Idaho Agricultural Experiment Station, Bulletin 101 (Moscow: University of Idaho, July 1917): 4–5; A. E. McClymonds, *Alfalfa Seed Production in Southern Idaho,* University of Idaho Agricultural Experiment Station, Bulletin 143 (Moscow: University of Idaho, April 1926), 16. For another illustration of bees as animals of the ecological commons, consider the comments of the farmer Fred Reddick, who, in August 1907 walked into the offices of the *Idaho Republican* in Blackfoot and proudly plunked down "a fine jar of honey" on the editor's desk. Reddick claimed to own thirty-eight hives producing 400 pounds of honey every week. Reddick said that he liked the insects because "they feed on the country at large, have no regard for fences or forest reserves, board themselves, ask for no pay, and know nothing of the employer's liability law." See *Idaho Republican,* 16 August 1907. A persistent tension in the irrigated landscape, however, one based on the intersection of private property and the ecological commons, was the concern that beekeepers had over the farmers' use of pesticides. Bees ranged across the landscape to pollinate crops, but the insects were vulnerable to the sprays that farmers applied to their fields. For examples of Idaho beekeepers' concern over chemical sprays, see *Idaho Republican,* 7 August 1919; *Idaho Farmer* 27 (7 April 1921): 364; 29 (6 April 1922): 316; Don B. Whelan, "Will the Alfalfa Weevil Run the Honey Bee Out of Idaho?" *Idaho Farmer* 30 (30 November 1922): 422. Beekeepers thought that sprays meant to destroy the alfalfa weevil killed bees, but agricultural scientists argued that it was the weevil, by destroying the blossoms, that was hurting the bees. See Whelan, "Will the Alfalfa Weevil Run the Honey Bee Out of Idaho?" and J. E. Nessly, "Growers Must Fight the Alfalfa Weevil," *Idaho Farmer* 32 (5 July 1923): 2.

45. Goulder, quoted in *Idaho Statesman,* 31 May 1877; Lowitt and Fabry, *Henry A. Wallace's Irrigation Frontier,* 127, 129; Robert S. Snyder et al., *Alkali Reclamation Investigations,* University of Idaho Agricultural Experiment Station, Bulletin 233 (Moscow: University of Idaho, October 1940), 3; Richard Calvin Montgomery, "Canyon County: The Economic Geography of a Southwestern Idaho Irrigated Area," M.A. thesis (University of Nebraska, 1951), 30; Siddik Atrushi, "Geographic Elements in Irrigation Agriculture in Boise Valley, Idaho," Ph.D. dissertation (Clark University, 1952), 183. Henry A. Wallace later became secretary of agriculture and vice-president of the United States.

46. Charles A. Jensen and B. A. Olshausen, "Soil Survey of the Boise Area, Idaho," in *Field Operations of the Bureau of Soils,* ed. Milton Whitney (Wash-

ington, D.C.: Government Printing Office, 1902), 429–445, and plates 65–68; Milton Fireman et al., *Characteristics of Saline and Alkaline Soils in the Emmett Valley Area, Idaho,* University of Idaho Agricultural Experiment Station, Research Bulletin 17 (Moscow: University of Idaho, May 1950); Osborne, *Weed Pests of Idaho and Methods of Eradication,* 12–13; George Stewart, "Pastures and Natural Meadows," in U.S. Department of Agriculture, *Grass: The Yearbook of Agriculture, 1948* (Washington, D.C.: Government Printing Office, 1948), 548; Lowitt and Fabry, *Henry A. Wallace's Irrigation Frontier,* 127. For another indication of the kinds of plants that thrived in seeped lands, consider the comment of the Idaho Supreme Court in its rendering of a judgment concerning drainage: "It is a well-recognized fact that under many of the irrigation systems of our State thousands of acres of land which were reclaimed from an arid condition and which for a time produced valuable crops have now become alkalined or water-logged, and thus ruined, and grow nothing but willows and tules because of the seepage of waters from canals and the irrigation of higher lands." See "Drainage for Irrigation Projects in Idaho," *Reclamation Record* 9 (February 1918): 68. When assessing the relative values placed on salinized landscapes, it is important to keep in mind that nature itself produces landscapes so salty that virtually nothing can grow in them. The Great Salt Lake and the Bonneville Salt Flats in Utah, which in one sense can be considered simplified, degraded ecosystems lacking in biodiversity, are now accorded the status of natural wonders and deemed worthy of preservation.

47. Holmes, *Swedish Homesteaders on the Minidoka Irrigation Project,* 60; *Caldwell Tribune,* 5 August 1910. See also *Idaho Republican,* 10 June, 28 October 1910, for a story of how settlers, near Springfield in the upper Snake River valley, created a pond that they stocked with ducks and fish for sporting purposes.

48. *Idaho Farmer* 39 (19 May 1927): 548; see also G. W. Grebe, "What Is or Should Be the Status of Wildlife as a Factor in Drainage and Reclamation Planning?" *Transactions of the Third North American Wildlife Conference* (1938): 119–124.

49. George W. Klontz and John G. King, *Aquaculture in Idaho and Nationwide* (Boise: Idaho Department of Water Resources, June 1975); *Idaho Statesman,* 30 April 1972, 19 February 1978, 9 March 1980; *Twin Falls Times-News,* 26 February 1978.

50. D. W. Fouch, "The Farmers' Unappreciated Friend," *Gem State Rural* 12 (28 February 1907): 1. Other people occasionally discussed the irrigated landscape in ecological terms and defended predators because of their usefulness in attacking species harmful to agriculture. T. G. Moore defended hawks; *Gem State Rural* 5 (October 1899): 2. A "Mr. Eiffe" defended coyotes at an Idaho State Horticultural Society meeting; *Gem State Rural* 8 (5 Febru-

ary 1903): 1. Julia M. Martin defended insectivorous birds in a poem titled "The Flying Police," which appeared in the *Idaho Farmer* 27 (16 June 1921): 593. "W. N." of Sugar City wrote to the *Idaho Farmer* asking for names of birds that destroyed rats, mice, and insects. The journal, quoting a book by Herrick, *Insects of Economic Importance,* responded: "Birds should be protected and encouraged to make their homes on the farm and about the orchards. They help to hold the balance of nature and certainly aid greatly in holding insects in check." *Idaho Farmer* 29 (13 April 1922): 353. See also "Farmers' Friends," *Idaho Register,* 7 June 1912. For an analysis of evolving attitudes toward predators, see Thomas Dunlap, *Saving America's Wildlife: Ecology and the American Mind, 1850–1990* (Princeton: Princeton University Press, 1988).

51. Julia M. Martin, "On Humby's Place," *Idaho Farmer* 28 (22 September 1921): 208.

52. For examples of these metaphors, see *Caldwell Tribune,* 22 September 1894; *Idaho Farmer* 24 (18 September 1919): 4; 37 (1 April 1926): 319; 39 (17 March 1927): 292; Henderson, *Twelve of Idaho's Worst Weeds,* 93 and passim; Osborne, *Weed Pests of Idaho and Methods of Eradication,* passim; Fred Schwendiman, "Battle With Germans," *Idaho Farmer* 23 (4 April 1918): 4. See also I. D. O'Donnell, "Weeds—Huns," *Reclamation Record* 9 (June 1918): 268: "To make our projects safe places in which to live, we must deal with weeds just as we would with the Huns. If possible, hold them at the first-line trenches. Get them with our long-range guns, and when closer use our machine guns (cultivators, hoes, etc.). Cut off their heads and cripple their feet. Mow them down in swathes. Smother them with gas. Take no prisoners and bury no dead. No Red Cross work is needed in this battle. Don't let them hold a single foot of ground. Take them while young, male and female alike. Don't trust them for a single minute, as they are as treacherous as a submarine. They must be drowned. So, while our boys are doing their duty at the front, let's get busy on the farm and do ours." Even D. W. Fouch's ecological concepts used a military metaphor: the balance of power.

53. *Idaho Farmer* 24 (16 January 1919): 4, and (6 March 1919): 21; "1928 Annual Report, Aberdeen-Springfield Canal Company," 11, ASCC.

54. For background on the weed fight on canal banks, see Dibble and Parry, "Control of Moss, Weeds, and Willows on the Minidoka Project"; "Annual Project History, Report of Construction and Operation and Maintenance, Minidoka, Volume XI, Year of 1917," p. 336, Roll 99, Microcopy M96, Minidoka Project, Idaho, 1903–1919, RG 115. On hunting, trapping, and otherwise guarding against burrowing mammals on canals, see D. W. Cole, "Patrol of Canals on Irrigation Projects," *Reclamation Record* 9 (January 1918): 24–25; "Annual Project History of Boise Project for 1918," pp. 147–148; "Annual Report of Canal Operations for Season of 1921," TFCC; *Idaho Farmer* 29 (26 January 1922): 71, and (11 May 1922): 439; E. H. Neal

to School Principal, Pingree School, 31 March 1932, File 4.07, Rodent Control, ASCC; New Sweden Irrigation District, "Minutes," 4 October 1932, New Sweden Irrigation District Records, New Sweden Irrigation District, Idaho Falls; E. L. Davis, "Memories of Early Aberdeen," MC 21-2-52, Oboler. For examples of inhabitants of the irrigated landscape who hunted and trapped along canals, see Holmes, *Swedish Homesteaders on the Minidoka Irrigation Project,* 23; *Idaho Farmer* 31 (8 February 1923): 120.

55. Dibble and Parry, "Control of Moss, Weeds, and Willows on the Minidoka Project," 193; *Idaho Farmer* 21 (30 November 1916): 6; 23 (9 May 1918): 5; 25 (4 March 1920): 362, and (22 April 1920): 661; R. K. Bonnett, "Pests Threaten Clover Seed Industry of Idaho," *Idaho Farmer* 26 (23 September 1920): 1383; R. K. Bonnett, "Red Clover Seed Production in Minidoka County," *Idaho Farmer* 26 (21 October 1920): 1515; Ayres et al., *Idaho Weeds: How to Know and Control Them,* 9; for Wheelon quotation, Twin Falls Canal Company, "Annual Report of Canal Operations for Season of 1922, " 5, TFCC.

56. "Report of Construction and Operation, Minidoka Project, Year of 1911," p. 86; Dibble and Parry, "Control of Moss, Weeds, and Willows on the Minidoka Project"; Adams, "Removal of Vegetation from Twin Falls Irrigation Canals"; Crawford, "Report on Cleaning Canals With Modified Disk Harrows, Minidoka Irrigation Project, Idaho"; "1929 Annual Report, Aberdeen-Springfield Canal Company," 2, 9, "1936 Annual Report, Aberdeen-Springfield Canal Company," 1, "1938 Annual Report, Aberdeen Springfield Canal Company," 8–9, all ASCC.

57. *Gem State Rural* 3 (October 1897): 1, 4; (November 1897): 6; (December 1897): 1, 2, 4; (January 1898): 1; *Idaho Farmer* 25 (15 July 1920): 1073; Claude Wakeland, "European Fly May Help Whip Alfalfa Weevil," *Idaho Farmer* 32 (19 July 1923): 39; Aldrich, *Grasshopper and Cricket Outbreaks,* 296–298. See also William H. Jordan, "The Weevil and the Wasp," *Natural History* 88 (December 1979): 37–42.

58. For examples of these techniques, see Claude Wakeland, "The Alfalfa Weevil," *Idaho Farmer* 27 (21 April 1921): 425, and (12 May 1921): 490; 28 (4 August 1921): 75; Bonnett, "Red Clover Seed Production in Minidoka County"; Byron Hunter and Samuel B. Nuckols, *An Economic Study of Irrigated Farming in Twin Falls County, Idaho,* U.S. Department of Agriculture, Bulletin 1421 (Washington, D.C.: Government Printing Office, October 1926), 12; Harold A. Vogel and Neil W. Johnson, *Types of Farming in Idaho,* Part 1, *Agricultural Resources and Factors Affecting Their Use,* University of Idaho Agricultural Experiment Station, Bulletin 207 (Moscow: University of Idaho, May 1934), 57; H. L. Spencer and H. W. Hulbert, *Idaho Perennial Weeds: Their Description and Control,* University of Idaho College of Agriculture, Extension Bulletin 98 (Moscow: n.p., June 1935), 24–25. To combat grasshoppers and crickets, irrigation farmers even scooped them up, sometimes with a

special device called a "hopper dozer," and destroyed them. See Aldrich, *Grasshopper and Cricket Outbreaks,* 298–299; and "Destroying Grasshoppers With Hopperdozer," *Gem State Rural* 13 (July 1907): 15.

59. See J. M. Aldrich, *Insecticides and Spraying,* University of Idaho Agricultural Experiment Station, Bulletin 7 (Moscow: Mirror Book Print, April 1894); W. C. Edmundson, *Insect Pests of the Orchards and Gardens of Idaho and Their Control,* University of Idaho Agricultural Experiment Station, Bulletin 87 (Moscow: University of Idaho, February 1916); Claude Wakeland and W. C. Hungerford, *Idaho Recommendation Chart for Plant Disease and Insect Control,* University of Idaho Agricultural Experiment Station, Bulletin 159 (Moscow: University of Idaho, May 1928); H. C. Manis and P. W. Portman, *Idaho Recommendations for Insect Control,* University of Idaho Agricultural Experiment Station, Bulletin 279 (n.p.: n.p., June 1950). On early poison programs for mammals, see *Idaho Falls Times,* 6 February 1916; *Idaho Republican,* 5 August 1919; *Idaho Farmer* 24 (13 November 1919): 5; *Aberdeen Times,* 30 August 1923, and undated *Aberdeen Times* clipping, both in file 4.07, Rodent Control, ASCC. For poison programs carried out within one irrigation system, see W. E. Crouch to Lester C. Walker, 29 August 1921, Melvin D. Smith to District Agent, Fish and Wildlife Service, 3 May 1949, and to Ival Goslin, 18 March 1948, all file 4.07, Rodent Control, ASCC; "1938 Annual Report, Aberdeen-Springfield Canal Company," 9, "Annual Report, Aberdeen-Springfield Canal Company, 1968," 5, both ASCC.

60. Benjamin, "Experimental and Historical Development of Weed Control in Idaho," 21; see also Ralph Scott Bristol, "Chemical Weed Control in Idaho," M.S. thesis (University of Idaho, 1932). On the biology of the bindweed, see Tom D. Whitson, ed., *Weeds of the West,* 5th ed. (Newark, Calif.: Western Society of Weed Science, 1996), 285.

61. Benjamin, "Experimental and Historical Development of Weed Control in Idaho"; Bristol, "Chemical Weed Control in Idaho"; *Idaho Farmer* 38 (22 July 1926): 64; 39 (7 April 1927): 391, and (5 May 1927): 501; 40 (14 July 1927): 24, and (8 September 1927): 187, and (22 December 1927): 552; V. F. Bruns et al., *The Use of Aromatic Solvents for Control of Submersed Aquatic Weeds in Irrigation Channels,* U.S. Department of Agriculture, Circular 971 (Washington, D.C.: Government Printing Office, 1955); *Idaho Statesman,* 17 October 1976. See also Aberdeen-Springfield Canal Company, "Manager's Report to Stockholders on Company's Operation in the 1949 Water Year," 11; "Annual Report, Aberdeen-Springfield Canal Company, 1961," 3; "Annual Report, Aberdeen-Springfield Canal Company, 1966," 3; "1984 Annual Report, Aberdeen-Springfield Canal Company," 2, all ASCC.

62. See, for example, Henderson, *Twelve of Idaho's Worst Weeds,* 98–101; *Idaho Farmer* 37 (1 April 1926): 319 (Bristol quoted); A. E. McClymonds, *Alfalfa*

Seed Production in Southern Idaho, University of Idaho Agricultural Experiment Station, Bulletin 143 (April 1926), 17.

63. On sagebrush fences and communal enclosures, see *Idaho Statesman,* 19 February 1870, 17, 22 May 1877; on the Mormon fence, see Norman Earl Ricks, "Mormon Settlement of Snake River Fork Country, 1883–1893," M.S. thesis (Brigham Young University, 1950), 71–73.

64. *Rigby Star,* 1 June, 6 July 1906; Wakeland and Shull, *The Mormon Cricket,* 3–5.

65. For a general description of rabbit drives, see Lovin, "Sage, Jacks, and Snake Plain Pioneers," 18–21. The account of the Market Lake drive is taken from the *Teton Peak,* 19 February 1908, republished in Martha B. Stoker, *Old Market Lake* (n.p.: n.p., 1983), 88–90. See also S. E. Piper, "The Jack Rabbit Problem," *Reclamation Record* 6 (March 1915): 123–124; and McAdoo and Young, "Jackrabbits." For brief descriptions of other rabbit drives, see, for example, *Idaho Farmer* 25 (5 February 1920): 176; (4 March 1920): 349; (15 April 1920): 615; 27 (3 February 1921): 144; (17 March 1921): 288; *Idaho Republican,* 3 January 1921; Carter, *Pioneer Irrigation,* 21, 63; Holmes, *Swedish Homesteaders on the Minidoka Irrigation Project,* 101.

66. See endnote 65 for examples of rabbit drives, which also occasionally mention the socializing that accompanied them. Shadduck is paraphrased in another newspaper article, the [Idaho Falls?] *Post-Register,* 8 August 1967, also republished in Stoker's *Old Market Lake,* 91–92.

67. Ray Erskine, "The Alfalfa Weevil in Southern Idaho," *Idaho Farmer* 28 (4 August 1921): 75; *Idaho Farmer* 27 (19 May 1921): 510.

68. For the development of these policies, see *General Laws of the State of Idaho* (Boise: Syms-York, 1907), 24–25; *Gem State Rural and Live Stock Journal* (April 1911): 21; *General Laws of the State of Idaho* (Boise: Syms-York, 1911), 381–383; *General Laws of the State of Idaho* (Boise: Syms-York, 1912), 46–47; *General Laws of the State of Idaho* (Boise: Syms-York, 1913), 476–477, 526; *General Laws of the State of Idaho* (n.p.: n.p., 1919), 86–88; Paul Wenger, *Pure Seed Law and Weed Control Act,* University of Idaho Agricultural Experiment Station, Circular 8 (Moscow: University of Idaho, April 1919); and Benjamin, "Experimental and Historical Development of Weed Control." For examples of their practical application, see *Idaho Farmer* 25 (29 January 1920): 144; 37 (25 March 1926): 296, and (1 April 1926): 319; 39 (17 March 1927): 292.

69. For examples of farmers working with various public agencies, see *Idaho Falls Times,* 6 February 1919; H. W. Hochbaum, "Working Plan of Farm Bureau," *Idaho Farmer* 23 (28 November 1918): 7; 24 (16 January 1919): 5, and (27 February 1919): 6, 13, and "Should Make War on Insect Pests," (6 March 1919): 21; 25 (13 March 1919): 5, and (15 January 1920): 95; 35 (1 January 1925): 4.

70. Consider the statement by the *Farm Bureau Monthly* of Glendale, California, reprinted in *Reclamation Record* 9 (September 1918): 426: "The man who refuses to stand up when The Star Spangled Banner is played, or to take off his hat when the flag passes, rightly should be made to feel the stigma of his neighbor's indignation. . . . But after all is he any worse than the farmer who refuses to exterminate the squirrels on his farm or to fight the grasshoppers and leaves his fields to become a breeding place to reinfest the surrounding country, or the man who kicks about 'personal' liberty when an ordinance is passed tending to reduce to a minimum the fire hazard of the grain fields?"

71. For examples of the irrigators' inability to drive weeds from the land, see E. H. Neal to Harold Ball, 18 August 1942, file 4.08, Weed Control, Inactive General Correspondence, ASCC; "General Manager's Annual Report for the Year 1935"; "Twin Falls Canal Company Manager's Annual Report," 31 December 1939; "Twin Falls Canal Company Annual Report, 1954," 2; "To the Board of Directors of the Twin Falls Canal Company, Twin Falls, Idaho," 10 January 1956, all TFCC. See also "1929 Annual Report, Aberdeen-Springfield Canal Company," 9; "1984 Annual Report, Aberdeen-Springfield Canal Company," 2, both ASCC.

72. "General Manager's Annual Report for the Year 1935," no page, "Twin Falls Canal Company Annual Report, 1954," 2, both TFCC.

73. For the unintended effects of weed control techniques, see U.S. Department of Health, Education, and Welfare, Public Health Service, "Upper Snake River Basin: Reconnaissance Report," in U.S. Bureau of Reclamation and U.S. Army Corps of Engineers, *Upper Snake River Basin,* Vol. 3, Part 1; K. E. Holte et al., *Environmental Surveys of the Teton River and Henry's Fork of the Snake River,* Idaho Water Resources Research Institute, Technical Completion Report WRRI Code USCE 44-418 (Moscow: Idaho Water Resources Research Institute, December 1972); U.S. Department of Agriculture, Soil Conservation Service, *Snake River Basin Cooperative Study: Main Report, Upper Snake River Basin* (n.p.: n.p., March 1979), 15–16; "Farmers View Aging Canal System," *Idaho State Journal,* 18 November 1983, in Irrigation vertical file, IHS; Bruns et al., *The Use of Aromatic Solvents for Aquatic Weeds in Irrigation Channels;* Charlotte V. Eberlein and William C. Schaffers, *Herbicide Carryover to Potatoes,* University of Idaho Agricultural Experiment Station, Current Information Series 864 (n.p.: n.p., April 1990).

74. On the persistence of DDT and other chemicals, see U.S. Department of Health, Education, and Welfare, Public Health Service, "Upper Snake River Basin: Reconnaissance Report."

75. In a sense, two systems were at work here: the economic resources of the irrigators and the reproductive and regenerative capacities of the animals. The irrigators' resources for killing could not overcome the biological

and ecological ability of muskrats, gophers, and other mammals to reestablish themselves in irrigation systems and restore their populations. Moreover, the irrigators' actions were somewhat paradoxical: they provided a habitat in the form of canals and ditches, and then they had to keep killing the mammals that would inevitably find those places attractive.

76. E. H. Neal to Chase Clark, 13 January 1923, Aberdeen-Springfield Canal Company to R. Mitchell et al., 19 November 1935, both file 4.07, Rodent Control, ASCC; "1942 Annual Report, Aberdeen-Springfield Canal Company," 5, "1945 Annual Report, Aberdeen-Springfield Canal Company," 2, "Manager's Report to Stockholders on Company's Operation in the 1948 Water Year, Aberdeen-Springfield Canal Company," 8, "Manager's Report to Stockholders on Company's Operation in the 1949 Water Year, Aberdeen-Springfield Canal Company," 10, all ASCC; "Annual Report of Canal Operations for Season of 1921," "Annual Report of General Manager for 1943," "Annual Report of the General Manager for 1944," all TFCC. For the decline of sheep on ditch banks: conversation with Jake Isaak, Aberdeen-Springfield Canal Company, Aberdeen, Idaho, 27 July 1990; conversation with Chuck Yost and Pam Muirbrook, Aberdeen-Springfield Canal Company, Aberdeen, Idaho, 5 October 1990.

77. E. H. Neal to Chase Clark, 13 January 1923, and to Owen W. Morris, 2 December 1941, Ival Goslin to Idaho State Fish and Game Warden, 17 May 1946, and to Melvin W. Smith, 6 June 1946, "Canal Official Blasts Game Bosses," undated newspaper clipping, all file 4.07, ASCC; "Twin Falls Canal Company Annual Report, 1953," TFCC; "1957 Annual Report to Stockholders, Aberdeen-Springfield Canal Company," 5, "Annual Report, Aberdeen-Springfield Canal Company, 1966," 3 (quotation), all ASCC. See also James O. Beck, *Twenty-first Biennial Report, Fish and Game Department of the State of Idaho* (n.p.: n.p., [1946?]), 37–39.

78. *Idaho Farmer* 29 (26 January 1922): 71, and (11 May 1922): 439; 31 (3 May 1923): 460; 32 (29 November 1923): 442; E. H. Neal to Chase Clark, 13 January 1923, Franklin Girard to I. J. Wenger, 24 March 1933, Idaho Department of Fish and Game, "Memorandum to All Canal Companies, Irrigation Districts, and Dyke Companies," 18 March 1943, and Idaho Department of Fish and Game, "Memorandum to Irrigationists," 17 March 1943, all file 4.07, Rodent Control, ASCC. Public opposition eventually led to a ban on the use of Compound 1080, an extremely effective poison. See Myron Dance and Elmer Isaak to Len B. Jordan, 14 March 1972, Milton B. Weston to Frank Church, 10 March 1972, file 4.07, Rodent Control, ASCC.

79. Wakeland, "The Alfalfa Weevil," 349; C. C. Vincent, "The Orchard Pests Are Always With Us," *Idaho Farmer* 31 (1 March 1923): 195.

80. "1936 Annual Report, Aberdeen-Springfield Canal Company," 9, ASCC; "Annual Report for 1938," and "Annual Report of the General Man-

ager of the Twin Falls Canal Company for 1945," both TFCC; Huntsinger quoted in Bob Lorimer, "Kuna Farmer Tells Secret of System for Keeping 'His Head above Water,'" *Idaho Statesman,* 21 July 1965.

3/ DIVIDING WATER

1. The story of Joe Koury's death (and Grover's eventual conviction of involuntary manslaughter) appeared in the *Idaho Republican* (Blackfoot), 8, 11, 15, 18 July, 21 November 1919. The Idaho writer, poet, and farmer Irene Welch Grissom penned a short story, "Thirsty Soil," that she probably based in part on the Koury-Grover incident. See her collection of writings, *We Harness a River* (Caldwell, Idaho: Caxton Printers, 1946), 51–70. In her memoir of life on an Idaho irrigated farm, the writer and farmer Annie Pike Greenwood discussed water stealing and the propensity of two neighbors, Baldy Parsons and Old Man Babcock, to divert water from the Greenwoods' fields:

"Water among us sagebrush folks was money. When a man stole your water, he committed grand larceny, no matter how much he himself might feel his crimes mitigated by the hymns he sang to Jesus a-Sundays. Water in the sagebrush country is not free, as the rain from heaven. Both the just and the unjust have to pay for it, or it is shut off. . . . If your head-gate is stuffed with weeds or gunny-sacks, it may mean the loss of all the money you can make that year, for your crop will die. . . .

"When another farmer steals your water, he takes the clothes from your children's backs, robs your wife of the medicine she probably needs, takes every penny out of your own overalls' pockets. It is no wonder that almost every year farmers are killed at the head-gates, one discovering another in the act of stealing water. There is sometimes, too, the murder of a ditch-rider who happens to catch the water racketeers, the answer to expostulation being a shot in the head. . . .

"Many times Charley [Greenwood's husband] came to the house for his gun and with it went walking away in the dusk up our main canal. I knew then that Bab or Baldy had stuffed our mutual head-gate when it was Charley's right to the water. He said he would fire a shot in the air to scare the guilty robber, but I was never easy until he came back. Then I stabbed him with my eyes to discover his dreadful secret. Fortunately, he never had any. He did not even fire his gun. The gunny-sacks or weeds had not been replaced during his absence, and no one appeared, though he sat there with his gun on the canal-bank, in the still of the long evening, for over an hour." See Greenwood, *We Sagebrush Folks* (New York: D. Appleton-Century, 1934; rpt. Moscow: University of Idaho Press, 1988), 213, 237, 300, 380–381 (quotation).

2. For the development of irrigation organizations, see Donald J. Pisani,

To Reclaim a Divided West: Water, Law, and Public Policy, 1848–1902 (Albuquerque: University of New Mexico Press, 1992); see also Robert G. Dunbar, *Forging New Rights in Western Waters* (Lincoln: University of Nebraska Press, 1983). Contemporary advocates of irrigation celebrated its cooperative basis. See, for example, Elwood Mead, "Rise and Future of Irrigation in the United States," in *Yearbook of the United States Department of Agriculture, 1899* (Washington, D.C.: Government Printing Office, 1900); William E. Smythe, *The Conquest of Arid America* (New York: Harper and Brothers, 1899, 1905; rpt. Seattle: University of Washington Press, 1985). The Corbett Slough irrigation system is briefly mentioned in Kate B. Carter, ed., *Pioneer Irrigation: Upper Snake River Valley* (Salt Lake City: Daughters of Utah Pioneers, 1955), 20. On mutualism in rural America, see, for example, the essays in Steven Hahn and Jonathan Prude, eds., *The Countryside in the Age of Capitalist Transformation: Essays in the Social History of Rural America* (Chapel Hill: University of North Carolina Press, 1985).

3. William Blackstone's famous adage about water and property was appropriate to the condition of Idaho irrigators: water, he said, "is a moveable, wandering thing, and must of necessity continue common by the law of nature; so that I can only have a temporary, transient, usufructuary right therein." Quoted in Theodore Steinberg, *Nature Incorporated: Industrialization and the Waters of New England* (Amherst: University of Massachusetts Press, 1991), 14. See also the frank discussion by J. B. Camp, manager of the King Hill irrigation project, on the problems surrounding the division of water among many irrigators. *Idaho Farmer* 27 (7 April 1921): 377.

4. On the Snake River and its hydrology, see John Wesley Powell, *Eleventh Annual Report of the United States Geological Survey to the Secretary of the Interior, 1889–'90,* Part 2, *Irrigation* (Washington, D.C.: Government Printing Office, 1891), 77–83; Israel C. Russell, *Geology and Water Resources of the Snake River Plains of Idaho,* U.S. Geological Survey, Bulletin 199 (Washington, D.C.: Government Printing Office, 1902), 25–26, 134; Harold T. Stearns et al., *Geology and Ground Water Resources of the Snake River Plain in Southeastern Idaho,* U.S. Geological Survey Water Supply Paper 774 (Washington, D.C.: Government Printing Office, 1938), 176–203. See also Lynn Crandall, "Snake River," [1958?], manuscript copy in Idaho–Reclamation of Land vertical file, Idaho Room, Idaho Falls Public Library (hereafter cited as Idaho Room).

5. For historical overviews of this settlement, see Carter, *Pioneer Irrigation;* Peter J. Valora, "A Historical Geography of Agriculture in the Upper Snake River Valley, Idaho," Ph.D. dissertation (University of Colorado, 1986); F. H. Newell, *Report on Agriculture by Irrigation in the Western Part of the United States at the Eleventh Census: 1890* (Washington, D.C.: Government Printing Office, 1894), 143–147; Russell, *Geology and Water Resources of the Snake River Plains,* 29; D. W. Meinig, "The Mormon Culture Region: Strategies and Patterns in

the Geography of the American West, 1847–1964," *Annals of the Association of American Geographers* 55 (June 1965): 204–205, 207–209; Davis Bitton, "Peopling the Upper Snake: The Second Wave of Mormon Settlement in Idaho," *Idaho Yesterdays* 23 (Summer 1979): 47–52. For information on expansion of irrigation, see D. W. Ross, *Biennial Report of the State Engineer to the Governor of Idaho for the Years 1899–1900* (Boise: Capital Printing Office, n.d.), 13; D. W. Ross, *Biennial Report of the State Engineer to the Governor of Idaho for the Years 1901–1902* (Boise: Statesman Print, n.d.), 54, 68, 77–78, 94; Wayne Darlington, *Biennial Report of the State Engineer to the Governor of Idaho, for the Years 1903–1904* (n.p.: n.p., n.d.), 39. The New Sweden Irrigation District, formed in 1899, was originally a private land and water development. The ethnic Swedes who settled the project subsequently purchased the land, canal system, and water rights and formed the irrigation district. See Charles E. Anderson et al., *After Fifty Years* (Caldwell, Idaho: Caxton Printers, 1941).

6. Newell, *Report on Agriculture by Irrigation in the United States,* 149, and Ross, *Biennial Report of the State Engineer, 1899–1900,* pp. 10–12.

7. "Water Law in Idaho," *Idaho Yesterdays* 25 (Spring 1981), provides a brief overview of Idaho water law, as does Paul L. Henry, "The Appropriation of Water," in R. W. Faris, *Seventh Annual Report of the Department of Reclamation, State of Idaho, 1931–1932* (Boise: n.p., n.d.), 13–16. For a lawyer's perspective on Idaho water law, see Wells A. Hutchins, "Idaho Law of Water Rights," *Idaho Law Review* 5 (Fall 1968): 1–129. For the national background of western water rights and the general features of prior appropriation, see Donald J. Pisani, *To Reclaim a Divided West;* and Donald Pisani, "Enterprise and Equity: A Critique of Western Water Law in the Nineteenth Century," *Western Historical Quarterly* 18 (January 1987): 15–37; and Hutchins, "Idaho Law of Water Rights." See also Dunbar, *Forging New Rights in Western Waters.* Pisani shows that through much of the nineteenth century, western territories and states combined the riparian and prior appropriation doctrines. Idaho's 1881 water law, for example, attempted to restrict the absolute title to water that prior appropriation conferred by making rights "correlative," or contingent on the quantity of water in a stream at any given time. See Pisani, *To Reclaim a Divided West,* 46–52.

8. Pisani, in "Enterprise and Equity," points out that prior appropriation was more a consequence of capitalist economic development than an adaptation to an arid environment. For a discussion of the relation between prior appropriation, capitalism, and the commodification of nature, see Donald Worster, *Rivers of Empire: Water, Aridity, and the Growth of the American West* (New York: Pantheon Books, 1985), 88–92. William Cronon's description of the transformation of nature into capital has influenced my characterization of water rights as an attempt to link nature flexibly to the market. See Cronon's *Nature's Metropolis: Chicago and the Great West* (New York: W. W.

Norton and Company, 1991), 97–259, especially chap. 3. In Idaho, water in a stream remained a form of common property; only when diverted did the water become an actual material possession. Still, Idaho law regarded a water right as real property or real estate. I think that the status of water rights as real property was another indication of how prior appropriation mediated between nature, the stream, and the capitalist market, which sought to commodify nature. Under prior appropriation, stream water, in a sense, was a sort of incipient private property. Although they recognized common property in the stream, irrigators could at least imagine and anticipate the ownership of part of that stream. See Hutchins, "Idaho Law of Water Rights," 6–7, and Ross, *Biennial Report of the State Engineer, 1899–1900,* p. 91.

9. Carter, *Pioneer Irrigation,* 6–7; Ross, *Biennial Report of the State Engineer, 1899–1900,* p. 90. For other examples of early irrigators' ignorance and lax record keeping, see Carter, *Pioneer Irrigation,* 13, 22, 25, 69–70, 99, 106, 169. In the late 1880s, James M. Cook of the Rexburg Agricultural and Horticultural Association testified that water shortage was not a problem for the Rexburg Irrigation Company. "We have always had a good supply of water," he stated. "This is distributed by the water master, who is allowed his own discretion and judgment in all cases. In no case have we had to measure out our water." See U.S. Congress, *Report of the Special Committee of the United States Senate on the Irrigation and Reclamation of Arid Lands,* Vol. 1: *The Northwest* (Washington, D.C.: Government Printing Office, 1890), 414.

10. For examples of the imprecision of recorded claims, see Ross, *Biennial Report of the State Engineer, 1899–1900,* 90–92; Jay D. Stannard, "The Use of Water From the Wood Rivers, Idaho," in *Report of Irrigation Investigations for 1902,* U.S. Department of Agriculture, Office of Experiment Stations, Bulletin 133 (Washington, D.C.: Government Printing Office, 1903), 91–92; and James Stephenson, Jr., *Sixth Biennial Report of the State Engineer to the Governor of Idaho, 1905–1906* (n.p.: n.p., n.d.), 7.

11. Ross, *Biennial Report of the State Engineer, 1901–1902,* pp. 54, 68, 77–78, 94; Ross, *Biennial Report of the State Engineer, 1899–1900,* p. 13; Darlington, *Biennial Report of the State Engineer, 1903–1904,* p. 39.

12. *Idaho Register* (Idaho Falls), 1 September 1905. Eventually, potatoes, peas, beans, and other crops requiring late-season irrigation further intensified the demand for water. See Lynn Crandall, "History of Irrigation Development in the Snake River Valley," [1946?], manuscript copy in Idaho–Reclamation of Land vertical file, Idaho Room.

13. *Idaho Register,* 25 August 1905, 7 September 1906 (quotation).

14. *Farmers' Friend* (Blackfoot), 27 March 1906.

15. *Idaho Republican* (Blackfoot), 7 August 1914.

16. Ibid.; Idaho, Sixth Judicial District Court, *In the District Court of the Sixth Judicial District of the State of Idaho, In and For the County of Fremont:*

Decree: Rexburg Irrigation Company et al., Plaintiffs, v. Teton Irrigation Canal Company et al., Defendants (n.p.: n.p., 1911).

17. *Idaho Republican,* 7 August 1914.

18. *Idaho Republican,* 7 August 1914; *Farmers' Friend,* 24 April 1906. The account of the Blackfoot irrigators' failed attempt to negotiate a cooperative arrangement with farmers upstream and their subsequent legal recourse come from these two newspapers.

19. *Farmers' Friend,* 27 March, 28 August 1906; *Idaho Register,* 24 August, 7 September 1906; *Idaho Republican,* 24 August 1906; *Rigby Star,* 17 August 1906.

20. See, for example, *Idaho Republican,* 18 August 1905.

21. *Farmers' Friend,* 27 March 1906.

22. *Idaho Register,* 7 September 1906, contended that Blackfoot irrigators were responsible for overcoming their hydrological plight; if they "would tap the river above the place where it sinks they would have no trouble in getting sufficient water."

23. On Jackson Lake as a storage site, see John Wesley Powell, *Tenth Annual Report of the United States Geological Survey to the Secretary of the Interior, 1888–'89,* Part 2, *Irrigation* (Washington, D.C.: Government Printing Office, 1890), 107. For information on the development of the dam and the Twin Falls and Minidoka projects, see W. G. Hoyt, *Water Utilization in the Snake River Basin,* U.S. Geological Survey Water-Supply Paper 657 (Washington, D.C.: Government Printing Office, 1935), 40, 124–127, and William Darrell Gertsch, "The Upper Snake River Project: A Historical Study of Reclamation and Regional Development, 1890–1930," Ph.D. dissertation (University of Washington, 1974), 43–168.

24. *Farmers' Friend,* 27 March 1906; *Idaho Falls Times,* 8 May 1906.

25. *Idaho Register,* 16 August 1910; *Rigby Star,* 18 August 1910. Newspapers sometimes got the group's name wrong; officially it was the Farmers Protective Irrigation Association; for the group's official letterhead, see John Empey to Fred Wilkey, 26 February 1916, Water District No. 1 Records (microfiche), Idaho Department of Water Resources, Boise (hereafter cited as IDWR).

26. Herbert Wing, *Tenth Biennial Report of the State Engineer to the Governor of Idaho, 1913–1914* (n.p.: n.p., n.d.), 19; *Idaho Republican,* 7, 14 August 1914.

27. For a discussion of stored water, natural flow, and consequent problems, see the articles in "Administrative Water Problems: A Symposium," Paper 1729, *American Society of Civil Engineers Transactions* 94 (1930), especially G. Clyde Baldwin, "Transmission and Delivery of Stored Water," 296–300, 324–325; Lynn Crandall, "Crandall on Administrative Water Problems," 318–321; and Thomas R. Newell, "Newell on Administrative Water Prob-

lems," 321–324. Although irrigators acquired rights to stored water at the reservoir, the law established no regulations for moving the water downstream. According to Baldwin, "There is no prescribed basis for converting a reservoir right into a right at the point of diversion many miles downstream." Baldwin, "Transmission and Delivery of Stored Water," 297.

28. Baldwin, "Transmission and Delivery of Stored Water"; Crandall, "Crandall on Administrative Water Problems"; Newell, "Newell on Administrative Water Problems"; D. G. Martin, *Eighth Biennial Report of the State Engineer to the Governor of Idaho, 1909–1910* (n.p.: n.p., n.d.), 292; A. E. Robinson, *Ninth Biennial Report of the State Engineer to the Governor of Idaho, 1911–1912* (n.p.: n.p., n.d.), 131–132.

29. Baldwin, "Transmission and Delivery of Stored Water"; Crandall, "Crandall on Administrative Water Problems"; Newell, "Newell on Administrative Water Problems."

30. See, for example, Robinson, *Ninth Biennial Report of the State Engineer, 1911–1912,* pp. 131–155, and Wing, *Tenth Biennial Report of the State Engineer, 1913–1914,* pp. 50–62. Astute observers even noted that the stored water actually raised the river's flow to the level of upper valley headgates, thus saving irrigators the trouble of having to build temporary wing dams into the riverbed. See, for example, *Idaho Republican,* 24 June 1919.

31. Baldwin, "Transmission and Delivery of Stored Water"; Crandall, "Crandall on Administrative Water Problems"; Newell, "Newell on Administrative Water Problems." More details on the Snake and its management appear in sources cited in subsequent notes.

32. For an example of estimated stored water loss, see Wing, *Tenth Biennial Report of the State Engineer, 1913–1914,* pp. 56–58.

33. See, for example, Wing, *Tenth Biennial Report of the State Engineer, 1913–1914,* pp. 59–62; *Idaho Register,* 26 August 1913; *Idaho Republican,* 31 July, 14 August 1914. Dates of water rights, and thus priority, are usually indicated in the capsule histories in Carter, *Pioneer Irrigation.* For a complete list of priority rights, see Idaho, Sixth Judicial District Court, *Rexburg v. Teton.* It is important to keep in mind that each canal or irrigator might have more than one water right, each with a different date.

34. For background on the administration of the upper Snake River, see *Idaho Republican,* 7 August 1914; Darlington, *Biennial Report of the State Engineer, 1903–1904,* pp. 15–16; Stephenson, *Sixth Biennial Report of the State Engineer, 1905–1906,* pp. 10–11; Martin, *Eighth Biennial Report of the State Engineer, 1909–1910,* pp. 124–125, 292; J. H. Smith, *Eleventh Biennial Report of the State Engineer to the Governor of Idaho, 1915–1916* (n.p.: n.p., n.d.), 17–21; Warren Swendsen, "Distribution of Snake River Water, District No. 36" (c. 1922), file 310.2, Minidoka Project Records, Record Group 115, Records of the Bureau of Reclamation, National Archives, Washington, D.C. (here-

after cited as RG 115); Baldwin, "Transmission and Delivery of Stored Water"; B. E. Stoutemyer to Commissioner, 30 October 1928, 4, 27 May 1929, 12 June 1929, all file 032, Minidoka Project Files, 1919–1929, General Administrative and Project Files, 1919–1945, RG 115. In 1903, Idaho divided the state into three sections, each under a water commissioner, who appointed watermasters to supervise the distribution of water. In 1909, a revision in the law provided that water users whose rights were covered by the same court decree should annually elect their own watermaster. Thus, 1911 was the first year that irrigators in the upper Snake River valley could elect their own watermaster, *Rexburg v. Teton* having been decided in late 1910. In 1915, Idaho established water districts that corresponded to river drainages; the upper Snake and its major tributaries constituted District 36. Eventually, District 36 encompassed the Minidoka–Twin Falls projects and the upper valley. This was a significant development for water politics because it joined, in the same administrative unit, the stored water users at Minidoka and Twin Falls and the natural flow users in the upper valley. Complicating the matter was the fact that each group had its own court decree; Minidoka–Twin Falls operated under a decree that decided the case of *Twin Falls Canal Co. et al. v. Charles N. Foster et al.,* 20 June 1913. The two decrees were finally combined under the 1929 Woodville stipulation, a joint agreement between the upper valley and the Minidoka–Twin Falls interests. Eventually, Water District 36 was redesignated District 1.

35. Robinson, *Ninth Biennial Report of the State Engineer, 1911–1912,* pp. 134–141.

36. See Wing, *Tenth Biennial Report of the State Engineer, 1913–1914,* pp. 59–62, for the account of the 1914 crisis.

37. Young, "Report of Watermaster, District No. 36, State of Idaho, Season 1915," Water District 1 Records (microfiche), IDWR; *Idaho Republican,* 30 July, 6, 13 August 1915; D. G. Martin, "Snake River Storage Project—Wyoming: Report on Distribution at Minidoka and Milner—Idaho, Season of 1916," and John Empey, "Watermaster's General Report for Water District No. 36, Season of 1917," both Water District 1 Records (microfiche), IDWR; Hoyt, *Water Utilization in the Snake River Basin,* 40–41; Barry Dibble, "Jackson Lake Reservoir," *Reclamation Record* 12 (September 1921): 423–424.

38. By 1914, twenty-six irrigation companies in the upper valley had joined the FPIA. "The board of directors," reported a local newspaper, "constitute a working unit to look after the rights of the greatest or [least] of the companies, and have prevented the undue waste of water, the unjust division of it, and in various ways have protected against water famine." See *Idaho Republican,* 14 August 1914. On the further consolidation of upper Snake River management, see John Empey, "Watermaster's General Report on Water District #36, Season of 1916," John Empey, "Watermaster's General Report

for Water District No. 36, Season of 1917," and John Empey, D. G. Martin, and C. F. Elford, "Report on Canal Deliveries of Natural and Stored Water from Snake River Between Jackson Lake, Wyoming and Milner, Idaho, for the Period July 1st to September 21st, Inclusive, 1918," all Water District 1 Records (microfiche), IDWR.

39. W. G. Swendsen, *First Biennial Report of the Department of Reclamation, State of Idaho, 1919–1920* (Boise: n.p., n.d.), 9–10; *Idaho Falls Times,* 1 May 1919; W. G. Swendsen, "Distribution of Snake River Water, District No. 36," (c. 1922), and G. Clyde Baldwin to Water Users' Committee of Nine, District 36, 16 February 1923, both file 310.2, Minidoka Project Records, RG 115.

40. *Idaho Falls Times,* 1 May 1919; *Idaho Republican,* 10, 24 June, 4 July 1919; Lynn Crandall, "Irrigation History Given," undated newspaper clipping in Idaho–Irrigation vertical file, Idaho Room. See also Stearns et al., *Geology and Ground Water Resources of the Snake River Plain,* 10–12; R. W. Faris, *Ninth Annual Report of the Department of Reclamation, State of Idaho, 1935–1936* (n.p.: n.p., n.d.), 17.

41. "Distribution of Snake River Water During Greatest Drought," *Engineering News-Record* 85 (11 November 1920): 927–931; *Idaho Republican,* 10, 24 June, 4, 8, 11, 15 July 1919.

42. "Distribution of Snake River Water During Greatest Drought"; *Idaho Republican,* 10, 24 June, 4, 8, 11, 15, 22 July 1919.

43. "Distribution of Snake River Water During Greatest Drought"; *Idaho Republican,* 8, 11, 22 July 1919.

44. *Idaho Republican,* 19 August 1919.

45. *Idaho Republican,* 6 August 1920.

46. *Idaho Republican,* 6 August 1920; *Rigby Star,* 24 November 1921; *Twin Falls Times-Register,* 12 July 1923, clipping in Water District 1 Records (microfiche), IDWR.

47. *Rigby Star,* 24 November 1921. At least one other organization formed in the upper Snake River valley around this time; the Water Users Association of the North Fork of Snake River represented more than two dozen canals on that stream. See Water Users Association of the North Fork of Snake River, Meeting Minutes, 29 July 1920, Water District 1 Records (microfiche), IDWR.

48. Warren Swendsen to F. J. Cowen, 28 March 1922, Water District 1 Records (microfiche), IDWR; *Rigby Star,* 21 November 1921, 6 April 1922. On Cotton's relationship to the Idaho Irrigation District, see *Idaho Register,* 21 June 1912, 25 April 1919. Cotton served as special deputy in charge of stored water during several seasons, including 1909, 1913, and 1914. See *Idaho Register,* 3 August 1909, and Wing, *Tenth Biennial Report of the State Engineer, 1913–1914.*

49. Cotton's essential points appeared in the following sources: "Water-Users Union Checks Up Distribution of Natural Flow Waters in Snake

River," *Idaho Falls Daily Post,* 4 May 1922, "Cotton Says Larger Per Cent Transmission Loss Should Be Charged Up to Storage Water," ibid. [c. May 1922?], "Natural Water Users to Start Establishing Some Rights in Big Question of Return Flow," ibid., 14 May 1922, "Says Natural Water Users Lose $80,000 Thru Evaporation," ibid., 2 June 1922, "Report of Meeting of Water-Users Held at St. Anthony Friday," [ibid.], 6 December 1922, "Water Users Review Accomplishments of the Past Year," *Times-Register,* 29 December 1922, all newspaper clippings in Water District 1 Records (microfiche), IDWR; Swendsen, "Distribution of Snake River Water, District No. 36"—the same document, titled "The Administration of Snake River," is located in Water District 1 Records (microfiche), IDWR; *Idaho Republican,* 25 January 1923.

50. Swendsen, "Distribution of Snake River Water, District No. 36," and G. Clyde Baldwin to Water Users' Committee of Nine, District 36, 16 February 1923, both file 310.2, Minidoka Project Records, RG 115.

51. Baldwin to Water Users' Committee of Nine, 16 February 1923.

52. W. G. Swendsen, "Re: District No. 36, " 19 January 1923, Water District 1 Records (microfiche), IDWR; *Idaho Republican,* 23 January 1923. See also the following documents in Water District 1 Records: John Lee, minutes of special meeting of District 36 water users, 30 January 1923; "Report of the Meeting of Water Users of District No. 36," [January 1923?], newspaper clipping; G. Clyde Baldwin to W. G. Swendsen, 31 January 1923. See also *Twin Falls Weekly News,* 3 February 1923, and *Idaho Falls Daily Post,* 3 February 1923. The plans for managing water for the 1923 season are mentioned in "Minutes of annual watermaster election of Water District No. 36," 3 March 1924, Water District 1 Records (microfiche), IDWR. Swendsen's meeting headed off the growing discontent of upper valley irrigators. See "Report of Meeting of Water-Users Held at St. Anthony Friday," [*Idaho Falls Daily Post?*], 6 December 1922, and "Water Users Review Accomplishments of the Past Year," *Times-Register,* 29 December 1922, both newspaper clippings in Water District 1 Records (microfiche), IDWR; *Rigby Star,* 14 December 1922.

53. *Idaho Republican,* 15 February 1923. See also "Minutes of Annual Water Master Election for Water District No. 36, Held . . . March 5 1923," and "A special meeting of the Committee of Nine . . . " (5 March 1923), both file 310.2, Minidoka Project Records, RG 115.

54. R. I. Meeker, "Report: Snake River Water Problems, Concerning Operation, Jackson Lake Reservoir, Transmission Losses Reservoir Water, Water District 36, Idaho," 9 February 1924, file 310.2, Minidoka Project Records, RG 115. Meeker's report is also summarized in *Rigby Star,* 13 March 1924, clipping, ibid. Cf. Daniel Tyler, "Delph E. Carpenter and the Principle of Equitable Apportionment," *Western Legal History* 9 (Winter/Spring 1996): 35–53.

55. Meeker, "Report: Snake River Water Problems"; B. E. Stoutemyer to Chief Counsel, 20 March 1924, E. B. Darlington to Commissioner [of Recla-

mation] (hereafter cited as CR), 3 August 1924, and Stoutemyer to CR, 20 December 1924, all file 310.2, Minidoka Project Records, RG 115. Cf. Tyler, "Carpenter and the Principle of Equitable Apportionment."

56. Darlington to CR, 3 August 1924; Stoutemyer to CR, 20 December 1924. For the controversies, politics, and compromises surrounding the agreements of 1924 and subsequent years, see also G. Clyde Baldwin to W. G. Swendsen, 6 June 1924, and "Minutes of a Special Meeting of Water Users Committee of Nine," 5 July 1924, both Water District 1 Records (microfiche), IDWR; Darlington to CR, 14, 23 August 1924, William Treichert to Hubert Work, 30 July 1924, H. A. Baker to Work, 9 August 1924, C. H. Burgher to Finney, 11 August 1924, S. J. Hawkins to Work, 8 August 1924, Stoutemyer to CR, 3 March 1925, Darlington to Chief Engineer, 7 March 1925, and to CR, 2 April 1926, Stoutemyer to CR, 24 May 1926, all file 310.2, Minidoka Project Records, RG 115. Although upper valley irrigators at first objected to Meeker's advice that they arrive at a compromise in water distribution, they were not displeased with the agreements that the Committee of Nine worked out. On the other hand, certain farmers in the Minidoka–Twin Falls area felt that the agreements slighted them, and the Committee of Nine then had to work out new arrangements to satisfy them.

57. E. J. Iddings, "Neighborliness Helps Idaho Over a Critical Year," *Idaho Farmer* 34 (14 August 1924): 103; *Idaho Farmer* 35 (19 March 1925): 268; Stoutemyer to CR, 24 May 1926.

58. For a subsequent example of successful negotiation, see Darlington to Chief Engineer, 7 March 1925. For droughts, dam construction, and future water management problems, see Gertsch, "The Upper Snake River Project"; Crandall, "Snake River"; and Hoyt, *Water Utilization in the Snake River Basin.* One of the worst years for water distribution would be 1934. See Crandall, "Irrigation History Given"; *Idaho Sunday Statesman,* 30 September 1934, in Irrigation vertical file, Idaho Historical Society Archives, Boise (hereafter cited as IHS); *Gooding Leader,* 11 April 1935, Snake River vertical file, IHS; "Annual Report of the Committee of Nine to the Water-Users of District No. 36," 4 March 1935, file 310.2, Minidoka Project Records, RG 115; Lynn Crandall, "History of Irrigation Development in the Snake River Valley." For an excellent history of the movement to build American Falls Dam, which grew out of the water shortages of the 1910s and 1920s, especially 1919, see Gertsch, "The Upper Snake River Project." For a brief analysis of the American Falls Reservoir complicating the hydrology and thus the distribution of Snake River water, see Lynn Crandall, "Ground Water Inflow to American Falls Reservoir," July 1959, in Box 2, Lynn Crandall Papers, American Heritage Center, University of Wyoming, Laramie.

59. For a critique of interpretations that emphasize the perpetual fragmentation of American society, see Thomas Bender, *Community and Social*

Change in America (New Brunswick, N.J.: Rutgers University Press, 1978; rpt. Baltimore: Johns Hopkins University Press, 1982). The notion of a "moral economy" is now widely used but derives from E. P. Thompson, "The Moral Economy of the English Crowd in the Eighteenth Century," *Past and Present* 50 (February 1971): 76–136. The practice of temporary transfers was an outgrowth of the social relationships among the people of the upper Snake River valley. "The practice of temporary transfers, or exchange of water between canals," wrote E. B. Darlington in 1926, "is one of long standing in the upper valley. It is practically a system of rotation, whereby at certain periods [owners of] some of the canals close their head gates and turn over . . . water to which they are entitled to other users. In a community like the upper valley, where there is a close relationship between many of the canal companies and districts, this practice results in a more general and beneficial use of water and higher crop returns." Darlington to CR, 2 April 1926. The study of water in the American West has placed too much emphasis on prior appropriation and too little on the sorts of cooperative, extralegal conventions that I have described in this chapter. For an example of a heavy emphasis on prior appropriation, see Charles Wilkinson, "West's Grand Old Water Doctrine Dies," *High Country News* 23 (12 August 1991): 1, 11–12.

60. Donald Worster contends that water conflict encouraged the growth of a centralized government agency, a Hobbesian Leviathan, that would end the anarchy and distribute the water. See Worster, *Rivers of Empire,* 92–96. On the expansion of Baldwin's power and his thoughts about the Committee of Nine and upper Snake River management, see G. Clyde Baldwin to CR, 31 January 1923, Water District 1 Records (microfiche), IDWR, and to Water Users' Committee of Nine, 16 February 1923; *Idaho Republican,* 8 March 1923; F. A. Banks, "Minutes of meetings held at Idaho Falls, District 36," 14 March 1923, file 310.2, Minidoka Project Records, RG 115.

61. For background on these kinds of cooperative organization, see Leonard Arrington, *Great Basin Kingdom: An Economic History of the Latter-day Saints, 1830–1900* (Cambridge, Mass.: Harvard University Press, 1958; rpt. Lincoln: University of Nebraska Press, 1966); Allan G. Bogue, "The Iowa Claims Clubs: Symbol and Substance," in *The Public Lands: Studies in the History of the Public Domain,* ed. Vernon Carstensen (Madison: University of Wisconsin Press, 1963), 47–69; Samuel P. Hays, *The Response to Industrialism, 1885–1914* (Chicago: University of Chicago Press, 1957); Ellis W. Hawley, *The Great War and the Search for a Modern Order: A History of the American People and Their Institutions, 1917–1933* (New York: St. Martin's Press, 1979). The *Idaho Republican,* 10 June 1919, appealed to cooperative values while recalling wartime conservation efforts: "During the war the American people, our valley included, conserved on everything and were surprised at themselves at

what they could do. This season it will probably be necessary to take the water out of our own ditches and give it to people who have no legal right to it."

4/ LABOR AND LANDSCAPE

1. *Idaho Statesman* (Boise), 17, 22, 29, 31 May 1877.

2. *Idaho Statesman,* 31 May 1877. See also *History of Idaho Territory, Showing Its Resources and Advantages; With Illustrations Descriptive of Its Scenery, Residences, Farms, Mines, Mills, etc.* (San Francisco: Wallace W. Elliot and Company, 1884), 191.

3. Few historians have examined the intersection of nature and work. Two important exceptions are Richard White, *The Organic Machine: The Remaking of the Columbia River* (New York: Hill and Wang, 1995), and Mart A. Stewart, *"What Nature Suffers to Groe": Life, Labor, and Landscape on the Georgia Coast, 1680–1920* (Athens: University of Georgia Press, 1996). Leonard Arrington and Dean May note the paucity of sources that discuss the skill and labor involved in small-scale irrigation in the American West; see "'A Different Mode of Life': Irrigation and Society in Nineteenth-Century Utah," *Agricultural History* 49 (1975): 2–20.

4. For capsule histories of Boise Valley canals, see H. H. Caldwell and Merle Wells, *Economic and Ecological History Support Study: A Case Study of Federal Expenditures on a Water and Related Land Resources Project: Boise Project, Idaho and Oregon* (Moscow: Idaho Water Resources Research Institute, 1974), 135–157.

5. Kate B. Carter, ed., *Pioneer Irrigation: Upper Snake River Valley* (Salt Lake City: Daughters of Utah Pioneers, 1955), 220–225. Carter's compilation is the single best source on upper Snake River valley irrigation.

6. For sources that provide general information on the development of the small companies, see Carter, *Pioneer Irrigation;* Paul Lloyd Murphy, "Irrigation in the Boise Valley, 1863–1903: A Case Study in Pre-federal Irrigation," M.A. thesis (University of California, 1947); and U.S. Congress, *Report of the Special Committee of the United States Senate on the Irrigation and Reclamation of Public Lands,* Vol 1: *The Northwest* (Washington, D.C.: Government Printing Office, 1890), 315–415. See also Robert G. Dunbar, *Forging New Rights in Western Waters* (Lincoln: University of Nebraska Press, 1983), and Wells A. Hutchins, *Mutual Irrigation Companies,* U.S. Department of Agriculture, Technical Bulletin 82 (Washington, D.C.: Government Printing Office, January 1929). For examples of the democratic basis of the home companies, see *Report of the Senate on Irrigation and Reclamation,* Vol. 1: *The Northwest,* 321–324; M. J. Hammond, "The Largest, the Best, and the Cheapest Irrigation System in the United States," *Rigby Star,* 26 August 1909. Goulder's state-

ment appeared in the *Idaho Statesman,* 17 May 1877. See also "Life Story of Michael Kimmer Hammer," MC 21-2-48, Special Collections Department, Eli M. Oboler Library, Idaho State University, Pocatello (hereafter cited as Oboler).

7. For examples of early irrigation construction in relation to natural topography, see Hammond, "The Largest, the Best, and the Cheapest Irrigation System in the United States"; Carter, *Pioneer Irrigation,* 6, 8, 10–11, 13, 21, 46, 52, 71, 82, 92, 98, 101, 122, 155, 170–171, 176, 177–178, 228, and passim; *Idaho Statesman,* 17, 22, 29, 31 May 1877; Caldwell and Wells, *Economic and Ecological History Support Study,* 135–157.

8. Few sources discuss in detail the planning and construction of small canals. For examples of folk design, see Carter, *Pioneer Irrigation,* 12, 25, 35–36, 100, 105 (quotation), 123, 161, 256–258, and passim.

9. Descriptions of the details of early ditch construction are meager. But see, for example, Carter, *Pioneer Irrigation,* 8–10, 11–14, 18–20, 67, 74–75, 78, 80, 106, 118–119, 168–169, 218, 223–224, 232, 247–248; and Merrill D. Beal and Merle W. Wells, *History of Idaho,* Vol. 2 (New York: Lewis Historical Publishing Company, 1959), 126–131.

10. Brief descriptions of these and other small canals can be found in Caldwell and Wells, *Economic and Ecological History Support Study,* 135–157, and Carter, *Pioneer Irrigation,* 17 (quotation) and passim.

11. Carter, *Pioneer Irrigation,* 81 and passim.

12. Names of early small canals can be found in Caldwell and Wells, *Economic and Ecological History Support Study,* 135–157; Carter, *Pioneer Irrigation;* and *Report of the Senate on Irrigation and Reclamation,* Vol. 1: *The Northwest,* 315–415.

13. Background on the transition from family to industrial mode can be gleaned from the many irrigation system histories in Carter, *Pioneer Irrigation.* On irrigation districts, see Wells A. Hutchins, *Irrigation Districts: Their Organization, Operation, and Financing,* Department of Agriculture, Technical Bulletin 254 (Washington, D.C.: Government Printing Office, June 1931). See also William F. Ringert, "Irrigation Districts: Purpose, History, Funding, and Problems," *Idaho Yesterdays* 30 (Spring/Summer 1986): 64–70.

14. O. K. Meservy, "Enterprise Irrigation District History" [ca. 1967], Enterprise Irrigation District Records, Enterprise Irrigation District, Teton, Idaho (hereafter cited as Enterprise); Carter, *Pioneer Irrigation,* 255.

15. Meservy, "Enterprise Irrigation District History"; "Circular Letter for the Benefit of Enquiring Parties," 8 June 1912, and Minute Book of Enterprise Irrigation District, 5 November 1928, both Enterprise; Carter, *Pioneer Irrigation,* 254–255; A. E. Robinson, *Ninth Biennial Report of the State Engineer to the Governor of Idaho, 1911–1912* (n.p.: n.p., n.d.), 76–77.

16. Meservy, "Enterprise Irrigation District History"; Minute Book of

Enterprise Irrigation District, 5 November 1928; Carter, *Pioneer Irrigation,* 254–255.

17. For overviews of large private, Carey Act, and federal government irrigation projects, see Neil H. Carlton, "A History of the Development of the Boise Irrigation Project," M.A. thesis (Brigham Young University, 1969); John Francis, ed., *Sagebrush to Cropland: A Western Water Project* (Moscow: Idaho Water Resources Research Institute, University of Idaho, March 1979); Murphy, "Irrigation in the Boise Valley"; William Darrell Gertsch, "The Upper Snake River Project: A Historical Study of Reclamation and Regional Development, 1890–1930," Ph.D. dissertation (University of Washington, 1974); Hugh Lovin, "Water, Arid Land, and Visions of Advancement on the Snake River Plain," *Idaho Yesterdays* 35 (Spring 1991): 3–18; Mikel H. Williams, *The History of Development and Current Status of the Carey Act in Idaho* (Boise: Idaho Department of Reclamation, 1970).

18. Few sources provide detailed descriptions of the labor conditions on irrigation construction projects. See Gladys R. Joslin, "Early Burley Pioneer Days Recalled: Minidoka Project History Is Related Here," *Burley Bulletin,* 1 May 1945 (Joslin was a Bureau of Reclamation clerk at Burley); Harold Rhodenbaugh, "Brawny Armies Built Canals That Conquered Desert," *Idaho Statesman,* 7 April 1929, and Alma Hanson, "Hell Raisers on the Payette Canal," *Frontier Times* (April-May 1965), both in Irrigation vertical file, IHS. For the sketchy details of a strike on the Idaho Canal and Irrigation Company system: *Idaho Falls Times,* 22 May 1906. Carter, *Pioneer Irrigation,* occasionally discusses work on the larger systems. On numbers of men and draft animals at work on the Twin Falls Land and Water Company's system, see "Work At Twin Falls Canal," [*Journal*? 29 May 1903?], in untitled large scrapbook at Twin Falls County Historical Society Museum, Twin Falls, Idaho; and *Twin Falls News,* 28 October 1904. According to these reports, Nelson Bennett and Company of Tacoma, the primary contractor on canal construction, was then employing between 400 and 500 men and 500 horse teams, while Feris and Kesl of Boise had 75–80 men at work on Milner Dam. See also Elliot Paul's *Desperate Scenery* (New York: Random House, 1954), 180 (quotation), a gritty but humorous memoir of one man's employment with the U.S. Reclamation Service on the construction of Jackson Lake dam. It is the single best source on large-scale construction associated with irrigated agriculture.

19. See Carlos Schwantes's superb account of transient laborers in the Pacific Northwest during the late nineteenth and early twentieth centuries: *Hard Traveling: A Portrait of Work Life in the New Northwest* (Lincoln: University of Nebraska Press, 1994), xi. For mention of Japanese and other ethnic laborers, see *Reports of the Immigration Commission: Immigrants in Industries,* Part 25, *Japanese and Other Immigrant Races in the Pacific Coast and Rocky Moun-*

tain States, Vol. 1: *Japanese and East Indians,* 61st Cong., 2d Sess., 1911, Senate Document 633, p. 319; Joslin, "Early Burley Pioneer Days Recalled"; and Paul, *Desperate Scenery,* 199, 202–203.

20. Alfred Peters, "Early History of the Twin Falls Canal Company and My Adult Life With the Company," file 1511.01, History of the Twin Falls Canal Company, Twin Falls Canal Company Records, Twin Falls, Idaho (hereafter cited as TFCC). Even the National Labor Relations Board recognized the nonagricultural character of canal work. In 1952, the NLRB ruled that the Twin Falls Canal Company workers could hold an election to decide upon their representation by the Teamsters. The union lost by only a few votes. See Frank L. Stephan to the Stockholders of Twin Falls Canal Company, 9 January 1951, 8 January 1952, both in "Minutes of Annual Meeting, 1951–1955," and "Minutes of the Annual Meeting of the Stockholders of the Twin Falls Canal Company," 13 January 1953, all TFCC.

21. Carter, *Pioneer Irrigation,* 15, 18, 55–57, passim.

22. Ibid., 42; Richard Lowitt and Judith Fabry, eds., *Henry A. Wallace's Irrigation Frontier: On the Trail of the Corn Belt Farmer, 1909* (Norman: University of Oklahoma Press, 1991), 151, 155, 160; Alvin C. Holmes, *Swedish Homesteaders in Idaho on the Minidoka Irrigation Project* (n.p.: n.p., 1976), 40, 41, 82, 83, 109. See also George H. Bliss, "Organization of Irrigation Operating and Maintenance Force and Its Training," *Engineering and Contracting* 41 (6 May 1914): 535; and P. M. Fogg, "A History of the Minidoka Project, Idaho, to 1912 Inclusive," August 1915, Vol. 1: 29, 50, Roll 96, Microcopy M96, Minidoka Project, Idaho, 1903–1919, Record Group 115, Records of the Bureau of Reclamation, National Archives, Washington, D.C. (hereafter cited as RG 115). On the Twin Falls Land and Water Company project, contracts stipulated that settlers be given preference in hiring. Also, the company claimed that it would allow settlers to bid on lateral construction projects. See "Twin Falls Land and Water Company," an informational brochure from the early 1900s located at TFCC.

23. D. W. Ross to Chief Engineer, U.S. Reclamation Service, 20 February 1907, F. E. Weymouth and H. M. Shilling to Reclamation Commission, 21 January 1915, both file 140, Minidoka Project Records, General Administrative and Project Files, 1902–1919, RG 115; Gerard H. Matthew to Ross, 11 March 1907, Ross to Chief Engineer, 20 March 1907, B. E. Stoutemyer to Supervising Engineer, 23 April 1910, C. H. Paul to Director, 2 May 1910, all file 30, ibid. See also Fogg, "A History of the Minidoka Project," 75–76, 136–137; F. E. Weymouth, "Methods of Operation and Management," in "Report of Conference of Operating Engineers for Irrigation Canal Systems Located in Idaho, Oregon, and Washington, November 1911," pp. 6–9, Roll 23, Microcopy M96, Project Histories and Reports of Reclamation Bureau Projects, 1905–1925, RG 115; and Charles Coates, "Federal-Local Relation-

ships on the Boise and Minidoka Projects, 1904–1926," *Idaho Yesterdays* 25 (Summer 1981): 2–9. By 1915, the Reclamation Service had taken over 191 miles of laterals on the north unit of the Minidoka project, leaving only 21 miles still under the farmers' control.

24. For examples of settlers who worked on railroads before or while they operated an Idaho irrigated farm, see Carter, *Pioneer Irrigation,* 23, 26–27, 57–58, 67, 75, 124, 147, 173. See also Leonard Arrington, "Recalling a Twin Falls Childhood," *Idaho Yesterdays* 25 (Winter 1981): 31–40, for the story of a Mormon farm family whose male members worked on the railroad and at a sugar factory near Twin Falls. On the Mormon agrarian impulse during the late nineteenth and early twentieth centuries, see Chas Peterson to Stanley L. Ponce, 28 December 1992, letter on file at the Utah State Historic Preservation Office, Salt Lake City.

25. E. B. Darlington, "Preparation of Land for Irrigation and Best Methods of Applying Water, Economical Heads, etc.," in "Report of Third Annual Conference of Operating Engineers, Held in Boise, Idaho, February 3, 4, and 5, 1914," p. 28, Roll 23, Microcopy M96, Project Histories and Reports of Reclamation Bureau Projects, 1905–1925, RG 115.

26. Ibid. For other discussions of field irrigation techniques, see *Idaho Statesman,* 23 June, 7 July 1874, 17, 29 May 1877; *Caldwell Tribune,* 6 September 1890; H. G. Raschbacher, *Irrigation From Snake River, Idaho,* U.S. Department of Agriculture, Office of Experiment Stations, Circular 65 (Washington, D.C.: Government Printing Office, 1906), 9–11; "Prize Winning Irrigator and His Methods," *Gem State Rural* (Caldwell) 11 (8 November 1906): 2; *Twin Falls News,* 25 November 1904, 18, 25 May, 7 September, 26 October 1906.

27. Darlington, "Preparation of Land for Irrigation," 28.

28. Carter, *Pioneer Irrigation,* 115–116. For another example, see *Twin Falls News,* 4 August 1905.

29. "Eilers Family History," MC 21-2-33, Oboler.

30. For McPherson on rotation, see *Twin Falls News,* 25 May 1906. See also "Annual Project History, Report of Construction and Operation, Minidoka Project—Idaho, Volume X, Year of 1916," p. 109, Roll 99, Microcopy No. M96, Minidoka Project, Idaho, 1903–1919, RG 115.

31. "Personal Journal of Delbert Fowler," MC 21-1-21, Oboler.

32. For one engineer's argument against continuous flow and for rotation, see D. W. Ross, *Biennial Report of the State Engineer to the Governor of Idaho for the Years 1899–1900* (Boise: Capital Printing Office, n.d.), 81–89. For an example of the efficient, timed, measured, industrial-style method of water distribution, see E. B. Darlington, "Use of Water on the Salmon River Tract," *Reclamation Record* 9 (May 1918): 225–228. Useful in thinking about this aspect of water management is E. P. Thompson, "Time, Work Discipline, and Industrial Capitalism," *Past and Present* 38 (December 1967): 56–97.

33. On the labor requirement of wheat and other crops, see Byron Hunter and S. B. Nuckols, *Farm Costs and Relative Profitableness of Seven Crops, Twin Falls County, Idaho, 1919 and 1920,* University of Idaho Agricultural Experiment Station, Research Bulletin 2 (Moscow: University of Idaho, March 1922), 7; Byron Hunter, *Preliminary Report on Farm Organization in Twin Falls and Latah Counties,* University of Idaho Agricultural Experiment Station, Bulletin 123 (Moscow: University of Idaho, February 1921), 7; Byron Hunter and Samuel B. Nuckols, *An Economic Study of Irrigated Farming in Twin Falls County, Idaho,* U.S. Department of Agriculture, Bulletin 1421 (Washington, D.C.: Government Printing Office, October 1926), 53–57.

34. For descriptions of the cooperative work arrangements of the wheat harvest, see "Eilers Family History," MC 21-2-33, and "The Life Story of Charles Clifford Robinson," MC 21-6-186, both Oboler; Paul A. Eke and Neil W. Johnson, *Profitable Systems of Farming for the Idaho Falls Area,* University of Idaho Agricultural Experiment Station, Bulletin 198 (Moscow: University of Idaho, June 1933), 25; Carter, *Pioneer Irrigation,* 59; Gwen Berrett Fillmore et al., *Menan, Idaho, 1879–1986* (n.p.: n.p., n.d.), 39; ZoAnn Simmons, *Iona Centennial History Book, 1883–1983* (n.p.: n.p., n.d.), 17. Newspapers occasionally mentioned farmer-owned or custom threshers. According to the *Rigby Star,* 3 October 1907, the East LaBelle Threshing Company took delivery of a horse-powered thresher. An account of a steam engine, owned by the threshermen "Ed. Jones and O. Sayer," collapsing a bridge over the Rigby Canal is found in *Rigby Star,* 17 October 1907. The *Idaho Republican* (Blackfoot), 30 September 1919, told the story of a farmer killed in a thresher accident while working on the farm of a neighbor. Thomas D. Isern, *Bull Threshers and Bindlestiffs: Harvesting and Threshing on the North American Plains* (Lawrence: University Press of Kansas, 1990), describes the organization and operation of threshing outfits on the Great Plains.

35. Annie Pike Greenwood's account of threshing appears in her memoir, *We Sagebrush Folks* (New York: D. Appleton-Century, 1934; rpt. Moscow: University of Idaho Press, 1988), 173–180. Isern, *Bull Threshers and Bindlestiffs,* includes discussion of cooperative threshing "rings" and female labor.

36. Isern discovered general but not absolute differences in threshing and harvesting techniques depending on farm sizes and conditions. Farms with smaller acreages, more level ground, and spring wheat tended to use binders to cut wheat. Large winter-wheat farms with uneven ground, usually the type of operation found on the far western Great Plains, used headers for the harvest. Similarly, farmers located close together and with small patches of grain usually resorted to some form of cooperative labor, while pure custom threshing predominated on the large, dispersed farms of the Far West. See *Bull Threshers and Bindlestiffs,* 26–30, 74–76. Harvesting of Idaho's irrigated wheat—small patches, often spring wheat, on level ground—

generally relied on the use of the binder and cooperative labor, while on nearby large, hilly, dry farms the header prevailed. See *Parker Centennial Chronicles, 1884–1984* (Parker, Idaho: Parker Centennial Committee, 1984), 324; and S. J. Rich, *Commissioner of Immigration, Labor, and Statistics, Seventh Biennial Report, 1911–1912* (Caldwell, Idaho: Caxton Printers, [1912?]), 106.

37. *Rigby Star,* 3 October 1907, 20 August 1908, 28 April 1909. An engineer surveyed Lee's farm and its irrigation ditches, according to the *Idaho Register* (Idaho Falls), 5 October 1906. The paper also reported that the work of building ditches and clearing land, directed by Lee, "goes on with as precise a systemization as if it were a piece of railroad construction." For examples of other big farmers and labor systems contemporary with Lee, see *Idaho Farmer* (Caldwell) 30 (3 August 1922): 68.

38. For overviews of sugar beet production in Idaho and the mountain West, see Leonard Arrington, *Beet Sugar in the West: A History of the Utah-Idaho Sugar Company* (Seattle: University of Washington Press, 1966); J. R. Bachman, *Story of Amalgamated Sugar Company, 1897–1961* (n.p.: Amalgamated Sugar Company, 1962); Fred G. Taylor, *A Saga of Sugar* (n.p.: Utah-Idaho Sugar Company / Deseret News Press, 1944). For details of how the sugar companies established themselves in Idaho, see Arrington, *Beet Sugar in the West,* 3–39, 54–73, 184–186; Charles F. Saylor, *Progress of the Beet-Sugar Industry in the United States in 1904,* U.S. Department of Agriculture, Report 80 (Washington, D.C.: Government Printing Office, 1905), 54–55; Charles F. Saylor, *Progress of the Beet-Sugar Industry in the United States in 1909,* U.S. Department of Agriculture, Report 92 (Washington, D.C.: Government Printing Office, 1910), 30, 37; F. S. Harris, *The Sugar Beet in America* (New York: Macmillan Company, 1919), 312. For sugar beet acreages, see U.S. Department of Commerce, Bureau of the Census, *Thirteenth Census of the United States,* 1910, Vol. 6: *Agriculture* (Washington, D.C.: Government Printing Office, 1913), 405, and *Fourteenth Census of the United States,* 1920, Vol. 6, Part 3: *Agriculture* (Washington, D.C.: Government Printing Office, 1922), 142.

39. For descriptions of the grueling labor necessary to thin, weed, and hoe sugar beets, see Harry Schwartz, *Seasonal Farm Labor in the United States* (New York: Columbia University Press, 1945), 103–104; Harris, *The Sugar Beet in America,* 45–46; and especially Lenora Braithwaite, "We Are of Yesterday," MC 21-5-156, Oboler.

40. Saylor, *Progress of the Beet-Sugar Industry, 1904,* pp. 17–20, 26 (quotation), 26–27, 42; Charles F. Saylor, *Progress of the Beet-Sugar Industry in the United States in 1907,* U.S. Department of Agriculture, Report 86 (Washington, D.C.: Government Printing Office, 1908), 11–12; Schwartz, *Seasonal Farm Labor in the United States,* 103–104; Harris, *The Sugar Beet in America,* 49–50, 52–53, 92–102; *Reports of the Immigration Commission,* Part 25, Vol. 1: *Japa-*

nese and East Indians, 94. For an example of farmers negotiating a contract, see *Reclamation Record* 12 (March 1921): 106.

41. Arrington, *Beet Sugar in the West,* 37–38, 60; *Reports of the Immigration Commission: Immigrants in Industries,* Part 25: *Japanese and Other Immigrant Races in the Pacific Coast and Rocky Mountain States,* Vol. 2: *Agriculture,* 61st Congress, 2d Session, 1911, Senate Document 633, 148. Saylor also pointed out the prosperity that sugar beet production seemed to bring: Saylor, *Progress of the Beet-Sugar Industry, 1904,* pp. 54–55, 110–112; Saylor, *Progress of the Beet-Sugar Industry, 1909,* pp. 30–31. See also *Idaho Republican,* 5, 22 September 1916, 5 January, 20 August 1917, 22 November 1918, and Harris, *The Sugar Beet in America,* 250–257. Of course, sugar factory workers (perhaps including some farmers), like other industrial laborers, sometimes called strikes; see *Idaho Republican,* 29 November 1918. See Leonard Arrington, "Recalling a Twin Falls Childhood," 31–32, for a story of railroad work, sugar factory work, and life on an Idaho irrigated farm.

42. Saylor, *Progress of the Beet-Sugar Industry, 1904,* pp. 110–112; Arrington, *Beet Sugar in the West,* 132–134.

43. *Gem State Rural* 8 (25 June 1903): 4.

44. Saylor, *Progress of the Beet-Sugar Industry, 1904,* pp. 111–112; *Reports of the Immigration Commission,* Part 25, Vol. 1: *Japanese and East Indians,* 70–71, 319, Vol. 2: *Agriculture,* 145–148; *Idaho Falls Times,* 25 May 1906. For a general overview of agricultural labor in Idaho, see Patricia K. Ourada, *Migrant Workers in Idaho* (Boise: Boise State University, 1980).

45. Schwartz, *Seasonal Farm Labor in the United States,* 107–110; *Reports of the Immigration Commission,* Part 25, Vol. 1: *Japanese and East Indians,* 70–71, Vol. 2: *Agriculture,* 145–148. Different labor contracts for each group reflected the effort on the part of Utah-Idaho Sugar to use the presence of the German Russians to control the Japanese workers. German Russians received a flat rate; Japanese received bonuses for greater effort but penalties for production below certain levels.

46. Schwartz, *Seasonal Farm Labor in the United States,* 110–111; Erasmo Gamboa, *Mexican Labor and World War II: Braceros in the Pacific Northwest, 1942–1947* (Austin: University of Texas Press, 1990), 7–8. See also *Idaho Farmer* 28 (13 October 1921): 274, and (20 October 1921): 291, for reports of Mexican laborers brought in by Amalgamated Sugar to harvest beets in the Twin Falls area.

47. *Reports of the Immigration Commission* Part 25, Vol. 2: *Agriculture,* 145–146, 298–310; *Idaho Republican,* 29 September 1916. For an excellent overview of Japanese and their involvement in Idaho sugar beet production, see Laurie Mercier and Carole Simon-Smolinski, eds., *Idaho's Ethnic Heritage: Historical Overviews,* 3 vols. (n.p.: Idaho Centennial Commission and National

Park Service, U.S. Department of the Interior, 1990), I: 44–66. For information on anti-Japanese laws and other particulars of Japanese settlement in Idaho, see also Robert C. Simms, "The Japanese-American Experience in Idaho," *Idaho Yesterdays* 22 (Spring 1978): 2–10. By 1920, 129 Japanese farmers were operating in Idaho; *Idaho Farmer* 27 (21 April 1921): 416.

48. *Reports of the Immigration Commission,* Part 25, Vol. 2: *Agriculture,* 145–148, 297–298; Schwartz, *Seasonal Farm Labor in the United States,* 107 (quotations); Greenwood, *We Sagebrush Folks,* 181.

49. Mercier and Simon-Smolinski, *Idaho's Ethnic Heritage,* I:63–66; Saylor, *Progress of the Beet-Sugar Industry, 1904,* pp. 36–38; *Reports of the Immigration Commission,* Part 25, Vol. 2: *Agriculture,* 145–147.

50. Arrington, *Beet Sugar in the West,* 169; *Idaho Farmer* 24 (3 July 1919): 7; 31 (23 August 1923): 123; Carter, *Pioneer Irrigation,* 120–122.

51. Schwartz, *Seasonal Farm Labor in the United States,* 103.

52. For an example of irrigation as the basis for regenerating the American republic, see William F. Smythe, *The Conquest of Arid America* (New York: Harper and Brothers, 1899, 1905; rpt. Seattle: University of Washington Press, 1969).

53. Isern, *Bull Threshers and Bindlestiffs,* 174–209; Idaho State Employment Service, *Newsletter,* 1 September 1937; Carter, *Pioneer Irrigation,* 235–237; Peters, "Early History and My Adult Life With the Company"; Simmons, *Iona Centennial History Book,* 25.

54. Arrington, *Beet Sugar in the West,* 141–155; Wayne D. Rasmussen, "Technological Change in Western Beet Sugar Production," *Agricultural History* 41 (1967): 31–35; *Idaho Register,* 5 June 1908; *Idaho Republican,* 24 September, 6 October, 5 November 1918, 7 October 1919. See also Hobart Beresford, *March of Mechanization of Sugar Beet Production in Idaho,* University of Idaho Agricultural Experiment Station, Circular 111 (n.p.: n.p., July 1946).

55. Rasmussen, "Technological Change in Western Beet Sugar Production," 33–35; Arrington, *Beet Sugar in the West,* 147–151.

56. For an insightful discussion of crop seed and agricultural labor, see Deborah Fitzgerald, "Farmers Deskilled: Hybrid Corn and Farmers' Work," *Technology and Culture* 34 (April 1993): 324–343.

5/ FROM FIELD TO MARKET

1. A. G. Seiler, "The Story of An Irrigated Farm," *Reclamation Record* 6 (April 1915): 192.

2. According to Gilbert Fite, "The years between the Spanish-American War and World War I became known as the Golden Age of American agri-

culture." Fite, *American Farmers: The New Minority* (Bloomington: Indiana University Press, 1981), 17. Although the Reclamation Service and its farmers were encountering various social, economic, and environmental problems by this time, crop values were rising, new settlers were still opening homesteads on Reclamation Service projects, and the amount of acreage that the agency served was still expanding. The Reclamation Service's greatest troubles lay in the future, with the farm depression of the 1920s. See Paul Wallace Gates, *History of Public Land Law Development* (Washington, D.C.: Government Printing Office, 1968), 654–698.

3. "Report of Construction and Operation (Annual Project History), Minidoka Project—Idaho, Vol IX, Year of 1915," p. 308, Roll 98, Microcopy M96, Minidoka Project, Idaho, 1903–1919, Record Group 115, Records of the Bureau of Reclamation, National Archives, Washington, D.C. (hereafter cited as RG 115); *Idaho Farmer* (Caldwell) 22 (8 March 1917): 12.

4. *Idaho Republican* (Blackfoot), 25 June 1917; *Idaho Farmer* 24 (12 December 1918): 6; Annie Pike Greenwood, *We Sagebrush Folks* (New York: D. Appleton-Century, 1934; rpt. Moscow: University of Idaho Press, 1988), 11.

5. On Idaho's early economy and agriculture, see: Carlos A. Schwantes, *In Mountain Shadows: A History of Idaho* (Lincoln: University of Nebraska Press, 1991), 49–52, 77–103; Dean L. May and Jenny Cornell, "Middleton's Agriminers: The Beginnings of an Agricultural Town," *Idaho Yesterdays* 28 (Winter 1985): 2–11; Paul L. Murphy, "Early Irrigation in the Boise Valley," *Pacific Northwest Quarterly* 44 (October 1953): 177–184; *History of Idaho Territory, Showing Its Resources and Advantages: With Illustrations Descriptive of Its Scenery, Residences, Farms, Mines, Mills, etc.* (San Francisco: Wallace W. Elliot and Company, 1884); "A Look at the Boise Valley," *Idaho Yesterdays* 28 (Winter 1985): 12–19; and the correspondence of W. A. Goulder in the *Idaho Statesman* (Boise), 17, 22, 29, 31 May 1877. On railroads and Idaho agriculture, see James L. Ehernberger and Francis G. Gschwind, *Smoke down the Canyon: Union Pacific, Idaho Division* (Callaway, Neb.: E. and G. Publications, 1966), 6–19; and William W. Kratville's classic study of the Union Pacific, *Golden Rails* (Omaha: Kratville Publications, 1965), 189–213. On the evolution of American dietary habits and the "beef and potatoes syndrome," see Harvey A. Levenstein, *Revolution at the Table: The Transformation of the American Diet* (New York: Oxford University Press, 1988).

6. U.S. Department of Commerce, Bureau of the Census, *Thirteenth Census of the United States,* 1910, Vol. 6: *Agriculture* (Washington, D.C.: Government Printing Office, 1913), 405.

7. F. D. Coburn, *The Book of Alfalfa* (New York: Orange Judd Company, 1918), discusses the utility and versatility of the alfalfa plant. Clothier is quoted on 11–12.

8. Coburn, *The Book of Alfalfa;* Stephen R. Chapman and Lark P. Carter,

Crop Production: Principles and Practices (San Francisco: W. H. Freeman and Company, 1976), 58.

9. Elias Nelson, *Alfalfa,* University of Idaho Agricultural Experiment Station, Bulletin 66 (Moscow: Idaho Post Print, 1909), 31–32; Richard Lowitt and Judith Fabry, eds., *Henry A. Wallace's Irrigation Frontier: On the Trail of the Corn Belt Farmer, 1909* (Norman: University of Oklahoma Press, 1991), 163.

10. On the usefulness of alfalfa as fodder, see Nelson, *Alfalfa,* 32. On the keeping of sheep and other livestock, see *Gem State Rural* 10 (16 March 1905): 4; *Idaho Republican,* 5 July 1918; *Idaho Farmer* 22 (15 March 1917): 12; 24 (3 July 1919): 7; 25 (4 March 1920): 362; 39 (31 March 1927): 351.

11. Siddik Atrushi, "Geographic Elements in Irrigation Agriculture in Boise Valley, Idaho," Ph.D. dissertation (Clark University, 1952), 156–171, provides a local analysis of the alfalfa-livestock relation. James A. Young and B. Abbott Sparks, *Cattle in the Cold Desert* (Logan: Utah State University Press, 1985), 121–180, places the phenomenon in the wider context of the Great Basin and the intermountain West. See also, for example, *Gem State Rural* 9 (16 March 1905): 4; and W. E. M'Lendon, "Soil Survey of the Blackfoot Area, Idaho," in U.S. Department of Agriculture, Bureau of Soils, *Field Operations of the Bureau of Soils, 1903,* Fifth Report (Washington, D.C.: Government Printing Office, 1904), 1043–1044.

12. U.S. Department of Commerce, Bureau of the Census, *Thirteenth Census of the United States,* 1910, Vol. 6: *Agriculture,* 396, 398; U.S. Department of Commerce, Bureau of the Census, *Fourteenth Census of the United States,* 1920, Vol. 6, Part 3: *Agriculture* (Washington, D.C.: Government Printing Office, 1922), 142, 145. On the postwar collapse of agricultural prices, see Gwynn Barrett and Leonard Arrington, "The 1921 Depression: Its Impact on Idaho," *Idaho Yesterdays* 15 (Summer 1971): 10–15.

13. On the alfalfa weevil and its impact on Idaho, see the multipart article by Claude Wakeland, "The Alfalfa Weevil," *Idaho Farmer* 27 (31 March 1921): 349; (7 April 1921): 375; (14 April 1921): 401; (21 April 1921): 438; (28 April 1921): 438. On the alfalfa quarantine, see *Idaho Republican,* 8 October 1917; *Idaho Farmer* 24 (12 December 1918): 6; "Official Report of the Proceedings Before the Fact Finding Commission Appointed by the Secretary of the Interior" (18 January 1924), 342, Exhibit No. 4 file, Exhibits Accompanying Final Reports, 1923–1924, Committee of Special Advisers on Reclamation, RG 115.

14. On the hay situation during the war, see, for example, *Idaho Republican,* 27 August, 14 September, 5, 8 October 1917, 19 July, 2 August, 24 September 1918. Stockmen themselves suffered from the recovery of European agriculture and from the expansion into world markets of animal products from places like Australia; Barrett and Arrington, "The 1921 Depression."

15. "Report of Conditions on the South Side Pumping Unit of the

Minidoka Project," 17 January 1924, Minidoka file, Exhibits Accompanying Final Reports, 1923–1924, Committee of Special Advisers on Reclamation, RG 115; Greenwood, *We Sagebrush Folks,* 17.

16. Shepherd's statement is in "Official Report of the Proceedings Before the Fact Finding Commission" (18 January 1924), 343. On the expansion of Idaho dairy production, see F. W. Atkeson et al., *The Dairy Situation in Idaho,* University of Idaho Agricultural Experiment Station, Bulletin 152 (Moscow: University of Idaho, July 1927). For examples of expanding dairy herds and cheese factories on the Minidoka irrigation project, see Barry Dibble to C. J. Blanchard, 2 April 1923, D. W. Davis to S. R. Winters, 29 March 1924, E. B. Darlington to Dorothy Lampen, 9 November 1928, all in file 555, Minidoka Project Records, RG 115; "Cheese Factories Succeed on the Minidoka Project," *Reclamation Record* 14 (November–December 1923): 319. While touring the Snake River valley in 1922, J. L. Kraft told local farmers that, of all agricultural commodities, only dairy products had not fallen below cost of production since the end of the world war. See J. E. Nessly, "Easterners 'Discover' Idaho's Possibilities," *Idaho Farmer* 30 (17 August 1922): 3, 14. See also Oren A. Fitzgerald, "The Milk Cow Brings Money into Idaho," *Idaho Farmer* 37 (17 June 1926): 3, 6.

17. George Stewart, *Alfalfa Growing in the United States and Canada* (New York: Macmillan, 1926), 1–6; R. A. Oakley, "The Seed Supply of the Nation," in *Yearbook of the United States Department of Agriculture, 1917* (Washington, D.C.: Government Printing Office, 1918), 497–536; W. A. Wheeler and G. C. Edler, "Some Effects of the War Upon the Seed Industry of the United States," in *Yearbook of the United States Department of Agriculture, 1918* (Washington, D.C.: Government Printing Office, 1919), 195–214; *Idaho Farmer* 25 (5 February 1920): 195, 27 (24 March 1921): 322, (31 March 1921): 346, and (5 May 1921): 458; *The Farming Business in Idaho,* University of Idaho Agricultural Experiment Station, Bulletin 151 (Moscow: University of Idaho, July 1927), 50–60.

18. C. B. Ahlson, *Report of the Seed Commissioner for the Biennium 1921–1922,* University of Idaho Agricultural Experiment Station, Circular 31 (Moscow: University of Idaho, January 1923), 14–15. Good color did not actually reveal the viability of seed: "To grow a well-colored seed," said Idaho's state seed commissioner in 1920, "has about the same value as for a fruit stand man to wipe an apple on his coat sleeve—it does not increase the value of an apple, but it increases his income." Bright-colored seed attracted the attention of buyers. See Edward F. Mason, "Idaho's High Grade Seeds," *Idaho Farmer* 25 (5 February 1920): 195. See also L. C. Aicher, *The Production of Alfalfa Seed in Southern Idaho,* University of Idaho Agricultural Experiment Station, Bulletin 101 (Moscow: University of Idaho, July 1917).

19. E. J. Iddings, *Idaho Pure Seed Law,* University of Idaho Agricultural

Experiment Station, Circular 65 (Moscow: University of Idaho, August 1931); Ahlson, *Report of the Seed Commissioner for the Biennium 1921–1922.*

20. Iddings, *Idaho Pure Seed Law;* Ahlson, *Report of the Seed Commissioner for the Biennium 1921–1922;* Aicher, *The Production of Alfalfa Seed in Southern Idaho.*

21. See Chapter 2 for details of the spread of weeds through the irrigated landscape; see also Mason, "Idaho's High Grade Seeds," 195; R. K. Bonnett, "Pests Threaten Clover Seed Industry of Idaho," *Idaho Farmer* 26 (23 September 1920): 1383; R. K. Bonnett, "Dodder, and the Means for Its Control," *Idaho Farmer* 26 (4 November 1920): 1583; *Idaho Farmer* 27 (24 March 1921): 322; B. F. Sheehan, "Dodder in Idaho," *Idaho Farmer* 27 (31 March 1921): 346. Dodder had appeared in Idaho alfalfa fields as early as 1895. See *Gem State Rural* 1 (November 1895): 5.

22. Irwin is quoted in *Idaho Farmer* 24 (12 December 1918): 6; *The Farming Business in Idaho,* 59; Aicher, "The Production of Alfalfa Seed in Southern Idaho," 3.

23. See Chapter 2 for the weed threat and efforts to control it. On the seed certification program, see *The Farming Business in Idaho,* 59; *Idaho Farmer* 26 (8 July 1920): 1057; Ahlson, *Report of the Seed Commissioner for the Biennium 1921–1922.*

24. Mason, "Idaho's High Grade Seeds." In 1926, Idaho alfalfa seed farmers, and American growers generally, received some help with the problem of seed identity from the Federal Seed Act. Under its terms, all foreign alfalfa seed had to be stained a shade of red, green, or violet as an indication of its adaptability to American climates. Once again, when nature failed to distinguish seed variety, humans compensated with their own color schemes. See H. W. Hulbert et al., *Adaptability of Alfalfa Strains and Varieties for Idaho,* University of Idaho Agricultural Experiment Station, Bulletin 199 (Moscow: University of Idaho, December 1933), 3.

25. *Idaho Republican,* 1 August 1919; *Idaho Farmer* 25 (1 January 1920): 6; 26 (8 July 1920): 1057, and (23 September 1920): 1385; 27 (10 February 1921): 169, and (17 February 1921): 190; 29 (9 February 1922): 123; B. F. Sheehan, "Greatest Seed Show Ever Held in West Scheduled," *Idaho Farmer* 26 (25 November 1920): 1699; O. A. Fitzgerald, "The Seed Industry in Idaho Makes Remarkable Progress," *Idaho Farmer* 35 (19 March 1925): 268, 270.

26. Some people in the American West tried to promote the human consumption of processed alfalfa in the form of medicine for nerves, breakfast food, buns, pies, pudding, and cigars. See *Idaho Register* (Idaho Falls), 10 August 1909, 9 August 1910, 14 June 1912. A few mills also ground alfalfa into meal for animal feed; see *The Farming Business in Idaho,* 74.

27. J. S. Welch, "Potato Culture Under Irrigation," in *Potato Culture,* Idaho Experiment Station Bulletin 79 (n.p.: n.p., April 1914); James W. Davis and

Nikki Balch Stilwell, *Aristocrat in Burlap: A History of the Potato in Idaho* (n.p.: Idaho Potato Commission, 1975), 16–17 (*American Falls Press* quoted), 26–29.

28. Davis and Stilwell, *Aristocrat in Burlap,* 16–19; U.S. Department of the Interior, Census Office, *Report on the Statistics of Agriculture in the United States at the Eleventh Census: 1890* (Washington, D.C.: Government Printing Office, 1895), 466; U.S. Department of the Interior, Census Office, *Twelfth Census of the United States,* 1900, Vol. 6: *Agriculture,* Part 2, *Crops and Irrigation* (Washington, D.C.: Census Office, 1902), 366. Markets for Idaho potatoes, like potato markets everywhere, were constantly changing. "The selection of markets depends upon conditions of demand, supply, and price, which change continually," two economists reported. George B. Fiske and Paul Froehlich, *Marketing the Early Potato Crop,* U.S. Department of Agriculture, Farmers' Bulletin 1316 (Washington, D.C.: Government Printing Office, 1923), 9.

29. *Idaho Register,* 12 November 1912. For more background on "spud alley" and the hoopla associated with potato production, see Davis and Stilwell, *Aristocrat in Burlap.*

30. *Idaho Register,* 23 June, 21 August 1908, 23 September, 15 November, 13, 23 December 1910, 4 October, 1, 12, 15, 19 November, 3 December 1912.

31. C. F. Wells and H. C. Dale, *The Potato Situation in Idaho,* University of Idaho Agricultural Experiment Station, Bulletin 153 (Moscow: University of Idaho, June 1927); Fiske and Froehlich, *Marketing the Early Potato Crop;* Wells A. Sherman et al., *Marketing Late-Crop Potatoes,* U.S. Department of Agriculture, Farmers' Bulletin 1578 (Washington, D.C.: Government Printing Office, 1929), 29–30; Davis and Stilwell, *Aristocrat in Burlap,* 64.

32. For a discussion of these problems, see Fiske and Froelich, *Marketing the Early Potato Crop,* and Sherman et al., *Marketing Late-Crop Potatoes.*

33. William Stuart, *Potato Storage and Storage Houses,* U.S. Department of Agriculture, Farmers' Bulletin 847 (Washington, D.C.: Government Printing Office, 1917; revised 1930); *Idaho Register,* 23 June 1908; *Idaho Republican,* 2 March, 24 August 1917, 7, 28 September 1922; "Annual Project History, Report of Construction and Operation and Maintenance, Minidoka, Volume XI, Year of 1917," p. 247, Roll 99, Microcopy M96, Minidoka Project, Idaho, 1903–1919, RG 115; E. R. Bennett, "Storing the Potato the Most Natural Way," *Idaho Farmer* 28 (1 September 1921): 1. See also *Idaho Register,* 12 November 1912, for more information on the storage systems of Joe Taylor and his neighbors.

34. Wells and Dale, *The Potato Situation in Idaho,* 5; *Idaho Republican,* 10 August, 28 September 1922.

35. C. E. Temple, "Potato Diseases and Insect Pests," in *Potato Culture,* Idaho Agricultural Experiment Station, Bulletin 79 (n.p.: n.p., April 1914), 40; Davis and Stilwell, *Aristocrat in Burlap,* 29–30; O. A. Pratt, "Experiments With Clean Seed Potatoes on New Land in Southern Idaho," *Journal of Agricultural Research* 6 (July 1916): 573–575; O. A. Pratt, "Soil Fungi in Relation to Diseases

of the Irish Potato in Southern Idaho," *Journal of Agricultural Research* 8 (April 1918): 73–99. Grubb is quoted in *Idaho Republican*, 25 July 1913.

36. Wells and Dale, *The Potato Situation in Idaho*, 35; Davis and Stilwell, *Aristocrat in Burlap*, 50, 53; Edward F. Mason, "Fremont Potatoes Are on the Map," *Idaho Farmer* 26 (7 October 1920): 1452; see also Chapter 2.

37. Davis and Stilwell, *Aristocrat in Burlap*, 30–31 (including quotation); *Idaho Register*, 13 December 1910; *Idaho Republican*, 18 July 1913; James Stephenson, Jr., *Irrigation in Idaho*, U.S. Department of Agriculture, Office of Experiment Stations, Bulletin 216 (Washington, D.C.; Government Printing Office, 1909), 29.

38. Davis and Stilwell, *Aristocrat in Burlap*, 31–32; Sherman et al., *Marketing Late-Crop Potatoes*, 29–30; Wells A. Sherman, et al., *Marketing Main-Crop Potatoes*, U.S. Department of Agriculture, Farmers' Bulletin 1317 (Washington, D.C.: Government Printing Office, 1923), 20; O. L. Mimms and George W. Woodbury, *Markets and Market Preferences for Idaho Potatoes*, University of Idaho Agricultural Experiment Station, Bulletin 231 (Moscow: University of Idaho, June 1939), 5; Stephenson, *Irrigation in Idaho*, 29. By 1920, when the irrigated landscape in Idaho's Snake River valley totaled some 2 million acres, about 43,000 of these were devoted to potatoes. See U.S. Department of Commerce, Bureau of the Census, *Fourteenth Census of the United States*, 1920, Vol. 6, Part 3: *Agriculture* (Washington, D.C.: Government Printing Office, 1922), 142.

39. C. T. More and C. R. Dorland, *Commercial Handling, Grading, and Marketing of Potatoes*, U.S. Department of Agriculture, Farmers' Bulletin 753 (Washington, D.C.: Government Printing Office, 1917), 16–23; *Idaho Republican*, 5 August 1910; *Idaho Register*, 1 October 1912.

40. More and Dorland, *Commercial Handling, Grading, and Marketing of Potatoes*, 16–21; Fiske and Froelich, *Marketing the Early Potato Crop*, 1516; Sherman et al., *Marketing Main-Crop Potatoes*, 14–15; Sherman et al., *Marketing Late-Crop Potatoes*, 18–19, 30; J. W. Park, *Marketing the Late Crop of Potatoes*, U.S. Department of Agriculture, Circular 475 (Washington, D.C.: Government Printing Office, June 1938), 34–35; Davis and Stilwell, *Aristocrat in Burlap*, 85–94; *Idaho Republican*, 27 August, 7, 21 September, 22 October 1917, 10 September 1918, 26 January 1921 (quotation). Byrd Trego "thought potato growers should be educated to the necessity of properly grading their spuds by sorting"; *Idaho Republican*, 5 August 1910. Wholesalers also advocated that farmers grade their produce; see *Idaho Register*, 1 October 1912.

41. *Idaho Republican*, 27 August, 7, 21 September, 22 October 1917, 10 September 1918; Harold W. Samson, *United States Grades for Potatoes*, U.S. Department of Agriculture, Department Circular 238 (Washington, D.C.: Government Printing Office, 1922). For background on Food Administration policies, see D. F. Houston, "Report of the Secretary of Agriculture," in *Yearbook*

of the United States Department of Agriculture, 1917 (Washington, D.C.: Government Printing Office, 1918).

42. *Idaho Farmer* 24 (12 December 1918): 6.

43. For accounts of conflict over grading, see *Idaho Republican,* 24 December 1918, 17, 19, 21, 26 January 1921; *Idaho Falls Daily Post,* 10 January 1922; and *Idaho Farmer* 39 (15 February 1927): 167. See also *Idaho Farmer* 24 (26 January 1922): 1.

44. Davis and Stilwell, *Aristocrat in Burlap,* 30–31; *Idaho Republican,* 10 August 1922.

45. *Idaho Register,* 1 October 1912.

46. For a description of "stovepiping," see Davis and Stilwell, *Aristocrat in Burlap,* 90, 93. "This practice," the authors state, "was perhaps the darkest day in the history of the Idaho shipping industry in customer relations and an appreciation for quality." According to John R. Stilgoe, ancient farmers believed that grain millers regularly adulterated flour with sand and other substances. See Stilgoe's *Common Landscape of America, 1580 to 1845* (New Haven: Yale University Press, 1982), 300.

47. *Idaho Register,* 21 August 1908, 23 September, 15 November 1910, 12 November 1912; *Idaho Republican,* 18 July 1913. The *American Falls Press* material is reported in *Idaho Register,* 21 August 1908.

48. On Limbert and potato promotion, see Schwantes, *In Mountain Shadows,* 102–103, 171–172; see also Todd Shallat, ed., *Snake: The Plain and Its People* (Boise: Boise State University, 1994), 226, for a nice reproduction of a photograph of Limbert's giant potato. For other examples of advertising and promotion, see *Idaho Register,* 23 December 1910, 1 November, 3 December 1912; *Idaho Republican,* 24 December 1918; *Idaho Farmer* 26 (October 1920): 1441. See also *Idaho Register,* 15 November 1912, for a report of an exhibit of Idaho potatoes at the Chicago Land Show: "The potatoes will form the Idaho Falls exhibit and will be distributed free to visitors at the show. Each potato will be wrapped, the wrapper bearing Idaho Falls advertising. The potatoes will be hand picked and selected and bought at 20 cents per hundred above the market to secure the very best."

49. *Idaho Republican,* 21 September 1922; *Idaho Farmer* 32 (31 January 1924): 95; 35 (1 January 1925): 4; Schwantes, *In Mountain Shadows,* 102. The state of Idaho institutionalized advertising during the Great Depression: in 1937, it established an advertising commission, later the Idaho Potato Commission, to promote the tubers. Davis and Stilwell, *Aristocrat in Burlap,* 54, 104–105.

50. Davis and Stilwell, *Aristocrat in Burlap,* 101, 108, 111.

51. Ibid.

52. More and Dorland, *Commercial Handling, Grading, and Marketing of Potatoes,* 23–24; *Idaho Register,* 23 September 1910, 16 September 1913; *Idaho*

Republican, 2 March, 18 June, 1 October 1917, 24 December 1918, 21 June 1921, 21 September 1922, 14 June, 18 July, 30 August 1923; *Idaho Farmer* 25 (22 April 1920): 671; 39 (16 June 1927): 643; *The Farming Business in Idaho,* 28, 30; Sherman et al., *Marketing Main-Crop Potatoes,* 20. The farmers' belief that deceitful potato wholesalers stole a portion of their crop had ancient origins. European peasants believed that grain millers robbed them of a part of their crop; see Stilgoe, *Common Landscape of America,* 300–309. Idaho potato growers also fumed over consignment shipping. With the rock-bottom prices of the early 1920s, some dealers quit buying the crop and instead marketed it on consignment, for a fee. Farmers then complained that they had to pay demurrage, freight, and consignment costs while the dealers, who had no stake in the final profit, were careless about getting the best price for the carloads. See *Idaho Republican,* 7 September 1922.

53. Davis and Stilwell, *Aristocrat in Burlap,* 62–84, 95.

54. On Aroostook and Snake River valley farmers, see Bret Wallach, *At Odds with Progress: Americans and Conservation* (Tucson: University of Arizona Press, 1991), 12–14.

55. Idahoans often used an imperial metaphor that connoted wealth and prosperity and that minimized the problems inherent in raising the crop. Idahoans passed Joe Taylor's potato king mantle to other farmers and promoters. Joe Marshall, an engineer, farmer, businessman, and potato booster, eventually inherited the title. Marshall sometimes appeared at trade conventions wearing a crown and enthusiastically promoting Idaho spuds. See Davis and Stilwell, *Aristocrat in Burlap,* 45–61. Idahoans also attached names to the land that seemed to imply the ease with which farmers created wealth. The Boise Valley became the "Treasure Valley"; the area around Twin Falls, the "Magic Valley." Idaho itself was the "Gem State," and the Russet Burbank the "Netted Gem." On Idaho place names, see Lalia Boone, *Idaho Place Names: A Geographical Dictionary* (Moscow: University of Idaho Press, 1988).

56. Greenwood's memoir, *We Sagebrush Folks,* and the historian Jo Ann Ruckman's foreword and afterword to the book, provide a portrait of Charles Greenwood and his farming skills.

6/ INDUSTRIAL EDEN

1. For the classic account of the garden myth, see Henry Nash Smith, *Virgin Land: The American West as Symbol and Myth* (Cambridge, Mass.: Harvard University Press, 1950; rpt. New York: Vintage Books, 1957). An important recent analysis of the garden myth that pays close attention to gender is Carolyn Merchant, "Reinventing Eden: Western Culture as a Recovery Narrative," in *Uncommon Ground: Toward Reinventing Nature,* ed. William Cronon (New York: W. W. Norton, 1995). See also John Rennie Short, *Imagined Coun-*

try: Environment, Culture, and Society (London and New York: Routledge, 1991), 5–39. The classic work on the impact of industrialization on the pastoral form is Leo Marx, *The Machine in the Garden: Technology and the Pastoral Ideal in America* (New York: Oxford University Press, 1964). My focus on organic, technological, and divine metaphors owes much to Daniel B. Botkin, *Discordant Harmonies: A New Ecology for the Twenty-first Century* (New York: Oxford University Press, 1990), and Raymond Williams, "Ideas of Nature," in his *Problems in Materialism and Culture: Selected Essays* (London and New York: Verso, 1980). In addition to these works, my understanding of metaphor, myth, and landscape has been shaped by Annette Kolodny, *The Lay of the Land: Metaphor as Experience and History in American Life and Letters* (Chapel Hill: University of North Carolina Press, 1975) and *The Land before Her: Fantasy and Experience of American Frontiers, 1630–1860* (Chapel Hill: University of North Carolina Press, 1984); Carolyn Merchant, *Ecological Revolutions: Nature, Gender, and Science in New England* (Chapel Hill: University of North Carolina Press, 1989); Jules David Prown et al., *Discovered Lands, Invented Pasts: Transforming Visions of the American West* (New Haven: Yale University Press, 1992); Simon Schama, *Landscape and Memory* (New York: Knopf, 1995), especially 3–19; Richard Slotkin, "Myth and the Prodution of History," in *Ideology and Classic American Literature,* ed. Sacvan Bercovitch and Myra Jehlen (New York: Cambridge University Press, 1986); William H. Truettner, ed., *The West as America: Reinterpreting Images of the Frontier, 1820–1920* (Washington, D.C.: Smithsonian Institution Press for the National Museum of American Art, 1991); and Richard White, *"It's Your Misfortune and None of My Own": A History of the American West* (Norman: University of Oklahoma Press, 1991), 613–631.

2. E. B. Darlington, "The Romance of A River: A Past, Present, and Future Survey of Irrigation in Southern Idaho, *Reclamation Record* 11 (March 1920): 123.

3. Photograph 70–95.12, "Emmett, Idaho—Fruit Culture," Idaho Historical Society Library and Archives, Boise, Idaho (hereafter cited as IHS).

4. For a discussion of Fisher's poetry, see Dorys C. Grover, *A Solitary Voice: Vardis Fisher* (New York: Revisionist Press, 1973). "Joe Hunter" (pp. 52–53) is from Fisher's "Antelope People" sonnets, which portray the people of eastern Idaho's Antelope Hills, Fisher's native home. Joe Hunter was probably a dry farmer, but the masculine image explains irrigators just as well.

5. For an overview of rabbit drives, see Chapter 2. For general accounts of rabbit drives, see chapter 7 of E. L. Davis's memoir in the *Aberdeen Times,* 8 June 1950; Kate B. Carter, ed., *Pioneer Irrigation: Upper Snake River Valley* (Salt Lake City: Daughters of Utah Pioneers, 1955), 63; *Idaho Falls Times,* 3 February 1905. Useful in understanding the rabbit drives as a cultural expression is Clifford Geertz, *The Interpretation of Cultures* (New York: Basic

Books, 1973), especially 412–453. Aldo Leopold viewed hunting as a ritual: "The duck hunter in his blind and the operatic singer on the stage, despite the disparity in their accoutrements, are doing the same thing. Each is reviving, in play, a drama formerly inherent in daily life. Both are, in the last analysis, aesthetic exercises." See Leopold, *A Sand County Almanac, With Essays on Conservation from Round River* (New York: Oxford University Press, 1966; rpt. New York: Ballantine Books, 1970), 283. One government official, S. E. Piper of the U.S. Bureau of Biological Survey, believed that rabbit drives were not especially effective counters to jackrabbit irruptions. "While drives and organized hunts are useful . . . they are unsuccessful from an economic standpoint and are not applicable to the problem generally." See Piper, "The Jack Rabbit Problem," *Reclamation Record* 6 (March 1915): 123–124.

6. Annie Pike Greenwood, *We Sagebrush Folks* (New York: D. Appleton-Century, 1934; rpt, Moscow: University of Idaho Press, 1988). See pages 245–255 for quotations in this and the following paragraph.

7. *Idaho Republican* (Blackfoot), 31 January 1921.

8. *Idaho Register* (Idaho Falls), 5 October 1906. The *Rigby Star,* 3 October 1907, described Lee's work under the headline, "Successfully Subduing the Native Wilds: From a Sagebrush Plain to Growing Fields of Golden Grain Within a Few Months—Energy and Irrigation Does the Work."

9. Darlington, "The Romance of A River," 122–123.

10. C. J. Blanchard, "The Minidoka Project, Idaho," *Reclamation Record* 6 (February 1915): 78.

11. Fannie G. Brunt, "The Saga of the Snake River, or Magic Valley's Yesterdays" (n.d.), typescript copy in "Idaho—History—Drama" Vertical File, Idaho Room, Idaho Falls Public Library.

12. American Falls Press, *Commemorating the Dedication of American Falls Dam* (Official Souvenir Edition) (n.p.: n.p., [1925?]), Manuscript Collection 544, Idaho Travel and Tourism, IHS.

13. D. W. Ross, "Home Making in Idaho Under the Carey Law," *Irrigation Age* 6 (December 1894): 255; Greenwood, *We Sagebrush Folks,* 57; Brunt, "The Saga of the Snake River." When speaking of the Snake's capacity to drown people, Greenwood referred to the river as female, "tamed . . . but treacherous still." See *We Sagebrush Folks,* 267.

14. Quoted in *Reclamation Record* 14 (September 1923): 276.

15. J. H. Lowell, "The Boise Project," *Reclamation Record* 6 (November 1915): 517–518; see also the following newspaper clippings in Dams—Arrowrock Vertical File, IHS: "Arrow Rock Dam, Finishing Touch to Great Boise Irrigation Project, Is Formally Dedicated Today" and "More People Go To Arrowrock To See Great Dam." Images of dams in many ways conformed to Leo Marx's concept of the "technological sublime"; see *Machine in the Garden,* 194–226.

16. *Idaho Republican,* 17 May 1923. Although this story carried no byline,

it is editorial in form and appeared under the masthead identifying Trego as editor and proprietor. I assume that he wrote the article. For an overview of the iconography of Indians and progress in western art, see Truettner, ed., *The West as America,* 168–175.

17. For a discussion of the Johnny Appleseed myth, see Robert Price, *Johnny Appleseed: Man and Myth* (Bloomington: Indiana University Press, 1954), especially 246–264. Richard Slotkin, *Regeneration through Violence: The Mythology of the American Frontier, 1600–1860* (Middletown, Conn.: Wesleyan University Press, 1973), is one of the standard treatments of frontier mythology.

18. *Idaho Statesman* (Boise), 3 September 1891.

19. *Idaho Republican,* 2 February 1922.

20. Irene Welch Grissom, "The Desert Reclaimed," *New West Magazine* 11 (December 1920): 15, also reprinted in *Idaho Falls Daily Post,* 16 June 1923. "The Man With the Shovel," *Salt Lake Tribune,* 11 June 1907, is quoted in Ray G. Newman, "An Early Historical Account of the North Side Canal Company, Jerome, Idaho," M.A. thesis (Utah State University, 1964), 1–2.

21. *Gem State Rural* (Caldwell) 10 (31 August 1905): 4.

22. C. J. Blanchard, "The End of the Rainbow Ranch," *Reclamation Record* 13 (October 1922): 250–252. For examples of pastoral farm names: *Reclamation Record* 7 (March 1916): 224–225, (August 1916): 346; 9 (August 1918): 363, (September 1918): 417, (November 1918): 516; 15 (February 1924): front page.

23. Charles J. Lisle, "The Book Farm," *Gem State Rural and Live Stock Journal* 21 (April 1916): 1–4, for quotations in this and next paragraph.

24. Mary Hallock Foote, "The Irrigating Ditch," *Century Magazine* 38 (June 1889): 299–300, reprinted in Barbara Cragg et al., *The Idaho Stories and Far West Illustrations of Mary Hallock Foote* (Pocatello: Idaho State University Press, 1988), 285–287.

25. Reid's poem is in Carter, *Pioneer Irrigation,* 4.

26. Greenwood, *We Sagebrush Folks,* 110.

27. For examples of the organic aesthetic, see *Caldwell Tribune,* 14 February 1891, 10 December 1892; *Report of the Special Committee of the United States Senate on the Irrigation and Reclamation of Arid Lands,* Vol. 1: *The Northwest* (Washington, D.C.: Government Printing Office, 1890), 381. See also the report in *Gem State Rural* 10 (23 February 1905): 5. "Senator Steele, of Bingham county, whose home is at Iona, sent some monster potatoes to the St. Louis Exposition. A single specimen weighed 7½ pounds and 80 potatoes filled a barrel." For the organic-industrial aesthetic, see the comments of J. R. Shinn, "Fruit For Exhibition," *Gem State Rural and Live Stock Journal* (September 1910): 15. On the gift potatoes sent to the Coolidges, see *Idaho Farmer* (Caldwell) 35 (1 January 1925): 4. Annie Pike Greenwood complained of some judges at a fair: "They were the kind of judges whose only criterion

of excellence was giantism. They had not learned that vegetables have the equivalent of hyperactivity of the thyroid gland and that oversize is always at the expense of some other important attribute." Greenwood, *We Sagebrush Folks,* 232.

28. *Idaho Farmer* 35 (5 October 1916): 6. For additional examples of the metaphorical association of children and plants, see Photograph 73-221.425, "Fruit Culture (Apples)," IHS; "Some Pointed Advice By An Expert," *Irrigation Age* 25 (January 1910): 128; Greenwood, *We Sagebrush Folks,* 110, illustration facing 232; *Gem State Rural* 1 (June 1896): 3; and 9 (14 January 1904): 1.

29. May Nessly, "Four Perfect Idaho Products," *Idaho Farmer* 34 (25 September 1924): 214; and see (16 October 1924): 269. See also *Idaho Farmer* 40 (22 September 1927): 239, and *Reclamation Record* 5 (November 1914): 411. For another view, see the noted plant breeder Luther Burbank's book, *The Training of the Human Plant* (New York: Century Company, 1907), in which Burbank analogized child rearing to plant breeding and employed an interesting combination of technical and organic metaphors. He believed that what children needed was country life, not an "artificial" urban environment, "so like that of the hot-house" (p. 16). He even characterized America as a sort of wild garden in which the joining of races produced a healthy natural hybridization.

30. *Idaho Farmer* 31 (1 February 1923): 115; 27 (17 March 1921): 290. Compare also the comments of Governor C. C. Moore, who, in 1923 "urged that as much attention be given to the children as to the poultry and live stock in order that Idaho may produce a good 'crop' of citizens for future years." *Idaho Farmer* 32 (16 August 1923): 107.

31. *Idaho Register,* 13 December 1910; *Idaho Farmer* 40 (27 October 1927): 359.

32. For Spud's poem, see *Idaho Republican,* 1 October 1917. For children playing in canals, including the remembrance of Amanda Clark, see Carter, *Pioneer Irrigation,* 37, 79. See also Greenwood, *We Sagebrush Folks,* 41–42. Of course, one of the greatest fears in Idaho's irrigated areas was that children would drown in canals. See, for example, Carter, *Pioneer Irrigation,* 79, and Greenwood, *We Sagebrush Folks,* 163–165.

33. Raymond Williams, *The Country and the City* (New York: Oxford University Press, 1973), shows that the pastoral form as used in English literature masked the capitalist economic connections between the country and the city. The countryside, while seemingly natural and free of the social problems of the city, rested on the very real exploitation of rural people by capitalist elites. In this chapter I have borrowed Williams's notion that the pastoral obscures disruptive and exploitative economic conditions.

34. *Gem State Rural* 1 (November 1895): 2–3.

35. L. F. Henderson, *Twelve of Idaho's Worst Weeds,* University of Idaho

Agricultural Experiment Station, Bulletin 14 (Moscow: Moscow Mirror, 1898), 106–107.

36. Henderson, *Twelve of Idaho's Worst Weeds,* 113–114; Don B. Whelan, "Who Owns the Farm, You or Mr. Grasshopper?" *Idaho Farmer* 31 (8 March 1923): 219. See also Fred Schwendiman, "Battle With Germans," *Idaho Farmer* 23 (4 April 1918): 4. For an insightful discussion of the interplay of chemical weapons, warfare, farm pesticides, the struggle against pests, and metaphor, see Edmund P. Russell III, "'Speaking of Annihilation': Mobilizing for War against Human and Insect Enemies, 1914–1945," *Journal of American History* 82 (March 1996): 1505–1529.

37. Warren Swendsen, *Report on American Falls Reservoir District* (n.p.: n.p., 1924), 22.

38. *Salt Lake Tribune* quoted in *Reclamation Record* 17 (May 1926): inside front cover.

39. For a discussion of the antipastoral, see Short, *Imagined Country,* 35–39.

40. Mary Hallock Foote, *The Chosen Valley* (Boston and New York: Houghton Mifflin and Company; Cambridge, Mass.; Riverside Press, 1892). Because the novel is out of print and not readily available, I have used the more accessible 1892 serialization: see *Century Magazine* 44 (October 1892): 829 (quotation).

41. See Foote's concluding paragraph to her novel, in *Century Magazine* 44 (October 1892): 833. For detail on Foote's life and her relationship to her husband and his career, see Rodman W. Paul, ed., *Victorian Gentlewoman in the Far West: The Reminiscences of Mary Hallock Foote* (San Marino, Calif.: Huntington Library, 1972), 1–44.

42. For brief mention of Greenwood's life, see Federal Writers' Project, Works Progress Administration, *The Idaho Encyclopedia* (Caldwell, Idaho: Caxton Printers, 1938), 171. And see Jo Ann Ruckman's perceptive foreword and afterword to the 1988 reprint of *We Sagebrush Folks.*

43. Greenwood, *We Sagebrush Folks,* 13–14. For quotations in the next paragraph, see 339, 353, 115, 158, 5; and the next, 78–80, 451–481. On women wanting to escape the farm, 281–90. The memoir is silent on Greenwood's breakdown and separation from her husband; see Ruckman's afterword to the 1988 edition.

44. *Idaho Farmer* 31 (5 July 1923): 20.

45. Peter Boag notes the connection between the organicism of the garden and "artifice." See his *Environment and Experience: Settlement Culture in Nineteenth-Century Oregon* (Berkeley: University of California Press, 1992), 91.

46. For "chemical laboratory" and "foliage laboratory," see S. J. Rich, ed., *Commissioner of Immigration, Labor, and Statistics, Seventh Biennial Report, 1911–1912* (Caldwell, Idaho: Caxton Printers, [1912?]), 107.

47. For an interesting commentary on the prevalence of the settlers' dream

of the garden landscape, see Randall R. Howard, "Irrigation As the Home-seeker Finds It," *Pacific Monthly* 24 (November 1911): 521–532. For a good example of the lush, colorful imagery of advertisements, see the cover of the pamphlet titled "Idaho Orchards" in Agriculture—Crops (Fruit) Vertical File, IHS.

48. See, for example, the chapter titled "Wilderness," in Greenwood's *We Sagebrush Folks.* For an example of the modern view, consider the observation of a Boise resident, Rob Lesser, that Milner Dam's interruption of the Snake River flow is "a sacrilege" or consider what the environmental writer Tim Palmer says, that river development in the West turns the "biological bonanza of the riparian zone . . . into a biological desert." Both comments are found in Palmer's *Snake River: Window to the West* (Washington, D.C., and Covello, Calif.: Island Press, 1991), 143, 185. And compare this assessment to the novelist, historian, and essayist Wallace Stegner: "He wrote often, sometimes caustically, about one of the culture's embedded attitudes shared by many river-bending Westerners who followed the Mormons: that the desert should blossom as the rose. On numerous occasions he referred to irrigating the desert as engineering the arid land into what it wasn't intended by God to be. That rankled him; the West's 'original sin,' he called it." See James Thalman, "One of Ours," *Continuum: The Magazine of the University of Utah* 5 (Winter 1995–1996): 27.

CONCLUSION

1. Arthur Chapman, *Out Where the West Begins and Other Western Verses* (Boston and New York: Houghton Mifflin Company, 1916); Sanborn Souvenir Company, Inc., Commerce City, Colorado, published the postcard. Chapman's lyrics were featured in a historical pageant staged in Twin Falls in 1921. For descriptions of the pageant, see *Twin Falls Daily News,* 11 October, 30 November, 1, 2, December 1921.

2. Yi-Fu Tuan, "Discrepancies between Environmental Attitude and Behavior: Examples from Europe and China," in *Ecology and Religion in History,* ed. David Spring and Eileen Spring (New York: Harper and Row, 1974); Yi-Fu Tuan, "Our Treatment of the Environment in Ideal and Actuality," *American Scientist* 58 (May–June 1970): 244, 247–249; Jeanne Kay, "Preconditions of Natural Resource Conservation," in *The History of Soil and Water Conservation,* ed. Douglas Helms and Susan L. Flader (Washington, D.C.: Agricultural History Society, 1985); Steven R. Simms, "Wilderness as a Human Landscape," in *Wilderness Tapestry: An Eclectic Approach to Preservation,* ed. Samuel I. Zeveloff, L. Mikel Vause, and William H. McVaugh (Reno and Las Vegas: University of Nevada Press, 1992).

Selected Bibliography

Aberdeen–Springfield Canal Company Records. Aberdeen–Springfield Canal Company. Aberdeen, Idaho.

Aberdeen Times. 1936, 1950.

Adams, R. M. "Removal of Vegetation from Twin Falls Irrigation Canals." *Engineering News-Record* 85 (12 August 1920): 319–321.

Adams, Roger M. "Tile Drainage of Irrigated Land for Seepage Relief at Twin Falls." *Engineering News-Record* 84 (13 May 1920): 961.

"Administrative Water Problems: A Symposium." Paper 1729. *American Society of Civil Engineers Transactions* 94 (1930): 295–325.

Ahlson, C. B. *Report of the Seed Commissioner for the Biennium 1921–1922*. University of Idaho Agricultural Experiment Station, Circular 31. Moscow: University of Idaho, January 1923.

Aicher, L. C. *The Production of Alfalfa Seed in Southern Idaho*. University of Idaho Agricultural Experiment Station, Bulletin 101. Moscow: University of Idaho, July 1917.

———. *The Production of Clover Seed Under Irrigation in Southern Idaho*. University of Idaho Agricultural Experiment Station, Bulletin 100. Moscow: University of Idaho, June 1917.

Aldrich, J. M. *Grasshopper and Cricket Outbreaks*. University of Idaho Agricultural Experiment Station, Bulletin 41. Moscow: Moscow Weekly Mirror Print, March 1904.

———. *Insecticides and Spraying*. University of Idaho Agricultural Experiment Station, Bulletin 7. Moscow: Mirror Book Print, April 1894.

Alt, David, and Donald W. Hyndman. *Northwest Exposures: A Geologic Story of the Northwest*. Missoula: Mountain Press Publishing Company, 1995.

Alt, David D., and Donald W. Hyndman. *Roadside Geology of Idaho*. Missoula: Mountain Press Publishing Company, 1989.

American Falls Canal and Power Company Papers. Department of Special Collections and Archives, Merrill Library, Utah State University. Logan, Utah.

American Falls Press. *Commemorating the Dedication of American Falls Dam* (Official Souvenir Edition). N.p.: n.p., [1925?]. In Manuscript Collection

544, Idaho Travel and Tourism, Idaho Historical Society Library and Archives, Boise.

Anderson, Charles E., Irene Welch Grissom, Grover Jensen, Axel B. Anderson, and S. A. Johnson. *After Fifty Years*. Caldwell, Idaho: Caxton Printers, 1941.

Anderson, L. W. J. "Aquatic Weed Problems and Management in North America." In *Aquatic Weeds: The Ecology and Management of Nuisance Aquatic Vegetation,* edited by Arnold H. Pieterse and Kevin J. Murphy. London: Oxford University Press, 1990.

Arrington, Leonard J. *Beet Sugar in the West: A History of the Utah-Idaho Sugar Company*. Seattle: University of Washington Press, 1966.

———. *Great Basin Kingdom: An Economic History of the Latter-day Saints*. Cambridge, Mass.: Harvard University Press, 1958; rpt. Lincoln: University of Nebraska Press, 1966.

———. *History of Idaho*. 2 vols. Moscow: University of Idaho Press; and Boise: Idaho State Historical Society, 1994.

———. "Idaho and the Great Depression." *Idaho Yesterdays* 13 (Spring 1969): 2–8.

———. "Irrigation in the Snake River Valley: An Historical Overview." *Idaho Yesterdays* 30 (Spring/Summer 1986): 3–11.

———. "Launching Idaho's Sugar Beet Industry." *Idaho Yesterdays* 9 (Fall 1965): 16–27.

———. "The Mormon Settlement of Cassia County, Idaho, 1873–1921." *Idaho Yesterdays* 23 (Summer 1979): 36–46.

———. "Recalling a Twin Falls Childhood." *Idaho Yesterdays* 25 (Winter 1981): 31–40.

———. "Science, Government, and Enterprise in Economic Development: The Western Beet Sugar Industry." *Agricultural History* 41 (December 1967): 1–18.

Arrington, Leonard, and Dean May. "'A Different Mode of Life': Irrigation and Society in Nineteenth-Century Utah." *Agricultural History* 49 (1975): 2–20.

Atkeson, F. W., D. L. Fourt, George L. Sulerud, and B. H. Critchfield. *The Dairy Situation in Idaho*. University of Idaho Agricultural Experiment Station, Bulletin 152. Moscow: University of Idaho, July 1927.

Atkinson, Alfred. "Eradicating Weeds." *Gem State Rural* 12 (2 May 1907): 1.

Atrushi, Siddik. "Geographic Elements in Irrigation Agriculture in Boise Valley, Idaho." Ph.D. dissertation, Clark University, 1952.

Attebery, Louie W. "From Littoral to Lateral." *Idaho Yesterdays* 30 (Spring/Summer 1986): 26–32.

Ayres, Jessie C., H. W. Hulbert, and C. B. Ahlson. *Idaho Weeds: How to Know and Control Them.* University of Idaho College of Agriculture, Extension Bulletin 65. Boise: n.p., April 1926.

Bailey, Vernon. *Harmful and Beneficial Mammals of the Arid Interior.* U.S. Department of Agriculture, Farmers' Bulletin 335. Washington, D.C.: Government Printing Office, 1908.

———. *Revision of the Pocket Gophers of the Genus Thomomys.* U.S. Department of Agriculture, Bureau of Biological Survey, North American Fauna 39. N.p.: n.p., November 1915.

Balcom, Robert. *Preliminary Survey of the Aquatic Weed Problem in Irrigation Systems.* Denver: U.S. Department of the Interior, Bureau of Reclamation, Operation and Maintenance Division, March 1943.

Baldwin, G. Clyde. "Transmission and Delivery of Stored Water." *American Society of Civil Engineers Transactions* 94 (1930): 296–300, 324–325.

Baldwin, M., and F. O. Youngs. "Soil Survey of the Twin Falls Area, Idaho." In *Field Operation of the Bureau of Soils, 1921.* U.S. Department of Agriculture, Bureau of Soils, Twenty-third Report. Washington, D.C.: Government Printing Office, 1926.

Bark, Don H. "Duty of Water Investigation." In Idaho. State Engineer. *Biennial Report . . . 1909–1910.*

———. "Duty of Water Investigation." In Idaho. State Engineer. *Biennial Report . . . 1913–1914.*

Barrett, Gwynn, and Leonard Arrington. "The 1921 Depression: Its Impact on Idaho." *Idaho Yesterdays* 15 (Summer 1971): 10–15.

Bartlett, William Francis. "Irrigation Conditions in Raft River Water District, Idaho, 1904." In *Annual Report of Irrigation and Drainage Investigations, 1904.* U.S. Department of Agriculture, Office of Experiment Stations, Bulletin 158. Washington, D.C.: Government Printing Office, 1905.

Beal, Merrill D., and Merle W. Wells. *History of Idaho,* Vol. 2. New York: Lewis Historical Publishing Company, 1959.

Beck, James O. See Idaho. Department of Fish and Game. *Biennial Report.*

Bender, Thomas. *Community and Social Change in America.* New Brunswick, N.J.: Rutgers University Press, 1978; rpt. Baltimore: Johns Hopkins University Press, 1982.

Benjamin, Lester Vance. "Experimental and Historical Development of Weed Control in Idaho." M.S. thesis, University of Idaho, 1932.

Bennett, E. R. "Storing the Potato the Most Natural Way." *Idaho Farmer* 28 (1 September 1921): 1.

Bentley, E. B., and Glenn Oakley. "A Climate of Change." In *Snake: The Plain and Its People,* edited by Todd Shallat. Boise: Boise State University, 1994.

Bitton, Davis. "Peopling the Upper Snake: The Second Wave of Mormon Settlement in Idaho." *Idaho Yesterdays* 23 (Summer 1979): 47–52.

Blanchard, C. J. "The End of the Rainbow Ranch." *Reclamation Record* 13 (October 1922): 250–252.

———. "The Minidoka Project, Idaho." *Reclamation Record* 6 (February 1915): 78.

———. "Playgrounds on Reclamation Projects." *Reclamation Record* 8 (August 1917): 375.

———. "Taming the Snake: Controlling a Mighty River to Reclaim a Million Acres of Idaho's Desert." *Reclamation Record* 10 (December 1919): 563–565.

Bliss, George H. "Organization of Irrigation Operating and Maintenance Force and Its Training." *Engineering and Contracting* 41 (6 May 1914): 535–537.

Boag, Peter. *Environment and Experience: Settlement Culture in Nineteenth-Century Oregon.* Berkeley: University of California Press, 1992.

Bogue, Allan G. "The Iowa Claims Clubs: Symbol and Substance." In *The Public Lands: Studies in the History of the Public Domain,* edited by Vernon Carstensen. Madison: University of Wisconsin Press, 1963.

Bonnett, R. K. "Dodder, and the Means for Its Control." *Idaho Farmer* 26 (4 November 1920): 1583.

———. "Pests Threaten Clover Seed Industry of Idaho." *Idaho Farmer* 26 (23 September 1920): 1383.

———. "Red Clover Seed Production in Minidoka County." *Idaho Farmer* 26 (21 October 1920): 1515.

Bonnett, R. K., and H. W. Hulbert. *Sweet Clover.* University of Idaho Agricultural Experiment Station, Circular 22. Moscow: University of Idaho, March 1922.

Boone, Lalia. *Idaho Place Names: A Geographical Dictionary.* Moscow: University of Idaho Press, 1988.

Borror, Donald J., and Richard E. White. *A Field Guide to the Insects, North America North of Mexico.* Boston: Houghton Mifflin Company, 1970.

Botkin, Daniel B. *Discordant Harmonies: A New Ecology for the Twenty-first Century.* New York: Oxford University Press, 1990.

Bowler, Peter. "Mollusks of the Middle Snake River." In *Snake: The Plain and Its People,* edited by Todd Shallat. Boise: Boise State University, 1994.

Boyce, Ronald R. "The Mormon Invasion and Settlement of the Upper Snake River Plain in the 1880s: The Case of Lewisville, Idaho." *Pacific Northwest Quarterly* 78 (January–April 1987): 50–58.

Braithwaite, Lenora. "We Are of Yesterday," MC 21-5-156. Special Collections Department, Eli M. Oboler Library, Idaho State University, Pocatello, Idaho.

Brandhorst, L. Carl. "The Panacea of Irrigation: Fact or Fancy." *Journal of the West* 7 (October 1968): 491–508.

Brindley, T. A., and Joseph C. Chamberlain. "The Pea Weevil." In *Insects: The*

Yearbook of Agriculture. 1952. Washington: U.S. Government Printing Office, n.d.

Bristol, Ralph Scott. "Chemical Weed Control in Idaho." M.S. thesis, University of Idaho, 1932.

Bruns, V. F., M. M. Hodgson, H. F. Arle, and F. L. Timmons. *The Use of Aromatic Solvents for Control of Submersed Aquatic Weeds in Irrigation Channels.* U.S. Department of Agriculture, Circular 971. Washington, D.C.: Government Printing Office, 1955.

Brunt, Fannie G. "The Saga of the Snake River, or Magic Valley's Yesterdays." (n.d.) In Idaho–History–Drama vertical file, Idaho Room, Idaho Falls Public Library, Idaho Falls.

Budiansky, Stephen. *The Covenant of the Wild: Why Animals Chose Domestication.* New York: William Morrow and Company, 1992.

Burbank, Luther. *The Training of the Human Plant.* New York: Century Company, 1907.

Burt, William H., and Richard P. Grossenheider. *A Field Guide to the Mammals, North America North of Mexico.* 3d ed. Boston: Houghton Mifflin Company, 1976.

Caldwell, H. H., and Merle Wells. *Economic and Ecological History Support Study: A Case Study of Federal Expenditures on a Water and Related Land Resources Project: Boise Project, Idaho and Oregon.* Moscow: Idaho Water Resources Research Institute, 1974.

Caldwell Tribune (Caldwell). 1884–1910.

Carlton, Neil H. "A History of the Development of the Boise Irrigation Project." M.A. thesis, Brigham Young University, 1969.

Carter, George N. See Idaho. Department of Reclamation. *Biennial Report.*

Carter, Kate B., ed. *Pioneer Irrigation: Upper Snake River Valley.* Salt Lake City: Daughters of Utah Pioneers, 1955.

Carter, Walter. *Ecological Studies of the Beet Leafhopper.* U.S. Department of Agriculture, Technical Bulletin 206. Washington, D.C.: U.S. Government Printing Office, November 1930.

Cates, H. R. "The Weed Problem in American Agriculture." In *Yearbook of the United States Department of Agriculture, 1917.* Washington, D.C.: Government Printing Office, 1918.

Chapman, Arthur. *Out Where the West Begins and Other Western Verses.* Boston and New York: Houghton Mifflin Company, 1916.

Chapman, Joseph A., and George A. Feldhamer, eds. *Wild Mammals of North America.* Baltimore: Johns Hopkins University Press, 1982.

Chapman, Stephen R., and Lark P. Carter. *Crop Production: Principles and Practices.* San Francisco: W. H. Freeman and Company, 1976.

"Cheese Factories Succeed on the Minidoka Project." *Reclamation Record* 14 (November–December 1923): 319.

Christ, J. H. *Reed Canary Grass*. University of Idaho Agricultural Experiment Station, Circular 71. Moscow: University of Idaho, June 1933.

Churcher, Peter B., and John H. Lawton. "Beware of Well-fed Felines." *Natural History* 98 (July 1989): 40–46.

Clements, Louis J., and Harold Forbush. *Pioneering the Snake River Fork Country*. Rexburg: Eastern Idaho Publishing Company, 1972.

Coates, Charles. "Federal-Local Relationships on the Boise and Minidoka Projects, 1904–1926." *Idaho Yesterdays* 25 (Summer 1981): 2–9.

Coburn, F. D. *The Book of Alfalfa*. New York: Orange Judd Company, 1918.

Cole, D. W. "Patrol of Canals on Irrigation Projects." *Reclamation Record* 9 (January 1918): 24–25.

Collins, Henry Hill. *Complete Field Guide to American Wildlife*. New York: Harper and Row, 1959.

Courtenay, Walter R., Dannie A. Hensley, Jeffrey N. Taylor, and James A. McCann. "Distribution of Exotic Fishes in the Continental United States." In *Distribution, Biology, and Management of Exotic Fishes*, edited by Walter R. Courtenay, Jr., and Jay R. Stauffer, Jr. Baltimore: Johns Hopkins University Press, 1984.

Cragg, Barbara, Dennis M. Walsh, and Mary Ellen Walsh, eds. *The Idaho Stories and Far West Illustrations of Mary Hallock Foote*. Pocatello: Idaho State University Press, 1988.

Crandall, Alice Fern Ravenscraft. "From Dawn to Dusk: An Autobiography," MC 21-2-45. Special Collections Department, Eli. M. Oboler Library, Idaho State University, Pocatello, Idaho.

Crandall, Lynn. "Crandall on Administrative Water Problems." *American Society of Civil Engineers Transactions* 94 (1930): 318–321.

———. "Ground Water Flows of the Snake River Plain." Address to Idaho State Reclamation Association. 31 March 1953. Copy in Idaho–Reclamation of Land vertical file, Idaho Room, Idaho Falls Public Library, Idaho Falls.

———. "Ground Water Inflow to American Falls Reservoir." July 1959. Box 2, Lynn Crandall Papers, American Heritage Center, University of Wyoming, Laramie.

———. "History of Irrigation Development in the Snake River Valley." [1946?]. Manuscript copy in Idaho–Reclamation of Land vertical file, Idaho Room, Idaho Falls Public Library, Idaho Falls.

———. "Remarks by Lynn Crandall to the Snake River Waterusers Upon the Occasion of His 25th Election as Watermaster, March 1954." Manuscript copy in Idaho Reclamation of Land vertical file, Idaho Room, Idaho Falls Public Library, Idaho Falls.

———. "Snake River." [1958?]. Manuscript copy in Idaho–Reclamation of Land vertical file, Idaho Room, Idaho Falls Public Library, Idaho Falls.

———. "Use of Water on Twin Falls North Side Project, Idaho" [1923]. George N. Carter Papers. American Heritage Center, University of Wyoming, Laramie.

Crawford, Hugh L. "Report On Cleaning Canals With Modified Disk Harrows, Minidoka Irrigation Project, Idaho. *Reclamation Record* 13 (February 1922): 22–24.

Cronholm, F. N. "Drainage System for the North Side Minidoka Irrigation Project." *Engineering News-Record* 69 (25 April 1914): 468–469.

Cronon, William. *Changes in the Land: Indians, Colonists, and the Ecology of New England.* New York: Hill and Wang, 1983.

———. *Nature's Metropolis: Chicago and the Great West.* New York: Norton, 1991.

———. "A Place for Stories: Nature, History, and Narrative." *Journal of American History* 78 (March 1998): 1347–1376.

———. "The Trouble With Wilderness; or, Getting Back to the Wrong Nature." *Environmental History* 1 (January 1996): 7–28.

———, ed. *Uncommon Ground: Toward Reinventing Nature.* New York: W. W. Norton, 1995.

Crosby, Alfred. *Ecological Imperialism: The Biological Expansion of Europe, 900–1900.* New York: Cambridge University Press, 1986.

Crowe, F. T. "Draining the Pioneer Irrigation District." *Reclamation Record* 5 (October 1914): 373–374.

Dams–Arrowrock Vertical File. Idaho Historical Society Library and Archives, Boise.

Darlington, E. B. "Construction of the Milner-Gooding Canal." *New Reclamation Era* 22 (November 1931): 242–243, 245.

———. "Greater Minidoka Project." *Burley Bulletin,* 6 September 1928.

———. "The Romance of A River: A Past, Present, and Future Survey of Irrigation in Southern Idaho." *Reclamation Record* 11 (March 1920): 122–123.

———. "Sources and Uses of the Snake." *Aberdeen Times,* 7 March 1935.

———. "Use of Water on the Salmon River Tract." *Reclamation Record* 9 (May 1918): 225–228.

Darlington, Wayne. See Idaho. State Engineer. *Biennial Report.*

Davis, A. P. "Why Some Irrigation Canals and Reservoirs Leak." *Engineering News-Record* 80 (4 April 1918): 663–665.

Davis, Belinda. *A Study of Irrigation and the Development of Ada County.* N.p.: Ada County Historic Preservation Council, August 1990.

Davis, E. L. "E. L. Davis Starts Series on History and Growth of Aberdeen Area," *Aberdeen Times,* 27 April, 4, 11, 18, 25 May, 8, 15, 22, 29 June 1950.

———. "Memories of Early Aberdeen," MC 21-2-52. Special Collections Department, Eli M. Oboler Library, Idaho State University, Pocatello.

Davis, James W., and Nicki Balch Stilwell. *Aristocrat in Burlap: A History of the Potato in Idaho.* N.p.: Idaho Potato Commission, 1975.

DeBerard, W. W. "Southern Idaho, an Engineers' Creation." *Engineering News-Record* 89 (21 September 1922): 473–475.

Deer Flat National Wildlife Refuge Records. Deer Flat National Wildlife Refuge. Nampa, Idaho.

"Destroying Grasshoppers With Hopperdozer." *Gem State Rural* 13 (July 1907): 15.

Dibble, Barry. "Jackson Lake Reservoir." *Reclamation Record* 12 (September 1921): 423–424.

Dibble, Barry, and T. W. Parry. "Control of Moss, Weeds, and Willows on the Minidoka Project." *Reclamation Record* 8 (April 1917): 192–193.

Dille, F. M. "The Minidoka National Bird Reservation, Idaho." *Reclamation Record* 7 (December 1916): 564–565.

"Distribution of Snake River Water During Greatest Drought." *Engineering News-Record* 85 (11 November 1920): 927–931.

"Drainage for Irrigation Projects in Idaho." *Reclamation Record* 9 (February 1918): 68.

Dunbar, Robert G. *Forging New Rights in Western Waters.* Lincoln: University of Nebraska Press, 1983.

Dunlap, Thomas R. *Saving America's Wildlife: Ecology and the American Mind, 1850–1990.* Princeton: Princeton University Press, 1988.

Eberlein, Charlotte V., and William C. Schaffers. *Herbicide Carryover to Potatoes.* University of Idaho, College of Agriculture, Agricultural Experiment Station, Current Information Series 864. N.p.: n.p., April 1990.

Edmundson, W. C. *Insect Pests of the Orchards and Gardens of Idaho and Their Control.* University of Idaho Agricultural Experiment Station, Bulletin 87. Moscow: University of Idaho, February 1916.

Ehernberger, James L., and Francis G. Gschwind. *Smoke down the Canyon: Union Pacific, Idaho Division.* Callaway, Nebr.: E. and G. Publications, 1966.

"Eilers Family History," MC 21-2-33. Special Collections Department, Eli M. Oboler Library, Idaho State University, Pocatello.

Eke, Paul A., and Neil W. Johnson. *Profitable Systems of Farming for the Idaho Falls Area.* University of Idaho Agricultural Experiment Station, Bulletin 198. Moscow: University of Idaho, June 1933.

Enterprise Irrigation District Records. Enterprise Irrigation District. Teton, Idaho.

Erskine, Ray. "The Alfalfa Weevil in Southern Idaho." *Idaho Farmer* 28 (4 August 1921): 75.

"Explosion in the Magic Valley." *American Heritage* 32 (April/May 1981): 26–37.

Farmers' Friend (Blackfoot). 1906.

The Farming Business in Idaho. University of Idaho Agricultural Experiment Station, Bulletin 151. Moscow: University of Idaho, July 1927.

Fiege, Mark. "Creating a Hybrid Landscape: Irrigated Agriculture in Idaho." *Illahee: Journal for the Northwest Environment* 11 (Spring/Summer 1995): 60–76.

———. "Wildlife and Irrigation Systems along the Snake River, Idaho." *Transactions of the Fifty-seventh North American Wildlife and Natural Resources Conference* (1992): 724–732.

Fillmore, Gwen Berrett, Elaine Brinton Poole, and Fontella Bitton Spelts. *Menan, Idaho, 1879–1986.* N.p.: n.p., n.d.

Fireman, Milton, and H. E. Hayward. "Irrigation Water and Saline and Alkaline Soils." In *Water: United States Department of Agriculture, Yearbook of Agriculture, 1955.* Washington, D.C.: Government Printing Office, n.d.

Fireman, Milton, C. A. Mogen, and G. Orion Baker. *Characteristics of Saline and Alkaline Soils in the Emmett Valley Area, Idaho.* University of Idaho Agricultural Experiment Station, Research Bulletin 17. Moscow: University of Idaho, May 1950.

Fiske, George B., and Paul Froelich. *Marketing the Early Potato Crop.* U.S. Department of Agriculture, Farmers' Bulletin 1316. Washington, D.C.: Government Printing Office, 1923.

Fite, Gilbert. *American Farmers: The New Minority.* Bloomington: Indiana University Press, 1981.

Fitzgerald, Deborah. "Farmers Deskilled: Hybrid Corn and Farmers' Work." *Technology and Culture* 34 (April 1993): 324–343.

Fitzgerald, Oren A. "The Milk Cow Brings Money into Idaho." *Idaho Farmer* 37 (17 June 1926): 3, 6.

———. "The Seed Industry in Idaho Makes Remarkable Progress." *Idaho Farmer* 35 (19 March 1925): 268, 270.

Flynn, P. J. *Irrigation Canals and Other Irrigation Works.* San Francisco: George Spaulding and Company, 1892.

Fogg, Percival M. "Drainage of Irrigated Lands on the Minidoka Project, Idaho." *Journal of the Idaho Society of Engineers* 2 (June 1912): 106–118.

Foote, Arthur Dewint. *Report on the Irrigating and Reclaiming of Certain Desert Lands in Idaho and Other Projects Connected Therewith.* N.p.: n.p., 1887.

Foote, Mary Hallock. *The Chosen Valley.* Boston and New York: Houghton Mifflin and Company; Cambridge, Mass.: Riverside Press, 1892.

———. "The Irrigating Ditch." *Century Magazine* 38 (June 1889): 299–300.

———. *A Victorian Gentlewoman in the Far West: The Reminiscences of Mary Hallock Foote,* edited by Rodman W. Paul. San Marino, Calif.: Huntington Library, 1972.

Forbush, Harold. "A Trilogy of Affliction upon Our Valley Pioneers." *Snake River Echoes* 11 (1982): 100–103.

Fortier, Samuel. "Design and Structural Details of Concrete Lining for Irrigation Canals." *Concrete* 18 (April 1916): 174–178.

Fouch, D. W. "The Farmers' Unappreciated Friend." *Gem State Rural* 12 (28 February 1907): 1.

Fowler, Delbert. "Personal Journal of Delbert Fowler," MC 21-1-21. Special Collections Department, Eli M. Oboler Library, Idaho State University, Pocatello, Idaho.

Francis, John, ed. *Sagebrush to Cropland: A Western Water Project.* Moscow: Idaho Water Resources Research Institute, University of Idaho, March 1979.

Gabrielson, Ira N. *Wildlife Refuges.* New York: Macmillan, 1943.

Gamboa, Erasmo. *Mexican Labor and World War II: Braceros in the Pacific Northwest, 1942–1947.* Austin: University of Texas Press, 1990.

Garland, Hamlin. *A Son of the Middle Border.* New York: Macmillan, 1935.

Gates, Paul Wallace. *History of Public Land Law Development.* Washington, D.C.: Government Printing Office, 1968.

Geertz, Clifford. *The Interpretation of Cultures.* New York: Basic Books, 1973.

Gem State Rural (Caldwell). 1895–1909.

Gem State Rural and Live Stock Journal (Caldwell). 1910–1916.

Gertsch, William Darrell. "The Upper Snake River Project: A Historical Study of Reclamation and Regional Development, 1890–1930." Ph.D. dissertation, University of Washington, 1974.

Gleisner, Donna M. *Ring-necked Pheasant.* Idaho Department of Fish and Game, Wildlife Leaflet 3. N.p.: n.p., 1983.

Gneiting, Gary Wayne. "An Economic History and Analysis of the Great Feeder Canal of Southeastern Idaho." M.S. thesis, Utah State University, 1972.

Goin, Peter. *Humanature.* Austin: University of Texas Press, 1996.

Gooding Leader. 1935.

Goudie, Andrew. *The Human Impact on the Natural Environment,* 4th ed. Cambridge, Mass.: MIT Press, 1994.

Grebe, G. W. "What Is or Should Be the Status of Wildlife as a Factor in Drainage and Reclamation Planning?" *Transactions of the Third North American Wildlife Conference* (1938): 119–124.

Greenwood, Annie Pike. *We Sagebrush Folks.* New York: D. Appleton-Century, 1934; rpt. Moscow: University of Idaho Press, 1988.

Grissom, Irene Welch. "The Desert Reclaimed." *New West Magazine* 11 (December 1920): 15.

———. *We Harness A River.* Caldwell, Idaho: Caxton Printers, 1946.

Grover, Dorys C. *A Solitary Voice: Vardis Fisher.* New York: Revisionist Press, 1973.

Hackett, Bill, and Bill Bonichsen. "Volcanic Crescent." In *Snake: The Plain and Its People,* edited by Todd Shallat. Boise: Boise State University, 1994.

Haegele, Rowland W. *The Beet Leaf-Hopper (Eutettix tennellus Baker): A Survey in Idaho.* University of Idaho Agricultural Experiment Station, Bulletin 156. Moscow: University of Idaho, July 1927.

———. *Field Studies of the Beet Leafhopper.* University of Idaho Agricultural Experiment Station, Bulletin 182. Moscow: University of Idaho, February 1932.

Hahn, Steven, and Jonathon Prude, eds. *The Countryside in the Age of Capitalist Transformation: Essays in the Social History of Rural America.* Chapel Hill: University of North Carolina Press, 1985.

Hammer, Michael Kimmer. "Life Story of Michael Kimmer Hammer," MC 21-2-48. Special Collections Department, Eli M. Oboler Library, Idaho State University, Pocatello, Idaho.

Hammond, M. J. "The Largest, the Best, and the Cheapest Irrigation System in the United States." *Ribgy Star,* 26 August 1909.

Hannah, F. W. "Losses in Canals and Methods of Prevention." *Journal of the Idaho Society of Engineers* 1 (May 1911): 29–36.

Hansen, Gary W., Floyd E. Oliver, and N. E. Otto. *Herbicide Manual.* Denver: U.S. Department of the Interior, Bureau of Reclamation, 1983.

Harper, K. T. "Historic Environments." In *Handbook of North American Indians,* Vol. 2: *Great Basin,* edited by W. L. D'Azevedo. Washington, D.C.: Smithsonian Institution Press, 1986.

Harris, F. S. *The Sugar Beet in America.* New York: Macmillan Company, 1919.

Harrison, Earl J., and Donald R. Johnson. *Mammals of Idaho.* Moscow: University of Idaho Press, 1981.

Harvey, W. A. "Plant Diseases, Insects, and Weeds as Affected by Irrigation." In *Proceedings: First Intersociety Conference on Irrigation and Drainage: Can Man Develop a Permanent Irrigation Agriculture?* edited by Robert M. Hagan. San Francisco: n.p., 1957.

Hawley, Ellis W. *The Great War and the Search for Modern Order: A History of the American People and Their Institutions, 1917–1933.* New York: St. Martin's Press, 1979.

Hays, Samuel P. *The Response to Industrialism, 1885–1914.* Chicago: University of Chicago Press, 1957.

Henderson, L. F. *Twelve of Idaho's Worst Weeds.* University of Idaho Agricultural Experiment Station, Bulletin 14. Moscow: Moscow Mirror, 1898.

Henry, Paul L. "The Appropriation of Water." In *Seventh Annual Report of the Department of Reclamation. State of Idaho, 1931–1932,* by R. W. Faris. Boise: n.p., n.d.

Hinton, Richard J. *Irrigation in the United States.* 49th Cong., 2d Sess., 1887, Senate Miscellaneous Document 15.

———. *A Report on Irrigation and the Cultivation of the Soil Thereby.* 52d Cong., 1st Sess., 1893, Senate Executive Document 41.

History of Idaho Territory, Showing Its Resources and Advantages; With Illustrations Descriptive of Its Scenery, Residences, Farms, Mines, Mills, etc. San Francisco: Wallace W. Elliot and Company, 1884.

Hobson, Karl. "Costs and Factors Influencing the Costs of Producing Sugar Beets on 69 Idaho Farms in 1937." M.A. thesis, University of Idaho, 1941.

Hochbaum, H. W. "Working Plan of Farm Bureau." *Idaho Farmer* 23 (28 November 1918): 7.

Holmes, Alvin C. *Swedish Homesteaders in Idaho on the Minidoka Irrigation Project.* N.p.: n.p., 1976.

Holte, K. E., B. L. Keller, J. H. Merriam, G. W. Minshall, M. G. Pavesic, F. L. Rose, and C. H. Trost. *Environmental Surveys of the Teton River and Henry's Fork of the Snake River.* Idaho Water Resources Research Institute Technical Completion Report, WRRI Code USCE 44–418. Moscow: Idaho Water Resources Research Institute, 1972.

Hornaday, William T. *The American Natural History,* Vol. 1: *Mammals.* New York: Charles Scribners Sons, 1914.

Hornbein, Julius. "Seepage Water, Formerly Wasted, Now Pumped for Irrigation." *Engineering News-Record* 78 (26 April 1917): 192–193.

Houston, D. F. "Report of the Secretary of Agriculture." In *Yearbook of the United States Department of Agriculture, 1917.* Washington, D.C.: Government Printing Office, 1918.

Hoyt, W. G. *Water Utilization in the Snake River Basin.* U.S. Geological Survey Water-Supply Paper 657. Washington, D.C.: Government Printing Office, 1935.

Hulbert, H. W., J. H. Christ, and John L. Toevs. *Adaptability of Alfalfa Strains and Varieties for Idaho.* University of Idaho Agricultural Experiment Station, Bulletin 199. Moscow: University of Idaho, December 1933.

Hunter, Byron. *Preliminary Report on Farm Organization in Twin Falls and Latah Counties.* University of Idaho Agricultural Experiment Station, Bulletin 123. Moscow: University of Idaho, February 1921.

Hunter, Byron, and Samuel B. Nuckols. *An Economic Study of Irrigated Farming in Twin Falls County, Idaho.* U.S. Department of Agriculture, Bulletin 1421. Washington, D.C.: Government Printing Office, October 1926.

———. *Farm Costs and Relative Profitableness of Seven Crops, Twin Falls County, Idaho, 1919 and 1920.* University of Idaho Agricultural Experiment Station, Research Bulletin 2. Moscow: University of Idaho, March 1922.

Hutchins, Wells A. "Idaho Law of Water Rights." *Idaho Law Review* 5 (Fall 1968): 1–129.

———. *Irrigation Districts: Their Organization, Operation, and Financing.* U.S. Department of Agriculture, Technical Bulletin 254. Washington, D.C.: Government Printing Office, June 1931.

———. *Mutual Irrigation Companies.* U.S. Department of Agriculture, Technical Bulletin 82. Washington, D.C.: Government Printing Office, January 1929.

"Idaho a Mammoth Seed Factory." *Idaho Club Woman* 4 (June 1914): 39.

Idaho. Commissioner of Immigration, Labor, and Statistics. *Seventh Biennial Report, 1911–1912.* By S. J. Rich. Caldwell, Idaho: Caxton Printers, [1912?].
Idaho. Department of Fish and Game. "Between Land and Water: The Wetlands of Idaho." *Nongame Wildlife Leaflet 9.* Boise: Idaho Department of Fish and Game, 1990.
———. *Biennial Report.* N.p.: n.p., n.d.
———. *Twenty-first . . .* [1946?] By James O. Beck.
———. *Biennial Report of the Fish and Game Warden of the State of Idaho.* N.p.: n.p., n.d.
———. *Second . . . [1909–1908].*
———. *Eighth . . . 1919–1920.* By Otto Jones.
Idaho. Department of Reclamation. *Annual Report.* Boise: n.p., n.d.
———. *Seventh . . . 1931–1932.* By R. W. Faris.
———. *Ninth . . . 1935–1936.* By R. W. Faris.
———. *Biennial Report.* Boise: n.p., n.d.
———. *First . . . 1919–1920.* By W. G. Swendsen.
———. *Fourth . . . 1925–1926.* By George N. Carter
———. *Sixth . . . 1929–1930.* By George N. Carter
———. *Twenty-first . . . 1959–1960.* By George N. Carter
Idaho. Department of Water Resources, Records of Water District 1 (microfiche). Idaho Department of Water Resources. Boise, Idaho.
Idaho Falls Daily Post. 1922, 1923.
Idaho Falls Post-Register. 1934.
Idaho Falls Times. 1905–1916.
Idaho Farmer (Caldwell). 1916–1927.
Idaho. *General Laws of the State of Idaho.* Boise: Syms-York, 1907, 1911, 1912, 1913; n.p., 1919.
Idaho–History–Drama vertical file, Idaho Room, Idaho Falls Public Library, Idaho Falls, Idaho.
Idaho–Reclamation of Land vertical files, Idaho Room, Idaho Falls Public Library, Idaho Falls, Idaho.
Idaho Register (Idaho Falls). 1906–1919.
Idaho Republican (Blackfoot). 1907–1923.
Idaho. Sixth Judicial District Court. *In the District Court of the Sixth Judicial District of the State of Idaho, In and For the County of Fremont: Decree: Rexburg Irrigation Company et al., Plaintiffs, v. Teton Irrigation Canal Company et al., Defendants.* N.p.: n.p., 1911.
Idaho. State Employment Service. *Newsletter,* 1 September 1937. Copy at Idaho State Library, Boise.
Idaho. State Engineer. *Biennial Report . . . to the Governor.* N.p.: n.p., n.d.
———. *1899–1900.* By D. W. Ross. Boise: Capital Printing Office, n.d.
———. 1901–1902. By E. W. Ross. Boise: Statesman Print, n.d.

———. *1903–1904.* By Wayne Darlington.
———. *Sixth . . . 1905–1906.* By James Stephenson, Jr.
———. *Eighth . . . 1909–1910.* By D. G. Martin.
———. *Ninth . . . 1911–1912.* By A. E. Robinson.
———. *Tenth . . . 1913–1914.* By Herbert Wing.
———. *Eleventh . . . 1915–1916.* By J. H. Smith.

Idaho State Journal (Pocatello). 1977.

Idaho Statesman (Boise). 1865–1980.

Iddings, E. J. *Idaho Pure Seed Law.* University of Idaho Agricultural Experiment Station, Circular 65. Moscow: University of Idaho, August 1931.

———. "Neighborliness Helps Idaho Over a Critical Year." *Idaho Farmer* 34 (14 August 1924): 103.

Igler, David. "When Is a River Not a River? Reclaiming Nature's Disorder in *Lux v. Haggin.*" *Environmental History* 1 (April 1996): 52–69.

Irrigation vertical files. Idaho Historical Society Library and Archives, Boise, Idaho.

Isern, Thomas D.. *Bull Threshers and Bindlestiffs: Harvesting and Threshing on the North American Plains.* Lawrence: University Press of Kansas, 1990.

Jackson, Donald C. "Engineering in the Progressive Era: A New Look at Frederick Haynes Newell and the U.S. Reclamation Service." *Technology and Culture* 34 (July 1993): 539–574.

———. *Great American Bridges and Dams.* Washington, D.C.: Preservation Press, 1988.

Jackson, John Brinckerhoff. *Discovering the Vernacular Landscape.* New Haven: Yale University Press, 1984.

———. *The Necessity for Ruins, and Other Topics.* Amherst: University of Massachusetts Press, 1980.

"Jackson Lake Dam, the Savior of the Snake River Valley." *Engineering News-Record* 83 (11–18 December 1919): 992–994.

Jensen, Charles A., and B. A. Olshausen. "Soil Survey of the Boise Area, Idaho." In *Field Operations of the Bureau of Soils, 1901,* edited by Milton Whitney. U.S. Department of Agriculture, Bureau of Soils, 3d Report. Washington, D.C.: Government Printing Office, 1902.

Johnson, Hildegard Binder. *Order upon the Land: The U.S. Rectangular Survey and the Upper Mississippi Country.* New York: Oxford University Press, 1976.

———. "Rational and Ecological Aspects of the Quarter Section: An Example From Minnesota." *Geographical Review* 47 (July 1957): 330–348.

Johnson, Neil W., and Harold A. Vogel. *Types of Farming in Idaho,* Part 2: *The Type of Farming Areas.* University of Idaho Agricultural Experiment Station, Bulletin 208. Moscow: University of Idaho, June 1934.

Jones, Otto. See Idaho. Department of Fish and Game. *Biennial Report of the . . . Warden.*

Jordan, William H., Jr. "The Weevil and the Wasp." *Natural History* 88 (December 1979): 37–42.

Joslin, Gladys R. "Early Burley Pioneer Days Recalled: Minidoka Project History Is Related Here." *Burley Bulletin*, 1 May 1945.

Kay, Jeanne. "Preconditions of Natural Resource Conservation." In *The History of Soil and Water Conservation*, edited by Douglas Helms and Susan L. Flader. Washington, D.C.: Agricultural History Society, 1985.

"Keep the Ditches Clean." *Gem State Rural* 13 (September 1908): 12.

King, Verl. *Egin Bench Sprinkler Irrigation Impact Study Report.* Boise: Idaho Department of Water Resources, 1988.

Kirkpatrick, E. M. "Farm Lands, Their Value, and What Makes Them." *Gem State Rural* 12 (28 February 1907): 3, 4, 9.

Klontz, George W., and John G. King. *Aquaculture in Idaho and Nationwide.* Boise: Idaho Department of Water Resources, June 1975.

Kolodny, Annette. *The Land before Her: Fantasy and Experience of American Frontiers, 1650–1860.* Chapel Hill: University of North Carolina Press, 1984.

———. *The Lay of the Land: Metaphor as Experience and History in American Life and Letters.* Chapel Hill: University of North Carolina Press, 1975.

Kratville, William W. *Golden Rails.* Omaha: Kratville Publications, 1965.

Lampen, Dorothy. *Economic and Social Aspects of Federal Reclamation.* Baltimore: Johns Hopkins University Press, 1930.

Langston, Nancy. *Forest Dreams, Forest Nightmares: The Paradox of Old Growth in the Inland West.* Seattle: University of Washington Press, 1995.

Larson, A. O., T. A. Brindley, and F. G. Hinman. *Biology of the Pea Weevil in the Pacific Northwest with Suggestions for its Control on Seed Peas.* U.S. Department of Agriculture, Technical Bulletin 599. Washington, D.C.: Government Printing Office, April 1938.

Lauckhart, J. Burton, and John W. McLean. "Chinese Pheasants in the Northwest." In *Pheasants in North America*, edited by Durward L. Allen. Harrisburg, Pa.: Stackpole Company; and Washington, D.C.: Wildlife Management Institute, 1956.

Leopold, Aldo. *A Sand County Almanac, With Essays on Conservation from Round River.* New York: Oxford University Press, 1966; rpt. New York: Ballantine Books, 1970.

Leopold, Luna B. *Water: A Primer.* San Francisco: W. H. Freeman and Company, 1974.

Levenstein, Harvey A. *Revolution at the Table: The Transformation of the American Diet.* New York: Oxford University Press, 1988.

Lever, Christopher. *Naturalized Fishes of the World.* San Diego: Academic Press, 1996.

Lewis, H. G., and P. P. Peterson. "Soil Survey of the Portneuf Area, Idaho." In *Field Operations of the Bureau of Soils, 1918.* U.S. Department of Agriculture, Bureau of Soils, 20th Report. Washington, D.C.: Government Printing Office, 1924.

Lewis, Mary Gunnell. "History of Irrigation Development in Idaho." M.A. thesis, University of Idaho, 1924.

"Life and Memories of Amos William, Theresa Ellen Gengler Fawcett and children," MC 21-5-162. Special Collections Department, Eli M. Oboler Library, Idaho State University, Pocatello, Idaho.

Linder, Allan D. "Idaho's Alien Fishes." *TEBIWA: The Journal of the Idaho State Museum* 6 (2) (1963): 12–15.

Lindholm, G. F. and S. A. Goodell. *Irrigated Acreage and Land Uses on the Snake River Plain, Idaho and Eastern Oregon.* Department of the Interior, U.S. Geological Survey, Atlas HA-691. N.p.: U.S. Geological Survey, 1986.

Lisle, Charles J. "The Book Farm." *Gem State Rural and Live Stock Journal* 21 (April 1916): 1–4.

"A Look at the Boise Valley." *Idaho Yesterdays* 28 (Winter 1985): 12–19.

Lovin, Hugh. "'Duty of Water' in Idaho: A 'New West' Irrigation Controversy, 1890–1920." *Arizona and the West* 23 (Spring 1981): 5–28.

———. "The Carey Act in Idaho, 1895–1925: An Experiment in Free Enterprise Reclamation." *Pacific Northwest Quarterly* 78 (October 1987): 122–133.

———. "Footnote to History: 'The Reservoir . . . Would Not Hold Water.'" *Idaho Yesterdays* 24 (Spring 1980): 12–19.

———. "Free Enterprise and Large-scale Reclamation on the Twin Falls–North Side Tract, 1907–1930." *Idaho Yesterdays* 29 (Spring 1985): 2–14.

———. "Sage, Jacks, and Snake Plain Pioneers." *Idaho Yesterdays* 22 (Winter 1979): 13–24.

———. "Water, Arid Land, and Visions of Advancement on the Snake River Plain." *Idaho Yesterdays* 35 (Spring 1991): 3–18.

Lowell, J. H. "The Boise Project." *Reclamation Record* 6 (November 1915): 517–518.

Lowitt, Richard, and Judith Fabry, eds. *Henry A. Wallace's Irrigation Frontier: On the Trail of the Corn Belt Farmer, 1909.* Norman: University of Oklahoma Press, 1991.

McAdoo, J. Kent, and James A. Young. "Jackrabbits." *Rangelands* 2(4) (August 1980): 135–138.

McClymonds, A. E. *Alfalfa Seed Production in Southern Idaho.* University of Idaho Agricultural Experiment Station, Bulletin 143. Moscow: University of Idaho, April 1926.

McPherson, Alex. "Orchard and Farm Problems." *Gem State Rural* 5 (2 November 1899): 1.

M'Lendon, W. E. "Soil Survey of the Blackfoot Area, Idaho." In *Field Operations of the Bureau of Soils, 1903.* U.S.Department of Agriculture, Bureau of Soils, 5th Report. Washington, D.C.: Government Printing Office, 1904.

Magistad, O. C., and J. E. Christiansen. *Saline Soils: Their Nature and Management.* U.S. Department of Agriculture, Circular 707. Washington, D.C.: Government Printing Office, September 1944.

Manis, H. C., and P. W. Portman. *Idaho Recommendations for Insect Control.* University of Idaho Agricultural Experiment Station, Bulletin 279. N.p.: n.p., June 1950.

Martin, D. G. See Idaho. State Engineer. *Biennial Report.*

Martin, Julia M. "The Flying Police." *Idaho Farmer* 27 (16 June 1921): 593.

———. "On Humby's Place." *Idaho Farmer* 28 (22 September 1921): 208.

Marx, Leo. *The Machine in the Garden: Technology and the Pastoral Ideal in America.* New York: Oxford University Press, 1964.

Mason, Edward F. "Fremont Potatoes Are on the Map." *Idaho Farmer* 26 (7 October 1920): 1452.

———. "Idaho's High Grade Seeds." *Idaho Farmer* 25 (5 February 1920): 195.

May, Dean L., and Jenny Cornell. "Middleton's Agriminers: The Beginnings of an Agricultural Town." *Idaho Yesterdays* 28 (Winter 1985): 2–11.

Mead, Elwood. "Rise and Future of Irrigation in the United States." In *Yearbook of the United States Department of Agriculture, 1899.* Washington, D.C.: Government Printing Office, 1900.

Mehringer, P. J. "Prehistoric Environments." In *Handbook of North American Indians,* Vol. 2, *Great Basin,* edited by W. L. D'Azevedo. Washington, D.C.: Smithsonian Institution Press, 1986.

Meinig, D. W. "The Mormon Culture Region: Strategies and Patterns in the Geography of the American West, 1847–1964." *Annals of the Association of American Geographers* 55 (June 1965): 191–220.

Merchant, Carolyn. *Ecological Revolutions: Nature, Gender, and Science in New England.* Chapel Hill: University of North Carolina Press, 1989.

———. "Reinventing Eden: Western Culture as a Recovery Narrative." In *Uncommon Ground: Toward Reinventing Nature,* edited by William Cronon. New York: W. W. Norton, 1995.

Mercier, Laurie, and Carole Simon-Smolinski. *Idaho's Ethnic Heritage: Historical Overviews.* 3 vols. N.p.: Idaho Centennial Commission and National Park Service, U.S. Department of the Interior, 1990.

Merriam, C. Hart. "The Zoology of the Snake Plains of Idaho." *American Naturalist* (March 1892): 218–222.

Mimms, O. L., and George W. Woodbury. *Markets and Market Preferences for Idaho Potatoes.* University of Idaho Agricultural Experiment Station, Bulletin 231. Moscow: University of Idaho, June 1939.

Minidoka Project, see U.S. Department of the Interior, Bureau of Reclamation, Record Group 115.

Mogen, C. A., E. N. Poulsen, A. E. Poulsen, E. J. Van Slyke, and W. E. Colwell. *Soil Survey of the Idaho Falls Area, Idaho.* U.S. Department of Agriculture, Agricultural Research Administration. Bureau of Plant Industry, Soils, and Agricultural Engineering, Series 1939, No. 8. N.p.: n.p., December 1950.

Montgomery, John Theodore, and James Warren Barber. "An Economic Study of the History, Present Situation, and Outlook of Agriculture on the Minidoka Irrigation Project." M.S. thesis, University of Idaho, 1927.

Montgomery, Richard Calvin. "Canyon County: The Economic Geography of a Southwestern Idaho Irrigated Area." M.A. thesis, University of Nebraska, 1951.

More, C. T., and C. R. Dorland. *Commercial Handling, Grading, and Marketing of Potatoes.* U.S. Department of Agriculture, Farmers' Bulletin 753. Washington, D.C.: Government Printing Office, 1917.

Mundorf, M. J., E. G. Crosthwaite, and Chabot Kilburn. "Ground Water for Irrigation in the Snake River Basin in Idaho," U.S. Department of the Interior, Open-file report, October 1960. In U.S. Department of the Interior, Bureau of Reclamation, and U.S. Army, Corps of Engineers, *Upper Snake River Basin,* Vol. 3, Part 2: *Coordination and Reports of Cooperating Agencies: U.S. Geological Surveys—Ground Water Branch.* N.p.: n.p., 1961.

Murphy, Paul Lloyd. "Irrigation in the Boise Valley, 1863–1903: A Case Study in Pre-federal Irrigation." M.A. thesis, University of California, 1947.

———. "Early Irrigation in the Boise Valley." *Pacific Northwest Quarterly* 44 (October 1953): 177–184.

Nampa Cactus Hedge Fence and Orchard Company, Articles of Incorporation, 13 October 1892. File 101, Articles of Incorporations Files, Canyon County Courthouse. Caldwell, Idaho.

Nash, Gerald D. "The Sugar Beet Industry and Economic Growth in the West." *Agricultural History* 41 (1967): 27–30.

Nelson, Elias. *Alfalfa.* University of Idaho Agricultural Experiment Station, Bulletin 66. Moscow: Idaho Post Print, 1909.

Nessly, J. E. "Growers Must Fight the Alfalfa Weevil." *Idaho Farmer* 32 (5 July 1923): 2.

Newell, F. H. *Report on Agriculture by Irrigation in the Western Part of the United States at the Eleventh Census: 1890.* Washington, D.C.: Government Printing Office, 1894.

———. *Survey of Reservoir Sites.* 55th Cong., 3d Sess., 1899, Senate Document 116.

Newell, F. H., and Daniel William Murphy. *Principles of Irrigation Engineering.* New York: McGraw-Hill, 1913.

Newell, R. J. "40 Years of O & M." *Idaho Yesterdays* 2 (Fall 1958): 2–7.

Newell, Thomas R. "Newell on Administrative Water Problems." *American Society of Civil Engineers Transactions* 94 (1930): 321–324.

New Sweden Irrigation District Records. New Sweden Irrigation District. Idaho Falls, Idaho.

Oakley, R. A. "The Seed Supply of the Nation." In *Yearbook of the United States Department of Agriculture, 1917*. Washington, D.C.: Government Printing Office, 1918.

O'Donnell, I. D. "Duty of Water." *Reclamation Record* 7 (September 1916): 406–407.

———. "Weeds—Huns." *Reclamation Record* 9 (June 1918): 268.

Ojala, J. C., S. L. Love, J. J. Pavek, and D. L. Corsini. *Potato Varieties for Idaho.* University of Idaho, Cooperative Extension System and Agricultural Experiment Station, Current Information Series 454. N.p.: n.p., September 1989.

Olin, W. H. *American Irrigation Farming.* Chicago: A. C. McClure and Company, 1913.

"Opening of the Great Reclamation Project." *Gem State Rural* 14 (March 1909): 1.

Osborne, O. M. *Weed Pests of Idaho and Methods of Eradication.* University of Idaho Agricultural Experiment Station, Bulletin 71. N.p.: n.p., July 1911.

"Our Reservoirs As Pleasure Resorts." *Reclamation Record* 8 (November 1917): 526–527.

Ourada, Patricia K. *Migrant Workers in Idaho.* Boise: Boise State University, 1980.

Palmer, Tim. *The Snake River: Window to the West.* Washington, D.C., and Covello, Calif.: Island Press, 1991.

Palmer, T. S. "The Deer Flat National Bird Reservation, Idaho." *Reclamation Record* 7 (March 1916): 221–223.

Park, J. W. *Marketing the Late Crop of Potatoes.* U.S. Department of the Interior, Circular 475. Washington, D.C.: Government Printing Office, June 1938.

Parker Centennial Chronicles, 1884–1984. Parker, Idaho: Parker Centennial Committee, 1984.

Paul, Elliot. *Desperate Scenery.* New York: Random House, 1954.

Perry, Barbara. "Arrowrock Dam Is Built." *Idaho Yesterdays* 29 (Spring 1985): 15–16.

Peters, Alfred. "Early History of the Twin Falls Canal Company and My Adult Life With the Company." File 1511.01, History of the Twin Falls Canal Company, Twin Falls Canal Company Records, Twin Falls Canal Company. Twin Falls, Idaho.

Peterson, F. Ross. "Confronting the Desert." In *Snake: The Plain and Its People,* edited by Todd Shallat. Boise: Boise State University, 1994.

Peterson, F. Ross, and W. Darrell Gertsch. "The Creation of Idaho's Lifeblood: The Politics of Irrigation." *Rendezvous* 11 (Fall 1976): 53–61.

Peterson, Roger Tory. *A Field Guide to Western Birds.* 3d ed. Boston: Houghton Mifflin, 1996.

Photograph Collection. Idaho Historical Society Library and Archives. Boise, Idaho.

Piemeisel, R. L. *Weedy Abandoned Lands and the Weed Hosts of the Beet Leaf Hopper.* U.S. Department of Agriculture, Circular 229. Washington, D.C.: Government Printing Office, July 1932.

Pioneer Irrigation District Records. Pioneer Irrigation District. Caldwell, Idaho.

Piper, S. E. "The Jack Rabbit Problem." *Reclamation Record* 6 (March 1915): 123–124.

Pisani, Donald J. "Enterprise and Equity: A Critique of Western Water Law in the Nineteenth Century." *Western Historical Quarterly* 18 (January 1987): 15–37.

———. *To Reclaim a Divided West: Water, Law, and Public Policy, 1848–1902.* Albuquerque: University of New Mexico Press, 1992.

Plew, Mark. "Native Traditions." In *Snake: The Plain and Its People,* edited by Todd Shallat. Boise: Boise State University, 1994.

Pollan, Michael. *Second Nature: A Gardener's Education.* New York: Dell Publishing/Delta, 1991.

Pollard, Herb. "William H. Ridenbaugh, Pioneer Wildlife Entrepreneur." *Idaho Wildlife Review* 27 (May–June 1975): 8–9.

Poulsen, Ezra J. "Too Many Squirrels." *Harper's Magazine* 178 (February 1939): 302–305.

Poulson, E. N., L. B. Nelson, and A. E. Poulson. *Soil Survey, Blackfoot-Aberdeen Area, Idaho.* U.S. Department of Agriculture, Bureau of Plant Industry, Series 1937, 6. Washington, D.C.: Government Printing Office, May 1943.

Poulson, E. N., and J. A. Thompson. *Soil Survey of the Jerome Area, Idaho.* U.S. Department of Agriculture, Bureau of Chemistry and Soils, Series 1927, 16. Washington, D.C.: Government Printing Office, 1931.

Powell, John Wesley. *Eleventh Annual Report of the United States Geological Survey to the Secretary of the Interior, 1889–'90,* Part 2, *Irrigation.* Washington, D.C.: Government Printing Office, 1891.

———. *Tenth Annual Report of the United States Geological Survey to the Secretary of the Interior, 1888–'89,* Part 2, *Irrigation.* Washington, D.C.: Government Printing Office, 1890.

Powell, William J. "I'll Take It If It's Legal." *Pacific Northwesterner* 13 (Summer 1969): 33–40, (Fall 1969): 55–64.

Powers, W. L.. "The Drainage of Irrigated Lands." *Reclamation Record* 9 (September 1918): 424–425.

Pratt, O. A. "Experiments With Clean Seed Potatoes on New Land in Southern Idaho." *Journal of Agricultural Research* 6 (July 1916): 573–575.

———. "Soil Fungi in Relation to Diseases of the Irish Potato in Southern Idaho." *Journal of Agricultural Research* 8 (April 1918): 73–99.

Preston, William L. *Vanishing Landscapes: Land and Life in the Tulare Lake Basin.* Berkeley: University of California Press, 1981.

Price, Robert. *Johnny Appleseed: Man and Myth.* Bloomington: Indiana University Press, 1954.

"Prize Winning Irrigator and His Methods." *Gem State Rural* 11 (8 November 1906): 2.

Project Histories and Reports of Reclamation Bureau Projects, see U.S. Department of the Interior, Bureau of Reclamation, Record Group 115.

Prown, Jules David, Nancy K. Anderson, William Cronon, Brian W. Dippie, Martha A. Sandweiss, Susan P. Schoelwer, Howard B. Lamar. *Discovered Lands, Invented Pasts: Transforming Visions of the American West.* New Haven: Yale University Press, 1992.

Pyne, Stephen. *Fire in America: A Cultural History of Wildland and Rural Fire.* 1982. Paperback ed., Seattle: University of Washington Press, 1997.

———. *World Fire: The Culture of Fire on Earth.* 1995. Paperback ed., Seattle: University of Washington Press, 1997.

Raschbacher, H. G. *Irrigation From Snake River, Idaho.* U.S. Department of Agriculture, Office of Experiment Stations, Circular 65. Washington, D.C.: Government Printing Office, 1906.

Rasmussen, Wayne D. "Technological Change in Western Beet Sugar Production." *Agricultural History* 41 (1967): 31–35.

Raup, Hugh Miller. *Forests in the Here and Now.* Missoula: Montana Forest and Conservation Experiment Station, School of Forestry, University of Montana, 1981.

Reclamation Record (Washington, D.C.). 1916–1926.

Reisner, Marc. *Cadillac Desert: The American West and Its Disappearing Water.* New York: Penguin Books, 1987.

Rich, S. J. See Idaho. Commissioner of Immigration, Labor, and Statistics, *Biennial Report, 1911–1912.*

Rickets, Virginia. *Greater Twin Falls Area Historical Guide.* Twin Falls, Idaho: Twin Falls County Historical Preservation Commission, May 1988.

Ricks, Norman Earl. "Mormon Settlement of Snake River Fork Country, 1883–1893." M.S. thesis, Brigham Young University, 1950.

Rigby Star. 1906–1924.

Robinson, A. E. See Idaho. State Engineer. *Biennial Report.*

Robinson, C. Clifford. C. Clifford Robinson Oral History, MC 21-6-186. Special Collections Department, Eli M. Oboler Library, Idaho State University. Pocatello, Idaho.

———. "The Life Story of Charles Clifford Robinson," MC 21-6-186. Special Collections Department, Eli M. Oboler Library, Idaho State University. Pocatello, Idaho.

Ross, D. W. See Idaho. State Engineer. *Biennial Report.*

———. "Home Making in Idaho Under the Carey Law." *Irrigation Age* 6 (December 1894): 254–257.

Russell, Edmund P., III. "'Speaking of Annihilation': Mobilizing for War against Human and Insect Enemies, 1914–1945." *Journal of American History* 82 (March 1996): 1505–1529.

Russell, Israel C. *Geology and Water Resources of the Snake River Plains of Idaho.* U.S. Geological Survey, Bulletin 199. Washington, D.C.: Government Printing Office, 1902.

Salinger, Herbert E. "Factors Affecting Pheasant Production on the Irrigated Lands in Southern Idaho." M.S. thesis, University of Idaho, 1950.

Samson, Harold W. *United States Grades for Potatoes.* U.S. Department of Agriculture, Department Circular 238. Washington, D.C.: Government Printing Office, 1922.

Saylor, Charles F. *Progress of the Beet-Sugar Industry in the United States in 1904.* U.S. Department of Agriculture, Report 80. Washington, D.C.: Government Printing Office, 1905.

———. *Progress of the Beet-Sugar Industry in the United States in 1907.* U.S. Department of Agriculture, Report 86. Washington, D.C.: Government Printing Office, 1908.

———. *Progress of the Beet-Sugar Industry in the United States in 1909.* U.S. Department of Agriculture, Report 92. Washington, D.C.: Government Printing Office, 1910.

Schama, Simon. *Landscape and Memory.* New York: Knopf, 1995.

Scheffer, Thomas B. *Habits and Economic Status of the Pocket Gopher.* U.S. Department of Agriculture, Technical Bulletin 224. Washington, D.C.: Government Printing Office, January 1931.

Schwantes, Carlos. *Hard Traveling: A Portrait of Work Life in the New Northwest.* Lincoln: University of Nebraska Press, 1994.

———. *In Mountain Shadows: A History of Idaho.* Lincoln: University of Nebraska Press, 1991.

Schwartz, Harry. *Seasonal Farm Labor in the United States.* New York: Columbia University Press, 1945.

Schwendiman, Fred. "Battle With Germans." *Idaho Farmer* 23 (4 April 1918): 4.

Scobey, Fred C. *The Flow of Water in Flumes.* U.S. Department of Agriculture,

Technical Bulletin 393. Washington, D.C.: Government Printing Office, December 1933.

Scott, Patricia Lyn. "Idaho and the Carey Act, 1894–1930." M.A. thesis, University of Utah, 1983.

Seiler, A. G. "The Story of An Irrigated Farm." *Reclamation Record* 6 (April 1915): 192.

Shallat, Todd, ed. *Snake: The Plain and Its People.* Boise: Boise State University, 1994.

Sheehan, B. F. "Dodder in Idaho." *Idaho Farmer* 27 (31 March 1921): 346.

———. "Greatest Seed Show Ever Held in West Scheduled." *Idaho Farmer* 26 (25 November 1920): 1699.

Sheriff, Carol. *The Artificial River: The Erie Canal and the Paradox of Progress, 1817–1862.* New York: Hill and Wang, 1996.

Sherman, Wells A., George B. Fiske, and O. D. Miller. *Marketing Main-Crop Potatoes.* U.S. Department of Agriculture, Farmers' Bulletin 1317. Washington, D.C.: Government Printing Office, 1923.

Sherman, Wells A., George B. Fiske, and J. W. Park. *Marketing Late-Crop Potatoes.* U.S. Department of Agriculture, Farmers' Bulletin 1578. Washington, D.C.: Government Printing Office, 1929.

Shinn, Dean A. " Historical Perspectives on Range Burning in the Inland Pacific Northwest." *Journal of Range Management* 33 (November 1980): 418–423.

Shinn, J. R. "Fruit For Exhibition." *Gem State Rural and Live Stock Journal* (September 1910): 15.

Short, John Rennie. *Imagined Country: Environment, Culture, and Society.* London and New York: Routledge, 1991.

Simmons, ZoAnn. *Iona Centennial History Book, 1883–1983.* N.p.: n.p., n.d.

Simms, Robert C. "The Japanese-American Experience in Idaho." *Idaho Yesterdays* 22 (Spring 1978): 2–10.

Simms, Steven R. "Wilderness as a Human Landscape." In *Wilderness Tapestry: An Eclectic Approach to Preservation,* edited by Samuel I. Zeveloff, L. Mikel Vause, and William H. McVaugh. Reno and Las Vegas: University of Nevada Press, 1992.

Simpson, James, and Richard Wallace. *Fishes of Idaho.* Moscow: University Press of Idaho, 1982.

Slotkin, Richard. "Myth and the Production of History." In *Ideology and Classic American Literature,* edited by Sacvan Bercovitch and Myra Jehlen. New York: Cambridge University Press, 1986.

———. *Regeneration through Violence: The Mythology of the American Frontier, 1600–1860.* Middletown, Conn.: Wesleyan University Press, 1973.

Smith, Henry Nash. *Virgin Land: The American West as Symbol and Myth.* Cambridge, Mass.: Harvard University Press, 1950; rpt. New York: Vintage Books, 1957.

Smith, J. H. See Idaho. State Engineer. *Biennial Report.*

Smith, Ray F. "Plant Diseases, Insects, and Weeds As Affected by Irrigation." In *Proceedings: First Intersociety Conference on Irrigation and Drainage: Can Man Develop a Permanent Irrigation Agriculture?* edited by Robert M. Hagan. San Francisco: n.p., 1957.

Smythe, William E. *The Conquest of Arid America.* New York: Harper and Brothers, 1899, 1905; rpt. Seattle: University of Washington Press, 1969.

Snyder, Robert S., Mark R. Kulp, G. Orion Baker, and James C. Marr. *Alkali Reclamation Investigations.* University of Idaho Agricultural Experiment Station, Bulletin 233. Moscow: University of Idaho, October 1940.

Spencer, H. L., and H. W. Hulbert. *Idaho Perennial Weeds: Their Description and Control.* University of Idaho, College of Agriculture, Extension Bulletin 98. Moscow: n.p., June 1935.

Stannard, Jay D. "The Use of Water From the Wood Rivers, Idaho." In *Report of Irrigation Investigations for 1902.* U.S. Department of Agriculture, Office of Experiment Stations, Bulletin 133. Washington, D.C.: Government Printing Office, 1903.

Stearns, Harold T., Lynn Crandall, Willard G. Steward. *Geology and Ground Water Resources of the Snake River Plain in Southeastern Idaho.* U.S. Geological Survey, Water Supply Paper 774. Washington, D.C.: Government Printing Office, 1938.

Stegner, Wallace. *The American West as Living Space.* Ann Arbor: University of Michigan Press, 1987.

Stephenson, James, Jr. See Idaho. State Engineer. *Biennial Report.*

———. *Irrigation in Idaho.* U.S. Department of Agriculture, Office of Experiment Stations, Bulletin 216. Washington, D.C.: Government Printing Office, 1909.

Stewart, George. *Alfalfa Growing in the United States and Canada.* New York: Macmillan, 1926.

———. "Pastures and Natural Meadows." In *Grass: The Yearbook of Agriculture, 1948.* Washington, D.C.: U.S. Government Printing Office, 1948.

Stewart, Mart A. "Rice, Water, and Power: Landscapes of Domination and Resistance in the Lowcountry, 1790–1880." *Environmental History Review* 15 (Fall 1991): 47–64.

———. *"What Nature Suffers to Groe": Life, Labor, and Landscape on the Georgia Coast, 1680–1920.* Athens: University of Georgia Press, 1996.

Stilgoe, John R. *Common Landscape of America, 1580 to 1845.* New Haven: Yale University Press, 1982.

Stockton, Robert S. "Management of Irrigation Systems." *Engineering and Contracting* 41 (28 January 1914): 141–150.

Stoker, Martha B. *Old Market Lake.* N.p.: n.p., 1983.

Stuart, William. *Potato Storage and Storage Houses.* U.S. Department of Agriculture, Farmers' Bulletin 847. Washington, D.C.: Government Printing Office, 1917; rev. 1930.

Swanson, Earl, Jr. "The Snake River Plain." *Idaho Yesterdays* 18 (Summer 1974): 211.

Swendsen, W. G. See Idaho. Department of Reclamation. *Biennial Report.*

———. *Report on American Falls Reservoir District.* N.p.: n.p., n.d.

Taylor, Paul. "Hand Laborers in the Western Sugar Beet Industry." *Agricultural History* 41 (1967): 19–26.

Temple, C. E. "Potato Diseases and Insect Pests." In *Potato Culture.* Idaho Agricultural Experiment Station, Bulletin 79. N.p.: n.p., April 1914.

Thacher, W. F. G. "Buhl, Idaho—The Land That Beckons." *Pacific Monthly* 24 (December 1911): 717–718.

Thalman, James. "One of Ours." *Continuum: The Magazine of the University of Utah* 5 (Winter 1995–1996): 25–29.

Thomas, H. A. "Carey Act Lands Surrounding Shoshone, Idaho." *Irrigation Age* 24 (March 1909): 151–152.

Thompson, E. P. "Time, Work Discipline, and Industrial Capitalism." *Past and Present* 38 (December 1967): 56–97.

Thorp, James, and C. S. Scofield. "Drainage in Arid Regions." In *Soils and Men: United States Department of Agriculture, Year Book of Agriculture, 1938.* Washington, D.C.: Government Printing Office, n.d.

Truettner, William H., ed. *The West as America: Reinterpreting Images of the Frontier, 1820–1920.* Washington, D.C.: Smithsonian Institution Press, for the National Museum of American Art, 1991.

Tuan, Yi-Fu. "Discrepancies between Environmental Attitude and Behavior: Examples from Europe and China." In *Ecology and Religion in History,* edited by David Spring and Eileen Spring. New York: Harper and Row, 1974.

———. "Our Treatment of the Environment in Ideal and Actuality." *American Scientist* 58 (May–June 1970): 244, 247–249.

Twin Falls Canal Company Records. Twin Falls Canal Company. Twin Falls, Idaho.

Twin Falls Daily News. 1921.

Twin Falls Land and Water Company Papers. Idaho State Historical Society Library and Archives, Boise.

Twin Falls News. 1904–1907.

Twin Falls Times-News. 1947, 1978.

Twin Falls Times-Register. 1978.

Twin Falls Weekly News. 1923.

Tyler, Daniel. "Delph E. Carpenter and the Principle of Equitable Apportionment." *Western Legal History* 9 (Winter/Spring 1996): 35–53.

United States. See U.S.

United States Geological Survey. See U.S. Department of the Interior, Geological Survey.

U.S. Bureau of the Census. *Report on the Productions of Agriculture as Returned at the Tenth Census,* 1880. Washington, D.C.: Government Printing Office, 1883.

———. *Report on the Statistics of Agriculture in the United States at the Eleventh Census,* 1890. Washington, D.C.: Government Printing Office, 1895.

———. *Twelfth Census of the United States,* 1900, Vol. 5: *Agriculture,* Part 1, *Farms, Live Stock, and Animal Products.* Washington, D.C.: Census Office, 1902.

———. *Twelfth Census of the United States,* 1900, Vol. 6: *Agriculture,* Part 2, *Crops and Irrigation.* Washington, D.C.: Census Office, 1902.

———. *Thirteenth Census of the United States,* 1910, Vol. 6: *Agriculture.* Washington, D.C.: Government Printing Office, 1913.

———. *Fourteenth Census of the United States,* 1920, Vol. 6, Part 3: *Agriculture.* Washington, D.C.: Government Printing Office, 1922.

———. *Fourteenth Census of the United States,* 1920, Vol. 7: *Irrigation and Drainage.* Washington, D.C.: Government Printing Office, 1922.

———. *Fifteenth Census of the United States,* 1930, *Agriculture,* Vol. 3: *Type of Farm,* Part 3, *The Western States.* Washington, D.C.: Government Printing Office, 1932.

———. *Fifteenth Census of the United States,* 1930, *Agriculture,* Vol. 2, Part 3: *The Western States.* Washington, D.C.: Government Printing Office, 1932.

———. *Fifteenth Census of the United States,* 1930, *Drainage of Agricultural Lands.* Washington, D.C.: Government Printing Office, 1932.

———. *Census of Agriculture,* 1925, Part 3: *The Western States.* Washington, D.C.: Government Printing Office, 1927.

U.S. Bureau of Reclamation. See U.S. Department of the Interior.

U.S. Department of Agriculture, Soil Conservation Service. *Snake River Basin Cooperative Study: Main Report, Upper Snake River Basin.* N.p.: n.p., March 1979.

U.S. Department of Health, Education, and Welfare, Public Health Service. "Upper Snake River Basin, Reconnaissance Report." In U.S. Department of the Interior, Bureau of Reclamation, and U.S. Army, Corps of Engineers. *Upper Snake River Basin,* Vol. 3, Part 1.

U.S. Department of the Interior, Bureau of Reclamation. Record Group 115, Records of the Bureau of Reclamation. National Archives, Washington, D.C. [Boise Project].

———. *Federal Irrigation Reservoirs As Pleasure Resorts.* Washington: U.S. Government Printing Office, 1928.

U.S. Department of the Interior, Bureau of Reclamation, and U.S. Army, Corps of Engineers. *Upper Snake River Basin.* 3 vols. Vol. 3, Part 1: *Coordination and Reports of Cooperating Agencies.* N.p.: n.p., 1961.

U.S. Department of the Interior, Fish and Wildlife Service. "A Preliminary Survey of Fish and Wildlife Resources, Upper Snake River Basin." March 1960. In U.S. Department of the Interior, Bureau of Reclamation, and U.S. Army, Corps of Engineers, *Upper Snake River Basin,* Vol. 3, Part 1.

U.S. Department of the Interior, General Land Office. *Territory of Idaho.* (Map scale: 1 inch = 16 miles.) 1876. 1879. 1883.

U.S. Department of the Interior, Geological Survey. *The Hydrologic Cycle.* Washington, D.C.: Government Printing Office, 1989.

———. *Ground Water.* Washington, D.C.: Government Printing Office, 1986.

U.S. Senate. *Report of the Special Committee of the United States Senate on the Irrigation and Reclamation of Arid Lands,* Vol. 1: *The Northwest.* Washington, D.C.: Government Printing Office, 1890.

———. *Reports of the Immigration Commission: Immigrants in Industries,* Part 25, *Japanese and Other Immigrant Races in the Pacific Coast and Rocky Mountain States.* 3 vols. 61st Cong., 2d Sess., 1911, Senate Document 633.

Valora, Peter J. "A Historical Geography of Agriculture in the Upper Snake River Valley, Idaho." Ph.D. dissertation, University of Colorado, 1986.

Vincent, C. C. "The Orchard Pests Are Always With Us." *Idaho Farmer* 31 (1 March 1923): 195.

Vogel, Harold A., and Neil W. Johnson. *Types of Farming in Idaho,* Part 1, *Agricultural Resources and Factors Affecting Their Use.* University of Idaho Agricultural Experiment Station, Bulletin 207. Moscow: University of Idaho, May 1934.

Wakeland, Claude. "The Alfalfa Weevil." *Idaho Farmer* 27 (31 March 1921): 349; (7 April 1921): 375; (14 April 1921): 401; (21 April 1921): 425; (28 April 1921): 438.

———. "European Fly May Help Whip Alfalfa Weevil." *Idaho Farmer* 32 (19 July 1923): 39.

Wakeland, Claude, and W. C. Hungerford. *Idaho Recommendation Chart for Plant Disease and Insect Control.* University of Idaho, Agricultural Experiment Station, Bulletin 159. Moscow: University of Idaho, May 1928.

Wakeland, Claude, and W. E. Shull. *The Mormon Cricket, With Suggestions For Its Control.* University of Idaho, College of Agriculture, Extension Division, Extension Bulletin 100. Moscow: n.p., February 1936.

Walker, Eugene H. "The Geologic History of the Snake River Country of Idaho." *Idaho Yesterdays* 7 (Summer 1963): 18–31.

Wallach, Bret. *At Odds with Progress: Americans and Conservation.* Tucson: University of Arizona Press, 1991.

"Wants Pests Exterminated." *Gem State Rural* 13 (October 1907): 25.

Warrington, Jessie. "The Wonderful Redemption of a Desert: A Story of the Past, Present, and Future of the Famous Twin Falls Country." *See Idaho First* 5 (September 1914): 3–21.

"The Waste of Water." *Gem State Rural and Live Stock Journal* (September 1910): 41.

"Water Law in Idaho." *Idaho Yesterdays* 25 (Spring 1981).

Weber, R. C. E. "Economy of Concrete for Irrigation Canal Linings." *Concrete* 32 (October 1923): 158–159.

Welch, J. S. "Potato Culture Under Irrigation." In *Potato Culture*. Idaho Experiment Station Bulletin 79. N.p.: n.p., April 1914.

Wells, C. F., and H. C. Dale. *The Potato Situation in Idaho*. University of Idaho Agricultural Experiment Station, Bulletin 153. Moscow: University of Idaho, June 1927.

Wenger, Paul. *Pure Seed Law and Weed Control Act*. University of Idaho Agricultural Experiment Station, Circular 8. Moscow: University of Idaho, April 1919.

Wescoat, James L. "Challenging the Desert." In *The Making of the American Landscape*, edited by Michael P. Conzen. Boston: Unwin Hyman, 1990.

Wheeler, W. A., and G. C. Edler. "Some Effects of the War Upon the Seed Industry of the United States." In *Yearbook of the United States Department of Agriculture, 1918*. Washington, D.C.: Government Printing Office, 1919.

Whelan, Don B. "Who Owns the Farm, You or Mr. Grasshopper?" *Idaho Farmer* 31 (8 March 1923): 219.

———. "Will the Alfalfa Weevil Run the Honey Bee Out of Idaho?" *Idaho Farmer* 30 (30 November 1922): 422.

White, Richard. *"It's Your Misfortune and None of My Own": A History of the American West*. Norman: University of Oklahoma Press, 1991.

———. *Land Use, Environment, and Social Change: The Shaping of Island County, Washington*. Seattle: University of Washington Press, 1980.

———. *The Organic Machine: The Remaking of the Columbia River*. New York: Hill and Wang, 1995.

Whitson, Tom D., ed. *Weeds of the West*. 5th ed. Newark, Calif.: Western Society of Weed Sciences, 1996.

Wilkinson, Charles. "West's Grand Old Water Doctrine Dies." *High Country News* 23 (12 August 1991): 1, 11–12.

Williams, Mikel H. *The History of Development and Current Status of the Carey Act in Idaho*. Boise: Idaho Department of Reclamation, 1970.

Williams, Raymond. *The Country and the City*. New York: Oxford University Press, 1973.

———. "Ideas of Nature." In *Problems in Materialism and Culture: Selected Essays*. London and New York: Verso, 1980.

Wing, Herbert. See Idaho. State Engineer. *Biennial Report*.

Worster, Donald. *Rivers of Empire: Water, Aridity, and the Growth of the American West.* New York: Pantheon Books, 1985.

Young, James A. "Mormon Crickets." *Rangeman's Journal* 5 (December 1978): 193–196.

Young, James A., and B. Abbott Sparks. *Cattle in the Cold Desert.* Logan: Utah State University Press, 1985.

Youngs, F. O., Glenn Trail, and B. L. Young. *Soil Survey of the Gooding Area, Idaho.* United States Department of Agriculture, Bureau of Chemistry and Soils, Series 1929, No. 10. Washington, D.C.: Government Printing Office, n.d.

Youngs, F. O., Mark Baldwin, A. J. Kern, and G. R. McDole. *Soil Survey: Minidoka Area, Idaho.* U.S. Department of Agriculture, Bureau of Chemistry and Soils, Series 1923, No. 27. Washington, D.C.: Government Printing Office, 1928.

Index

Throughout the text, I refer to most plants and animals by their common names. In the index I also include scientific names for most organisms. See subentries under *Birds; Fish; Insects; Mammals; Weeds.*